NEW DIMENSIONS IN AGRICULTURAL GEOGRAPHY

VOLUME ONE

HISTORICAL DIMENSIONS OF AGRICULTURE

Concept's International Series in Geography

Volume 1 **Historical Dimensions of Agriculture**
Volume 2 **The Ecology of Agricultural System**
Volume 3 **Socio-economic Dimensions of Agriculture**
Volume 4 **Landuse and Agricultural Planning**
Volume 5 **Spatial Dimensions of Agriculture**
Volume 6 **Human Perception and Technological Change in Agriculture**
Volume 7 **Dynamics of Agricultural Development**
Volume 8 **Anthropogenic Dimensions in Agriculture**

Concept's International Series in Geography No. 4

NEW DIMENSIONS IN AGRICULTURAL GEOGRAPHY

VOLUME ONE

HISTORICAL DIMENSIONS OF AGRICULTURE

NOOR MOHAMMAD
Editor

CONCEPT PUBLISHING COMPANY, NEW DELHI 110059

ISBN 81-7022-400-4 (Set)
ISBN 81-7022-401-2 (Vol. 1)

First Published 1992

Published by
Ashok Kumar Mittal
Concept Publishing Company
A/15-16, Commercial Block, Mohan Garden
NEW DELHI 110059 (India)

Lasertypeset by
Microtech Advance Printing Systems (Pvt.) Ltd.
H-13, Bali Nagar
NEW DELHI 110015

Printed at
Sunil Printers
CB 386, Ring Road, Naraina
NEW DELHI 110028

PREFACE

AGRICULTURE CONTINUES to be the key industry. It is not only the dominant form of employment but also forms the basis for a way of life for, perhaps, two-thirds of the world's households. Agriculture being the function of physical, socio-institutional and techno-economic factors which are dynamic in nature, keeps on changing with the basic objective of increasing production and generation of foodgrains surplus. The traditional system of farming is in the process of transformation to modern agriculture which itself is undergoing a set of equally and generally even in more rapid changes. So it has added a new dimension in agriculture which needs to be studied and analysed.

A project was, therefore, taken up to accomplish this task. Original research papers dealing with various facets of agriculture were invited from scholars of different discipline from all over the world. The response was very encouraging and a large number of original papers were received. Besides these, some papers which have classical and everlasting value have also been picked up from my earlier publication – *Perspectives in Agricultural Geography*. The papers have been classified according to their basic themes and have been put into following eight volumes so as to give a systematic picture of various themes. The articles dealing with concepts, theories and models have been kept first followed by those dealing with the same theme at macro, meso and micro levels respectively.

The First Volume – *Historical Dimensions of Agriculture* – containing 19 articles deals with the origin, diffusion and development of agriculture, research trends in agricultural geography, world agricultural systems and methods and techniques in agricultural geography.

The Second Volume – *The Ecology of Agricultural System* – deals with the selected parameters of physical environment specially the climatic elements which affect the agricultural landscape in spite of technological advancement in agriculture.

The Third Volume – *Socio-economic Dimensions of Agriculture* –

containing 19 articles deals with socio-economic environment which have bearings on agricultural systems. The first 8 articles deal with social parameters specially labour, caste, women etc., while the last 11 articles examine the impact of techno-economic factors specially the various facets of irrigation which is a basic agricultural input.

The Fourth Volume – *Landuse and Agricultural Planning* – contains 19 articles dealing with concepts, theories, models of landuse, land capability and suitability, wastelands and dynamics of landuse.

The Fifth Volume – *Spatial Dimensions of Agriculture* – includes articles on agricultural systems of specific areas such as Himalayas, Bastar, Uttara Khand etc. It also deals with dynamics of crop landuse, cropping pattern and its diversification, crop ecology and crop-combination analysis.

The Sixth Volume – *Human Perception and Technological Change in Agriculture* – contains only 15 articles and covers four basic themes. The first set of 3 articles deals with human perception and his consequential response, while the second set includes articles on correlates of technological change and adoption of agricultural innovations. The third set of articles deals with the assessment of socio-economic transformation in rural India. The last set deals with agriculture and human settlements.

The Seventh Volume – *Dynamics of Agricultural Development* – includes articles on concept of development and its various facets. Besides assessing the trends of levels of agricultural development, it also examines and evaluates its impact on environment and ecological parameters.

The Eighth Volume – *Anthropogenic Dimensions of Agriculture* – containing 17 articles deals with regional imbalances in agricultural productivity, methods of its measurement, and also its correlates and determinant. It also focuses on food availability and food security at regional and continental level.

The collection of papers in eight volumes of *New Dimensions in Agricultural Geography* is designed to encapsule the major themes and recent trends and concepts in the field of agricultural geography with particular reference to conceptual revolution in the discipline. All the volumes contain a brief introduction which gives a glimpse into the topics dealt with in each of them.

The variety of articles contributed by the renowned scholars of different disciplines has made the character of these volumes multi-disciplinary and has enhanced their value. These volumes have been designed to help not only the students of a particular discipline but also to general students who wish to keep abreast of the latest developments in the field of agricultural geography. From these points of view these volumes hold a promise of rich fare and in addition to being heuristic, will provide a mine of information for discerning scholars. It is hoped that this unique collection of research papers will not only be valuable to the researchers in the fields of agricultural geography, economics, sociology, history, and agricultural science, but will also provide guidelines to decision-makers and planners.

The completion of this project has been possible due to the constant inspiration and unstinting co-operation of a host of people too numerous to acknowledge individually. However, I owe a debt of gratitude to my mentor — Dr. Mohammad Shafi, Professor Emeritus, Department of Geography, A.M.U., Aligarh who gave me valuable comments and suggestions ever since the inception of this project. Prof. R. Ramachandran and my other colleagues who offered sincere comments on the title of volumes and the collection of articles deserve special thanks. I would also like to thank my father, wife and children who provided a favourable working atmosphere to me for successful completion of this project. Last but not the least, I would like to put a special word of appreciation, love and affection to Mohammad Imran Ali Khan who is a source of hope and inspiration.

Department of Geography — NOOR MOHAMMAD
Delhi School of Economics
University of Delhi
Delhi-110007
January 26, 1992

PUBLISHER'S NOTE

When Concept's International Series in Geography No. 1 — *Perspectives in Agricultural Geography* edited by Dr. Noor Mohammad in five volumes was issued over a decade back, we expressed our apprehensions that 'Posterity will be the best judge of our efforts in this direction. The response to this set of volumes will provide the impetus for hastening the publication of further series that are already on the anvil'.

Perspectives in Agricultural Geography received such an overwhelming response that it became out of print soon after its publication. Its success provided the necessary encouragement to issue thereafter Concept's International Series in Geography No. 2 — *Perspectives in Urban Geography* edited by Dr. C.S. Yadav in 16 volumes and No. 3 — *Perspectives in Geomorphology* by Dr. H.S. Sharma in four volumes.

We are glad to offer now Concept's International Series in Geography No. 4 — *New Dimensions in Agricultural Geography* edited by Dr. Noor Mohammad in eight volumes. In view of the new emerging trends, concept, theories and models in the field of agricultural geography, the set presents a wide panoramic view of the latest advances in this field. Hope this set too would be equally welcomed by the students, researchers, scholars and planners in various disciplines.

February 11, 1992 NAURANG RAI

CONTENTS

LIST OF FIGURES

LIST OF PLATES

LIST OF TABLES

LIST OF CONTRIBUTORS

Allan, J.A.
School of Oriential and African Studies, Malt Street, University of London, London

Armstrong, Patrick
Department of Geography, University of Western Australia, NedLands, Australia

Bielecka, K.
Institute of Geography and Spatial Organization, Polish Academy of Sciences, Warsaw, Poland

Coppock, J.T.
Department of Geography, University of Edinburg, Drummond St. Edinburg EHS 9 XP, England

Dayal, Edison
Department of Geography, University of Wollongong N.S.W. Australia

Dhillon, S.S.
Department of Geography, University of Kurukshetra, Kurukshetra

Garg, S.P.
Department of Geography, D.A.V. (P.G.) College, Dehradun

Ghosh, Sanjib K.
Professor of Photogrammetry, Laval University, Quebec (QUEBEC) G1K 7P4, Canada

Jedrusik, Maciej
Faculty of Geography and Regional Studies, University of Warsaw, Warsaw, Poland

Kostrowicki, Jerzy
Institute of Geography and Spatial Organization, Polish Academy of Sciences, Warsaw, Poland

Mohammad, Noor
Department of Geography, Delhi School of Economics, University of Delhi, Delhi

Mukhopadhyay, Jyotirmoy
Vidyasagar College, Calcutta

Obara, Dunstan A.
Department of Geography, Nairobi University, Nairobi, Kenya

Paprzycki, M.
Institute of Geography and Spatial Organization, Polish Academy of Sciences, Warsaw, Poland

Piasecki, Z.
Institute of Geography and Spatial Organization, Polish Academy of Sciences, Warsaw, Poland

Raza, Mehdi
Department of Geography, Aligarh Muslim University, Aligarh

Sharma, B.L.
P.G. Department of Geography, Government College, Beawar

Shukla, R.K.
Lodhu Thok, ATARRA-210201, Banda, U.P

Singh, B.B.
Principal, M.M.H. College, Ghaziabad

Singh, Inderjit
Scientist, Indian Institute of Remote Sensing (NRSA), Dehradun

Szyrmer, Jacek H.
Institute of Geography and Spatial Organization, Polish Academy of Sciences, Warsaw, Poland

INTRODUCTION

THE PRESENT volume containing 19 research papers deals with origin, diffusion and development of agriculture, agricultural regionalisation and world agricultural systems, research trends in agricultural geography, and finally methods and techniques of its analysis.

Agriculture is usually understood to mean both – the cultivation of plants and the domestication of animals. The origin of agriculture, therefore, dates back to the beginning of raising of plants and animals. Man, animal and plants have been living together since antiquity and are inseparably interconnected with one another, but how, when and where man started the raising of plants and animals is still a challenge to the scholars of various disciplines. Archaeologists, anthropologists, botanists, palaeontologists, historians, geologists and scholars of the other branches of physical sciences have made their efforts to unravel this mystery but have not come to any common conclusion so far. Whether domestication of plants and animals was started simultaneously or at different points of time is in dispute but the majority supports their simultaneous origin.

Several theories have been put forth to explain as to why and when man started agriculture but there are three major theories viz.: (i) environmental change and origin of agriculture; (ii) cultural change and origin of agriculture; and (iii) population pressure and origin of agriculture.

Similar to dates and circumstances leading to its origin, the exact place where agriculture originated first is also highly debated. Some consider river valleys as probable places of its origins (Childe, 1935; and Braidwood, 1958), while others reject it on the ground of its being liable to floods and drought conditions (Sauer, 1952; and Wissaman, *et al.*, 1956). However, Sauer, after giving several premises has considered hilly upland of tropical regions specially South east Asia as regions of the origin of agriculture. A Russian biogeographer, Vavilov (1949-50) on the basis of his eighteen years expeditions of the world along with other scientists, and the analysis of thus collected material, has identified six major gene centres in the Old World and two gene centres in the New World.

Agriculture with its long history of at least twelve thousand years has diffused and developed in various time periods and phases which could be given as the pre-historic period; early historic period; pre-medieval period; medieval period, neo-modern period; and modern period. In fact up to the medieval period agriculture has diffused in various phases at various points of time. During this period more and more areas had been brought under cultivation and minimum efforts were made to intensify it. But gradually it has developed both horizontally and vertically. In the modern period a revolution in agriculture has come, although not with equal vigour all over the world. **Mohammad** in his paper has given a reasonable account of the origin, diffusion and development of agriculture.

Obara in his paper has, firstly, discussed the theories of origin of agriculture with special reference to principles governing the domestication of plants and animals. Secondly, he has examined the evolution, diffusion and probable dates of the origin of agriculture. Thirdly, the author has evaluated the significance of the theory of plant and animal introduction.

Dhillon has given a critical review of studies made from time to time by non-geographers such as anthropologists, botanists, paleobotanists, archaeologists, historians, ethnologists, geneticists, etc. in order to assess the origin and development of agriculture. In fact the origin and dispersal of agriculture in different regions of the world has great potential value. The transition from non-agricultural to agricultural ways of life is not revolutionary historic event rather it is an evolutionary complex process involving an interaction between man and the environmental conditions over a long period of time. Evolution of agriculture, in other words, actually took place independently and in isolation in very small ecologically favourable area of the world. It is believed that about ten or twelve thousand years ago, man began, in some places, to cultivate some plants and to domesticate some animals. This was the first quiet explosion of agriculture. However, the author has discussed in detail the origin and gradual development of agriculture in the Old World and New World in the light of studies made by scholars of other disciplines. Agriculture indeed has a history of at least twelve thousand years, but it is of great interest as to how a variety of changes in agriculture took place. Since the inception of agriculture and right up its present form, each successive step was an improvement over the preceding one. However, the author has briefly traced the development of agriculture during Neolithic period, proto-historic period, early historic period, pre-Middle Ages, Middle Ages, post-Middle Ages, and the present times.

Although, the origin of agriculture in India has taken place around 4500 BP (Mohammad, 1981), it has diffused in various parts of the country at different points of time, and has seen several ups and downs during the course of its development. **Shukla** in his paper has given an account of agriculture during Ramayanic age. He has drawn inferences and facts from the verses of *Balmiki Ramayan* out of the ritual procedure, discussions, and mythologies by way of interpretation from geographical view point. The author has covered almost all parametres of agriculture such as factors affecting agriculture, irrigation, agricultural implements, general landuse pattern, major crops – paddy, barley, wheat, cash crop, natural calamities, social factors and agricultural prosperity.

The concept of agricultural region, like concept of region in geography, is a controversial one and yet no precise definition and method of demarcation is evolved. **Mukhopadhyaya** has made an approach to the problem of understanding the nature of an agricultural region inductively. The paper does not commence with any preconceived idea regarding the nature and reality of agricultural region. Instead, it reviews critically some representative works with the aim of revealing the nature of the ideas behind them. At least four sets of ideas about agricultural region can be found which are: (1) separate areas of agricultural activity that correspond to the natural regions or physical divisions of a land; (2) areas distinguished according to any prominent feature of the agricultural landscape, where discrimination is not usually based on a particular attribute; (3) areas distinguished in terms of a few attributes that intersect variously to give rise to unique agricultural landscape, though the latter cannot always be distinguished on the basis of one or all the involved attributes; and (4) agricultural regions distinguished on the basis of cluster analysis of the similarity co-efficients between the different parts of an agricultural landscape involving the most significant variable. However, none of the above sets of ideas is completely acceptable, and any attempt to synthesise them is likely to fail. It is, therefore, very difficult to develop a concrete concept of agricultural region from the mess of disagreement and confusion. The paper concludes with inclination to the nominalists' view regarding the nature of a region in general.

In the second section conceptual and empirical studies done at macro, meso and micro levels are included. The section opens up with **Kostrowicki's** article. Realising the significance of identification of agricultural types, a separate Commission on Agricultural Typology was set up in International Geography Union in 1964 held at London. Since then the Commission was working hard on principles, criteria, methods and techniques to evolve a scheme of world type of agriculture. As a result of various discussions in Commission and Congress meetings, it was decided to base agricultural types on 4 principal groups of agricultural attributes viz. social, operation and production – each having 7 variables, and structural which was described by 6 variables only. To make the 27 such selected variables fully comparable, they have been classified into 5 groups representing world range of agricultural phenomena. To arrive at a proper method of grouping such multifeature units, a number of cartographic and mathematical methods have been tested by the Commission.

Finally, 5 types of first order – Traditional Extensive Agriculture, Traditional Intensive Agriculture, Market Oriented Agriculture, Socialised Agriculture and Highly Specialised Livestock Breeding, 20 types of 2nd order and 55 types of 3rd agricultural typologies were identified. The author has given a brief account of each type highlighting its attributes and spatial location.

The monumental works of Kostrowicki on agricultural typology does not put an end to it but there is plenty of scope for improvement in its various facets specially in criteria, procedure and techniques. **Szyrimer'** paper is an attempt in this direction, wherein the author has given adequate thinking on diagnostic variables, their normalisation, determination of types, and calculation of basic variables. The 27 variables selected by Kostrowicki have been converted into diagnostic variables which are ratios, i.e., their values are obtained by dividing one by an other one. The diagnostic variables have been normalised by dividing all their possible values into 5 classes, i.e., very low, low, medium, high and very high. The type of agriculture is determined in this hierarchical order represented as M, Mm, Mmn which are expressed in five code comprising 27 digits. Finally he has described and explained the calculation of all basic variables. The article on the whole is an improvement over the existing methods and practices.

Kostrowicki has applied the general concept criteria and method of typology accepted 1976 IGU Commission with certain modifications for mapping the agricultural types of Europe. After giving a little more elaboration of the concept of agriculture typology and agricultural regions, he has selected 27 variables grouped into four i.e., social, operational, production and structural attributes. After making an elaborate comments on each variable he has put them into 5 classes on the basis of its percentage strength. Since the units of observation in Western countries are not uniform, the author has tried to bring them at uniform level by changing their order from one to the other. For all the units thus obtained material was collected and transformed into model codes which was sent to the computer for getting agricultural types of different orders. Finally 6 types of the first order, 24 types of the second order and 66 types of the third order agricultural types have been identified and mapped. The method used by the author is not free from shortcomings but it could be standardised and used in any scale from the lowest to the highest.

In the empirical study **Jedrusik** has applied Kostrowicki's concept method and procedure for identifying agricultural types of 16 small tropical islands. Out of which 5 were Pacific Islands, 6 Atlantic Islands, and 5 were chosen from Indian Ocean. On the basis of the characteristics of agriculture of these islands the author has commented and streamlined the various elements of social, operational, production and structural attributes of farming in order to make them comparable. The study shows the absence of agricultural type of the first order. In these Islands three types of agriculture on hierarchical level of second order have been identified which have further been divided into agricultural types of the third order.

Prior to Kostrowicki's (1981) monumental work on agricultural typology no suitable method was evolved to measure multi-level regional development of agriculture which may absorb types, systems and scales simultaneously. **Sharma** has successfully applied the same technique with slight modification in India by taking examples from specific areas of different states. The agricultural types speak very well of their regional development, regional imbalance and regional disparity from the planning point of view, which will take into account both the scales – horizontal and vertical. However, this method may not be very suitable at national level planning.

Agricultural geography is one of the recent branches of geography. It had its origin in the late twenties when agricultural regionalisation and regional description were its common themes. **Singh** in his paper has presented the evolution of the discipline and has discussed the evolution of the subject in three arbitrary phases i.e. pre-1950,

1950-60, and post-1960. The first phase is characterised by regionalisation and regional description with inductive method, whereas the second phase is marked with systematic approach together with quantitative methods for in depth and systematic analysis of the phenomena. The third phase may be treated as the golden age in the field of agricultural geography. Many conceptual and thought provoking researches were conducted. The salient trends of the phase are: more emphasis on theoretical and systematic approach and less concern with geographical description and regional approach, retreat from deterministic interpretation of phenomena to a probabilistic and behavioural approach, greater stress on micro-regional and representative studies, primacy of statistical analysis and synthesis, inter-disciplinary approach and regional to typological analysis. The author after reviewing other view points, has pointed out three basic principles of ecological adjustment, functional relationship and areal differentiation and organisation which lead to the seven fundamental concepts of agricultural geography viz. concept of agricultural landscape; agricultural landscape is dynamic which experiences continuous change and constant evolution; agricultural landscape is a function of resource structure, process and stage; location of agricultural activities; agricultural region, areal differentiation and organisation; inter-regional equilibrium, and spatial-functional interaction and regional agricultural development and planning.

Ecosystem is a segment of nature with its included organisms and their environment. It enjoins the enquirer to make a whole complex of nature as his unit of study. Basically it is a biological idea but the notion has enthusiastically been adopted by geographers, particularly those interested in landuse, and by practical resource managers. **Armstrong** in his paper has highlighted the ecosystem concept as a framework for studies in agricultural geography. In fact landuse should be seen as ecosystem modification and landuse ecology is, perhaps, the central theme in geography. The agricultural geographers have frequently borrowed two concepts from ecologists — conceptual treasure *vs.* energy flow, and bio-chemical circulation models. The author has illustrated as to how these two pieces of intellectual hardware have been utilised in attracting certain problems of land resource use. The author himself has prepared flow chart and discussed it with citing examples from others. A very useful concept in the description of the flow of energy in agricultural system is the energy subsidy which may be used to arrive at a convenient classification of landuse and agricultural system: low energy culture, medium energy culture and high energy culture.

The agriculture has the greatest potential for ecosystem destabilisation specially in the light of the recent development which has taken place in it. **Raza** in his article highlights that if the trend of agricultural development beyond the tolerance limit of ecosystem is not checked in time, the function of biosphere as a viable global life support system may seriously be geopardised. The author has listed and described the spatial occurrence of a number of serious ecological consequences of agricultural development, and has suggested suitable measures to mitigate them. The article concludes that if long-term man environment symbiosis is to be maintained the development of a new ecologically compatible food strategy for mankind must be immediately evolved without any loss of time.

As agriculture is the most common and widespread way of getting a living, considerable proportion of geographical research should be devoted to agricultural geography. But this is not true in both developed and developing countries. The major cause of this is the non-availability of required data. The reasons for the deficiency vary. Sometimes they arise because the acquisition of data by conventional methods would be too expensive to justify with respect to the likely returns from further development; on others there may be institutional impediments and a shortage of professional skills preventing data collection or there may be no local tradition for gathering such materials. Finally, the deficiency may occur because the variables of concern are ephemeral and they change from season to season and from year to year. **Coppock** has highlighted this aspect in his paper with special reference to tropical Africa. Ideally each investigation in agricultural geography should involve four stages : the identification of problems, the formulation of hypothesis, the collection of relevant data and the testing and modification of the hypothesis to provide an adequate explanation but unfortunately information required for such studies is not available in the tropics and hence, the objective must be limited. Either the reconnaissance study may be made on the basis of the data collected in general surveys or in depth study of select communities covering a range of environment may be taken. Between these two extremes the range of possible problems is large depending on information that is available or that can be collected. The author has critically analysed merits and limitations of various possible sources of data, and has advocated that the geographer can contribute to agricultural development in four ways : by provision of broad studies of major crops and livestock which will facilitate the evaluation of existing resources, by reconnaissance studies to identify and classify the nature of particular problems and problem areas, by focusing attention on and evaluating the spatial links between farmers and their markets, and by service as a member of a team on particular development projects. Tropical regions provide an original field to agricultural geographers to study spatial arrangement of agriculture around a settlement, areal diffusion of new crops and new techniques, change in the values of a single variable and the introduction of some new factors, and regional variations in the method of grain shortage. Hence, the more valuable service that agricultural geographers can perform is essentially educational, bringing together a range of material from a wide variety of sources to provide a new perspective and possibly even a new dimension for technical experts and for the administrators.

The Von Thunen theory has been studied, applied, modified, appreciated and criticised right from its inception. In fact Von Thunen developed two different theories in his model viz. an intensity theory and a landuse theory (Hall, 1966), although the later is indirectly concerned with intensity. The intensity of agriculture decreases with distance from market due to transport cost, land value, labour cost and other inputs. The output per unit of land is more near market centre as compared to the distant areas. The crop theory is concerned with spatial organisation of agricultural systems around market centre which is governed by the rationale of maximisation of profit as in any economic venture. Consequently, spatial organisation of landuse takes place in accordance with per unit area production, per unit value and per unit transport cost.

Von Thunen has propounded a normative or behavioural model based on several assumptions arising out of his experience and observations in his own farm. Later on he relaxed some of the simplifying assumptions by adding complexities in order to bring the model closer to reality. There are four major areas in which these assumptions could be relaxed: the single method of transport, the uniform plain, the single central market, and the behaviour of the people. However, various scholars have discussed, analysed, modified and tested it in a variety of environmental context and applied it at international, national, regional and micro-scales. Peet (1969) using the data of imports into Britain in 19th century, analysed the world-wide zone of agricultural production especially the expansion of commercial agriculture into continental interior of the New World. He presumed Western Europe and Eastern America as the world's urban-industrial nucleus and proved the existence of Thunen type spatial organisation of agriculture around the world city. Several studies at national level (Crossley, 1976; Muller, 1973; and Griffin, 1973) have demonstrated the occurrence of Thunen type of landuse pattern. Jonasson (1925), Valkenburg and Held (1952), and Horvath (1967) have tested the theory at regional level in different parts of the world and have found its relevance. Chisholm (1966) has cited several studies which have been conducted at micro level, and have found Von Thunen type of zoning in landuse pattern around the settlements. In addition, several recent studies Blaikie, 1971 could also be cited in support of Von Thunen theory even today when physical, social, economic and technological conditions have changed to a considerable extent. **Dayal** concludes that Von Thunen theory is based on solid empirical foundation and logical reasoning that few would dispute.

But there are contrary views with regard to the relevance of Von Thunen theory. The influence of urban sprawl on spatial organisation of agriculture is as such that it tends to reverse the formation of Von Thunen type of agricultural zones. Sinclair (1976) argues that since the land near the city has a high potential for urban use, a farmer is reluctant to make investment in it with the fear that in near future he may have to sell his land to urban developers. Consequently, intensity in terms of inputs is less near the city and it increases with the distance from the city which leads to reversal of farming intensity zones. Sinclair's views are, however, not based on any measured attribute of agriculture, but are purely conjectural. Mattingly (1972) supported his views and arrived at some concrete conclusion while measuring agricultural intensity in USA. Grossman (1976) has also arrived at the same conclusion while studying the spatial organisation of landuse around a city settlement in the developing countries.

The third quarter of the 20th century is labelled as the period of quantitative revolution in geography. It is mainly because of two reasons: firstly, the research problems are increasingly becoming complex and they can often be solved only by handling observation materials which are huge and hard to grasp and interpret by simple description methods by quantitative techniques; and secondly, there is the old drive in science towards employing methods as objective as possible and producing results as accurate as possible. Hence, in all branches of geography statistical formulae have been applied to solve and analyse various associated problems. **Bielecka** *et al.* have examined the applicability of quantitative method in spatial analysis of agriculture. The utilisation of quantitative methods is usually a hard and complex task. It could be effectively employed if it is done in collaboration with the branch specialists — geographer, economist, planner etc. with the mathematicians in formulating and solving research problems. The former has to define the subject and aim of research together with the question the study is to answer, while the latter has to select the matching quantitative method to provide an adequate answer to the question. The approach would be more successful if the experts of both the branches have basic knowledge of each other's discipline. The authors have given a critical overview of the collaboration of research being carried out at the Polish Academy of Sciences. They have pointed out as to how the subject specialists have precisely outlined the various attributes and problems of their research and in response the mathematicians have recommended matching quantitative techniques to be employed. This has further been elaborated by taking the example of delineation of agricultural types on the basis of 27 criteria selected by Kostrowicki. They have critically examined the merits and limitations of *Ward, Orline, Farell* (Mad), *Gravity* and *Idwer* methods in this connection.

Landuse planning should be dynamic, flexible and capable of adopting the changing conditions. It requires up to date data which is only possible with landsat imageries. **Garg** and **Singh** have tried to analyse the landuse through digital image processing of landsat thematic mapper data in the Ranikhet Tehsil of Almora district. The authors have identified three major types of landuse i.e. land not available for cultivation, cultivable land, and land under cultivation. It would have been better if the authors have worked out landuse categories according to the scheme of National Remote Sensing Agency or Directorate of Economic and Statistics.

The study of agriculture and the disposition of the resources upon which agricultural development depends requires the deployment of a wide variety of methodologies of data acquisition and analysis. Rarely does one have enough data for a comprehensive review of renewable resources and very frequently, while some aspects are well covered others are deficient. However, new methodology becomes available through the development of a new technology which provides something better than is available to date. Remote sensing is an innovation which can be done in various ways. There are two distinct aspects of it viz. *qualitative* which deals with the recognition and interpretation of details in the area; and *qualitative* which involves precision measurements to obtain direct information related to shape, size and other dimensional features as well as derived information such as change statistical or associated parameters of inference. Although several systems have been developed during the past two decades for remote sensing, LANDSAT system is typical and best known amongst these. The latest of the series LANDSAT-3 has a resolution of µ 30 m which dictates the limit on its use for the work of man. **Ghosh** has discussed the geometry of this as one typical remote sensing system, and has presented its general data handling concepts. In addition to human elements, mapping involves in sensor data with the inherent problems: Ground with its diversities, and Map with its variation possibilities. However, the author has discussed in detail the process and techniques of Multispectoral Scanner System (MSS), Sensor-Ground relationship with modified collinearity condition, computational operations and Ground-map relationship with modified Universal Transverse Mercator's (UTM) projections.

Coppock has rightly remarked that absence of detailed authentic data about all the variables of agriculture is one of the major causes of the neglect of this branch. It is more so according to **Allan** in the developing countries like India, although here agriculture accounts for 50 per cent of the national income and engages three-fourths of the total population. The reasons for this deficiency are varied and complex. Besides these, the difficulties of obtaining available topographic maps, for example soil maps, can make the preliminary stages of a regional study in India protracted and on occasions unsatisfactory. Therefore, the availability of inexpensive satellite imagery can be of immense value to those carrying out such studies. The imagery is of such low resolution that it does not pose a security threat, yet at the same time it provides sufficient information to allow the delineation of major topographic features and the identification of natural resource disposition. A case study of Bulandshahr District in Western Uttar Pradesh is described which shows that satellite imagery permits the delineation of soil and other features much more quickly and effectively than by a protracted period of field work.

REFERENCES

Braidwood, R. J., 1958 : "Near East Pre-history", *Science*, Vol. 127, pp. 1419-1430.

Blaikie, P. M. 1971 : Organisation of Indian Villages, *Transaction of Institute of British Geographer*, No. 52, p. 15.

Childe, V. G., 1935 : *New Light on the Most Ancient East.*

Chisolin, M., 1966 : *Rural Settlement and Landuse.*

Crossley, J. C., 1976 : "Location of Beef Processing Industry", *Annals of the Association of American Geographers*, Vol. 66, pp. 60-75.

Griffin, E., 1973 : "Testing Von Thunen Theory in Uruguay", *Geographical Review*, Vol. 63, pp. 500-516.

Grossman, D., 1976 : "Von Thunen Rings and Contemporary Settlements : The Case of Nickland (Nigeria), in Gerasimov, I.P. (Ed), *Regional Geography.*

Horvath, R. J., 1967 : "Von Thunen an Urban Sprawl", *Annals of the Association of American Geographers*, Vol. 57, pp. 811-812.

Jonasson, O., 1924 : "Agricultural Regions of Europe", *Economic Geography*, pp. 277-314.

Kostrowicki, J., 1981 : 'A Hierarchy of World Type of Agriculture" in N. Mohammad (Ed) *Perspectives in Agricultural Geography*, Vol. I, Concept Publishing Co., Delhi.

Mattingly, P.F., 1972 : "Intensity of Agricultural Landuse Near Cities", *The Professional Geographer*, Vol. 24, pp. 7-10.

Muller, P.O., 1973 : "Trend Surface Analysis of American Agricultural Patterns : A Macro Thunen Analysis", *Economic Geography*, Vol. 49, pp. 228-242.

Prothero, R.M., 1957 : Land Use of Soba, Zaria Provinces, Northern Nigeria, *Economic Geography*, pp. 72-86.

Sauer, C. O., 1952 : *Agricultural Origin and Dispersal.*

Shafi, M., 1974 : "Assessment of Von Thunen's Land Use Analysis in India", *The Geographer*, pp. 1-13.

Sinclair, R., 1967 : "Von Thunen and Urban Sprawl", *Annals of the Association of American Geographers*, Vol. 57, pp. 72-87.

Van Valkenberg, S. and Held, C. C., 1952 : Europe.

Vavilov, N. I., 1949-50 : "The Origin, Variation, Immunity and Breeding of Cultivated Plants", *Chronica Botanica*, Vol. XIII, pp. 14-54.

Wissaman, H. V. et al., 1956 : "On the Role of Nature and Man on Changing the Face of Dry Belt of Asia" in Thomas, W.L. (Ed), *Man's Role in Changing the Face of the Earth*, pp. 278-303.

1

ORIGIN, DIFFUSION AND DEVELOPMENT OF AGRICULTURE

NOOR MOHAMMAD

"Nimble thoughts can jump both sea and land."

—Shakespeare

AGRICULTURE IS one of the oldest and most important occupations of man. It is generally understood to mean both the cultivation of plants and the raising of livestock. Hence, the origin of agriculture goes back to the beginning of domestication of plants and animals. Although man, plants and animals have always lived together — and have remained inseparably inter-connected with one another, the question of how, when and where man started domestication of plants and animals first has not been answered with a high degree of authenticity. Several scholars of different disciplines especially from anthropology, archaeology, biology, history, palaeobotany, genetics, ethnology and other branches of physical sciences have made efforts to unravel this mystery but have not reached a common conclusion as yet. They hold different views regarding the time and place of origin of agriculture. This paper is an attempt to ascertain the conditions which made man to domesticate animals and plants, the time of origin of agriculture and place or places of its origin with the help of the studies made from time to time by geographers and non-geographers in this connection. The paper also traces the evolution, diffusion and development of agriculture.

The history of mankind dates back to the hoary antiquity of more than a million years. Anatomically, modern man is the direct descendant of *Homo sapiens*. According to the time the earliest specialised plant-food processing tools (metates) and 10,000 years for the first appearance of agriculture (Carter, 1977). However, during the early periods of history, man's progress was at a snail's pace due to several physical and social factors but it was in the Stone Age (Palaeolithic) that man became an expert hunter and gatherer. Nevertheless, his life was largely a search for food; his camp sites were temporary and starvation was a constant threat (Brown and Finsterbusch, 1972). Hence, farming was, perhaps, developed to make the supply of food more regular and assured by selecting certain plant and animal species an taking special care of them. Agriculture is, therefore, an effort by man to go beyond the natural boundaries and to modify environment to better suit man's needs.

Man, perhaps, having experienced the better taste of certain wild plants, especially those which reproduce themselves by tuber or root stalk, might have tried to grow them near his dwelling (Candolle, 1967). The success of this experiment might have convinced him of the usefulness of their cultivation and naturally led to the repetition and subsequently to the improvement of its methods. Flannery (1968) termed this process as one of positive feedback in contrast to that of negative feedback where the seeds were merely gathered. This experiment was most probably done by societies after achieving a degree of sedentary life by intensive collecting and gathering of wild food (Candolle, 1967). As regards the first domestication of animals, various views are put forth but most probably women-folk and children living in the camp sites first started the domestication of those animals which either approached them in search of food, or were cubs captured by men-folk in hunting as pets or future decoys to be used in hunting. Sheep were domesticated first, followed by goats, pigs and cattle.

Several Theories

There are several theories which examine the circumstances leading to the origin of agriculture. Some of these theories have been examined thoroughly (Darlington, 1956; Wilsie, 1962; Janick et al., 1974). The concept of the origin of agriculture is an old phenomenon, but it seems to be a relatively recent innovation when viewed in terms of the historical development of mankind because man has been a hunter and gatherer for the greater part of his existence. Towards the end of the Pleistocene period, the palaeolithic man had a parasite economy, living on the bounty of nature as simple hunter and food-gatherer. Although he was quite capable of great achievements in technological and

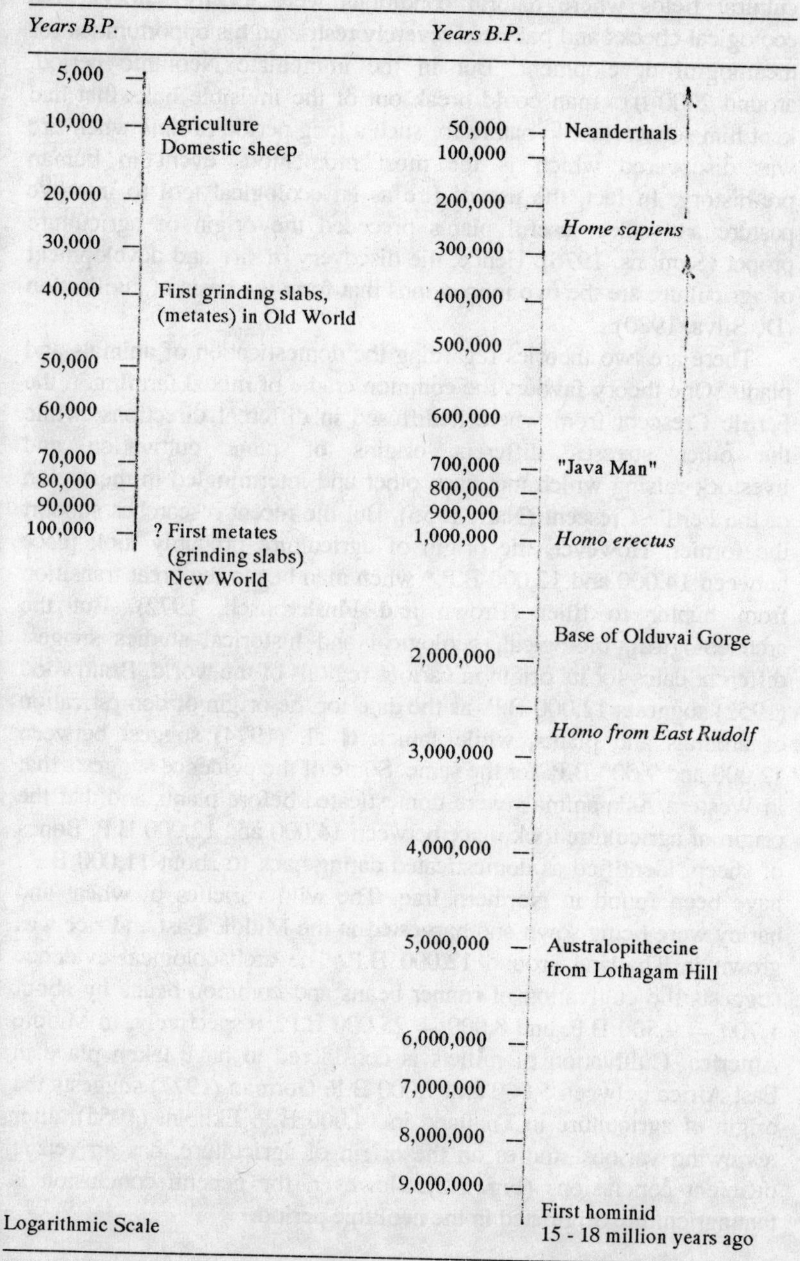

FIG. 1.1: Time scale of domestication.

cultural fields where natural conditions were locally suitable, the ecological checks and balances severely restricted his opportunities for meaningful development. But in the immediate Neolithic period, around 2000 B.C. man could break out of the invisible links that had kept him subservient to nature for such a long period of time when fire was discovered which is the most momentous event in human pre-history. In fact, the use of fire as an ecological tool to improve pasture and select useful plants preceded the origin of agriculture proper (Symons, 1976). Hence, the discovery of fire and development of agriculture are the two innovations that form the basis of civilisation (De Silva, 1980).

There are two theories regarding the domestication of animals and plants. One theory favours the common cradle of mixed farming in the Fertile Crescent from where it diffused in different directions; while the other stresses different origins of plant cultivation and livestock-raising which met each other and intermingled in the region of the Fertile Crescent (Narr, 1956). But the recent researches support the former. However, the origin of agriculture probably took place between 14,000 and 12,000 B.P.* when man began the great transition from hunter to tiller (Brown and Finsterbusch, 1972). But the archaeological, biological, geological and historical studies suggest different dates for its origin in various regions of the world. Braidwood (1952) suggests 12,000 B.P. as the date for the origin of domestication of animals and plants, while Janick et al. (1974) suggest between 12,000 and 9,000 B.P. for the same. Some of the evidence suggests that in Western Asia animals were domesticated before plants and that the origin of agriculture took place between 14,000 and 12,000 B.P. Bones of sheep, identified as domesticated dating back to about 11,000 B.P., have been found in Northern Iraq. The wild varieties of wheat and barley were being sown and harvested in the Middle East and rice was grown in Thailand around 12,000 B.P. The archaeological evidence suggests the cultivation of runner beans and common beans by about 1,700 — 9,500 B.P. and 8,000 — 25,000 B.P., respectively, in Middle America. Cultivation of millets is considered to have taken place in East Africa between 5,000 and 7,000 B.P. Gorman (1977) suggests the origin of agriculture in Thailand in 14,000 B.P. Ekholm (1955), after reviewing various studies on the origin of agriculture, has arrived at different conclusions (Fig. 1.2). However, the general conclusion is that agriculture originated in the neolithic period.

B.P. — Before Present (1950 as counting year).

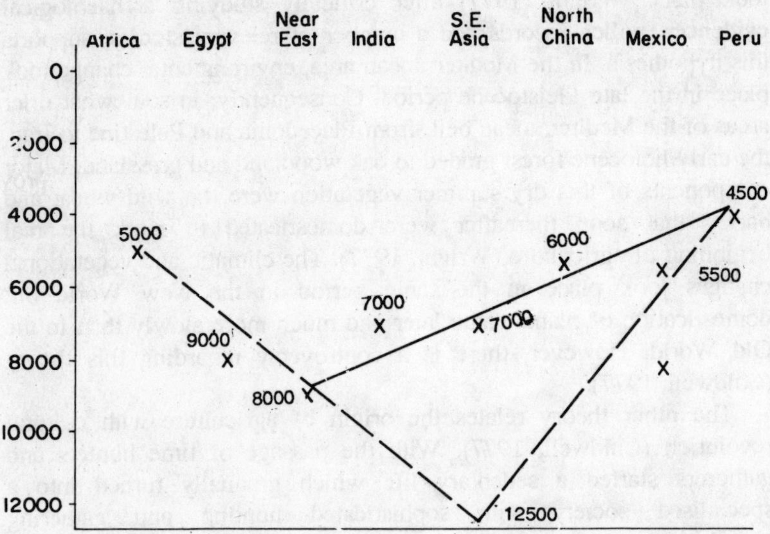

FIG. 1.2: **Agricultural origins in time and space.**

But the question arises as to why the origin of agriculture took place during this period particularly when man, animals and plants had been living together on the surface of the earth for 250,000 years. Three theories are put forth in this connection:

(i) Environmental change and origin of agriculture,
(ii) Cultural changes and origin of agriculture, and
(iii) Population pressure and origin of agriculture.

The natural environment undoubtedly exerts great influence on human life and conditions every social, cultural and economic situation. The history and substance of various forms of environmental determinism have been discussed by various scholars (Thomas, 1925; Sorokin, 1928; Hartshorne, 1939; Freeman, 1961; Brookfield, 1969). However, Childe (1929) presented his unified theory of environmental determinism in cultural history with particular reference to the origin of agriculture. This theory was based on inferences derived from glacial and pollen sequences, indicating the change of climate in the Near East from the moist to the semi-arid climate of today, forcing animals, plants and man to congregate at oases where domestication took place. Wright (1977) after critically studying archaeological evidences, pollen records and a number of relevant studies, supports this hypothesis. In the Mediterranean area, environmental change took place in the late Pleistocene period. Consequently, in somewhat drier areas of the Mediterranean belt, from Placedonia and Palestine to Iran, the early holocene forest graded to oak woodland and grassland. Major components of this dry summer vegetation were the wild wheat and barley that soon thereafter were domesticated to mark the real beginning of agriculture (Wright, 1977). The climatic and vegetational changes took place in the same period in the New World but domestication of plants came later and much more slowly than in the Old World. However, there is a controversy regarding this theory (Caldwell, 1977).

The other theory relates the origin of agriculture with cultural revolution (Caldwell, 1977). With the passage of time hunters and gatherers started a sedentary life which gradually turned into a specialised society with sophisticated hunting and gathering techniques. These societies underwent various changes with environmental change and gradually developed an intensive and pragmatic interest in wild plants and their properties. Eventually someone would have developed the technique of food production. This might have been followed by others and thus the agriculture diffused in other areas. Archaeological sites of Tamanlipas and Tehuacan (Mac Neish, 1958), Jarmo and Deh Luran, and Cayonu show the combination of wild and domesticated plants of similar species.

The third theory relates the origin of agriculture with the pressure of population. Man with his age-old experience of hunting and gathering was aware of the characteristics of wild plants and animals. He had some technical know-how particularly regarding harvesting of wild crops and hunting of animals. He was aware of the value of fire in promoting the growth of preferred species. When the pressure of population on his existing resources increased and his means of sustenance started decreasing quantitatively and qualitatively, he had to look for other alternatives. Among the various alternatives, domestication of animals and plants seems to be the most suitable, given his knowledge of the characteristics of wild plants and animals and the level of technology he had at that point in time. This hypothesis is supported by the evidence obtained from the village sites of Wade-un-Natuf and Kabaran. Recent discoveries at Spirit Cave

(Thailand) also support this hypothesis (Gorman, 1977).

However, the origin of agriculture is an evolutionary process rather than an event. It has taken thousands of years to evolve. There is spatio-temporal variation in its evolution. Therefore, it is quite likely that all these theories are correct and have relevance to the spatio-temporal evolution of agriculture.

Like the date, precisely where agriculture originated first is also a matter of opinion. There are several views regarding the place of origin of agriculture. The cultural historian Childe (1935) and the anthropologist Braidwood (1952, 1958) are of the view that river valleys and loess lands were the primary hearths of agriculture, and such regions are in China, South-West Asia and inter-tropical America. This hypothesis has a number of archaeological and botanical evidences in its support. But this view has been rejected by the geographers especially by Sauer (1952) and Wissman et al. (1956) on the ground that such areas are always liable to flooding and drought conditions.

Sauer considered the following six basic premises regarding the place of origin of agriculture:

(i) Agriculture did not originate in communities desperately short of food but where there was sufficient freedom from want to experiment,

(ii) The hearths of domestication are to be sought in regions of marked diversity of plants and animals,

(iii) Primitive agriculture began not in the large river valleys subject to lengthy floods and requiring protective dams, drainage or irrigation, but in the hilly lands,

(iv) Agriculture originated in woodlands which have soils easy to dig,

(v) The pioneers of agriculture had previously acquired special skills, but hunters would be least inclined towards domestication, and

(vi) The founders of agriculture were sedentary folks. Crop cultivation requires attention. Unless constantly guarded, the crop will be lost.

Hence, according to Sauer, agriculture originated in the tropical regions, most probably in South-East Asia as here most of his basic premises are fulfilled. No other area is equally well suited or equally well furnished for the rise of a fishing-farming culture (Harris, 1969). The earliest domestication of animals — dog, pig, fowl, duck and goose — took place as animals of the household. From here agriculture diffused northward in China and Japan, north-west in India, through the Indian Ocean to Africa and through the Mediterranean lands to Europe (Fig. 1.3). Similar gene-centres may be considered in tropical America, especially in Mexican-Central American border, where some of his premises are fulfilled. From this gene-centre also agriculture diffused towards north in the USA and towards south in several Latin American countries. Although, earlier, there was little evidence in support of this hypothesis, recent archaeological researches fully support his views.

Vavilov a Soviet biogeographer (1949-50), has made a significant contribution to the modern knowledge of the main centres of the origin of agriculture. He, along with other Russian scientists, made expeditions throughout the world during 1916–1934 and collected a lot of material regarding plant species which was analysed later. Finally, on the basis of differential phyto-geographic technique involving a number of logical steps and procedures, he listed six and two independent centres of origin of agriculture in the Old World and New World, respectively (Fig. 1.4). This research is duly supported by archaeological findings. On the basis of their nature and characteristics he has classified the important plant species of each gene-centre into various categories as given in Table 1.1.

Gene-centres of the Old World

Although there is a dispute regarding the first gene-centre in the Old-World, there is a common agreement among the scholars of various disciplines till date that the domestication of animals and plants started first in one of the six gene-centres of the Old World discussed below:

The Near East Gene-centre

This gene-centre extending from east to west includes parts of Greece, Crete, Turkey, Iran and Turkistan (Fig. 1.4). Vavilov has listed 83 varieties of plant species in this region, out of which 24 belong to cultivated fruits, 18 to cereals and other grains, 15 to melon plants and the remaining to oil plants, forage plants, spice and stimulant plants,

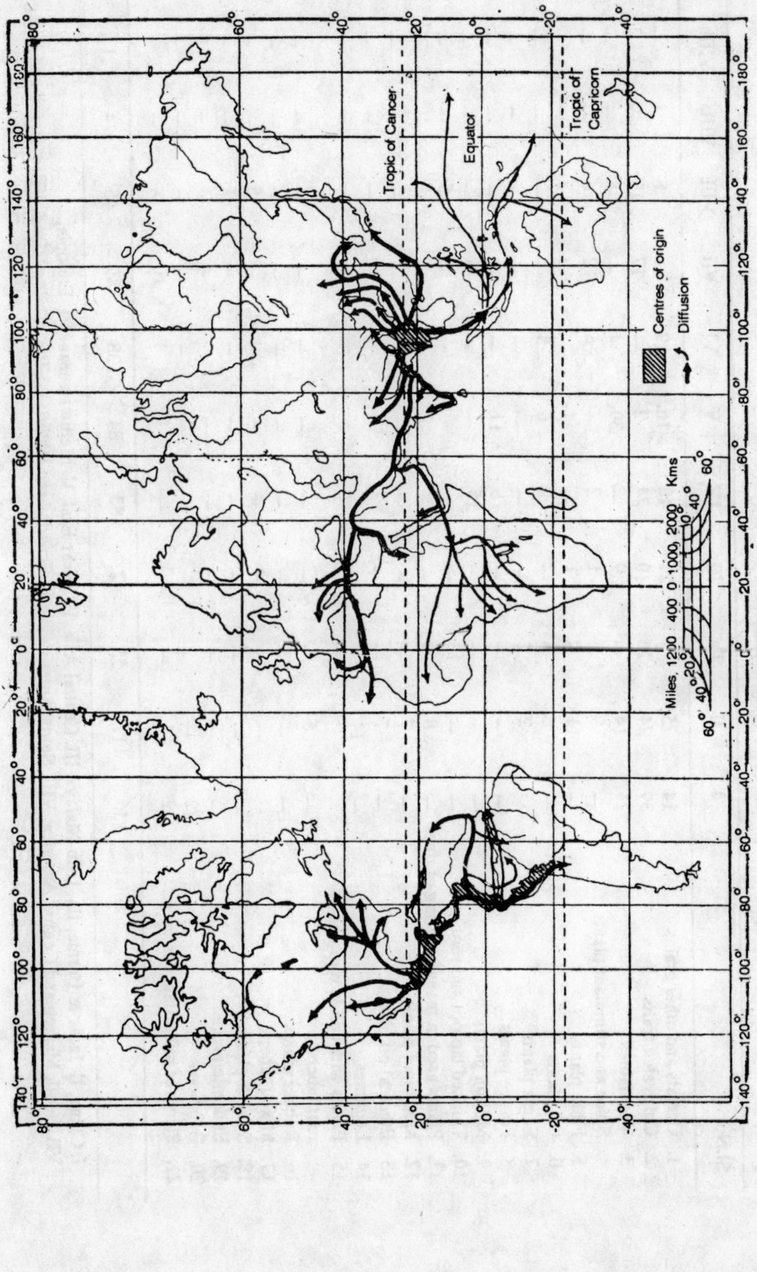

FIG. 1.3: Dispersal of agriculture (after Sauer).

TABLE 1.1: Number of plant species in gene-centres of agriculture

Sl.No.	I	II	IIa	III	IV	V	VI	VII	VIII	VIIIa	VIIIb	Total
1. Cereals and other grains	14	5	1	21	18	16	19	8	8	4	—	114
2. Cultivated fruits	39	24	26	10	24	2	—	17	13	—	7	162
3. Vegetables	24	14	—	9	—	30	4	—	5	—	—	86
4. Spice and stimulant plants	—	9	5	1	2	—	6	2	4	—	—	29
5. Fibre plants	—	11	2	1	—	1	—	3	2	—	—	20
6. Oil plants	—	6	4	—	9	6	5	—	—	—	—	30
7. Sugar plants	35	2	2	—	—	—	—	—	—	—	—	39
8. Forage plants	—	—	—	—	8	11	—	—	—	—	—	19
9. Million plants	—	—	—	—	15	—	—	6	—	—	—	21
10. Thicked taproot and root tuber	—	—	10	—	—	—	—	3	5	—	—	18
11. Plants used in manufacturing of dyes	—	8	1	—	—	2	—	—	—	—	—	11
12. Medicinal plants	—	7	—	—	—	—	—	2	—	—	—	9
13. Ethereal oil plants	—	7	—	—	7	15	—	—	—	—	—	29
14. Legumes	—	14	1	—	—	—	—	—	—	—	—	15
15. Plants with edible root and root tubers	—	6	—	—	—	—	—	—	—	—	—	6
16. Rubber plants	—	—	1	—	—	—	—	—	—	—	—	1
17. Mixed tubers	—	—	—	—	—	—	—	—	5	—	—	5
18. Squash plants	—	—	—	—	—	—	—	—	1	—	—	1
19. Endemic plants	—	—	—	—	—	—	—	—	—	—	6	6
20. Bamboos	22	—	2	—	—	—	—	—	—	—	—	24
21. Plants of miscellaneous uses	2	4	—	—	—	1	4	10	—	—	—	21
Total	136	117	55	42	83	84	38	49	45	4	13	666

I. China, II. India & Burma, IIa. Indo-Malaya, III. Central Asia, IV. Near East, V. Mediterranean, VI. Ethiopia, VII. South Mexico and Central America, VIII. South America (Peru, Ecuador and Bolivia), VIIIa. Chiloe Island, VIIIb. Brazil and Paraguay.

ethereal oil plants, resin and tannin plants. It was the pulsating heart of the Neolithic Revolution when the origin of agriculture is believed to have taken place first. This and the Mediterranean region are considered to be the most likely places of man's maiden experiment with agriculture and it goes back to about 12,000 B.P. All other agricultural beginnings are not only later in time but are progressively later as one goes away from this hearth area (Ekholm, 1955). The only possible challenge is by the South-East Asian centre proposed by Sauer on philosophic grounds and possibly to be supported by the finds from the Spirit Cave (Thailand).

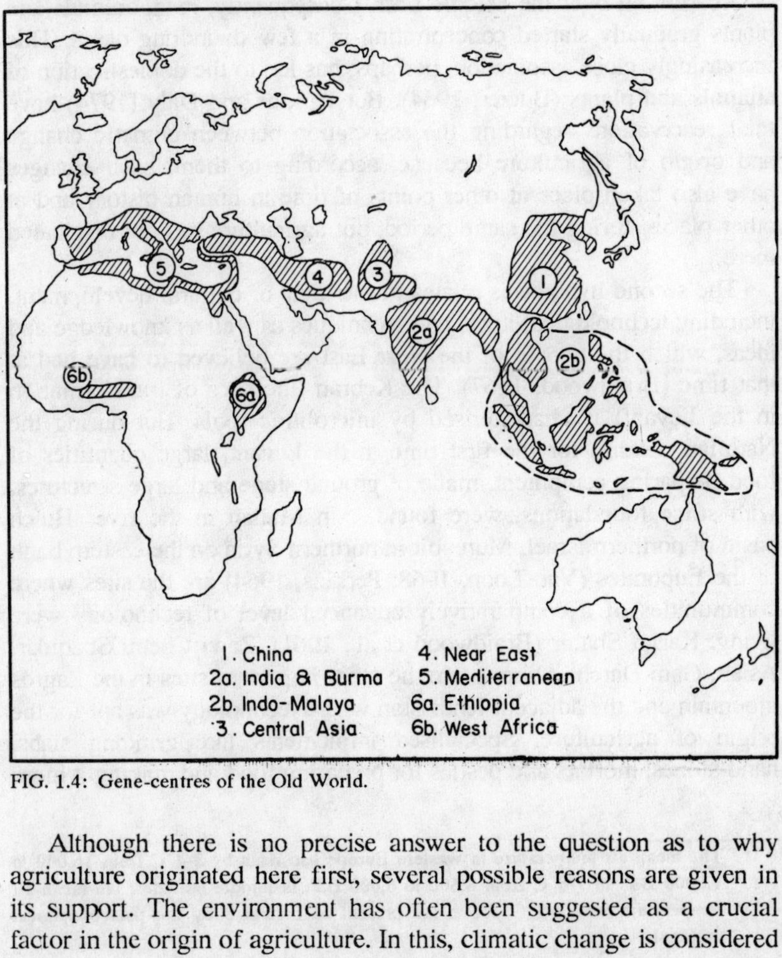

FIG. 1.4: Gene-centres of the Old World.

Although there is no precise answer to the question as to why agriculture originated here first, several possible reasons are given in its support. The environment has often been suggested as a crucial factor in the origin of agriculture. In this, climatic change is considered to be a primary force in the introduction of agriculture here. The Carbon-14 datings and palaeotemperature indicators quantifying climatic phenomena indicate a rise in air and sea water temperature during the late Pleistocene period* (Fairbridge, 1967).

During the glacial period, the present rainbelts were in the south. The cyclonic storms then were traversing the Mediterranean areas, northern Sahara, Lebanon, Mesopotamia, Saudi Arabia, Iran and India, with the result these area had humid climate suitable for growth of plants and development of animal species. But the retreat of glaciers shifted the rainbelts to almost their present position resulting in desiccation all over the Middle East. Consequently, men, animals and plants gradually started concentrating in a few dwindling oases. This increasingly close association, perhaps, has led to the domestication of animals and plants (Butzer, 1964). But Parkins and Daly (1974) have their reservations regarding the association between climatic change and origin of agriculture because, according to them, such changes have also taken place at other points of time in human history and at other places during the same period, but agriculture has not originated there.

The second hypothesis relates to the level of cultural development, including technological items and techniques as well as knowledge and ideas, which the people of the Near East are believed to have had at that time (Braidwood, 1967). The Kebran (the sites of transformation in the Levant) is characterised by microlithic tools. But during the Natufian culture, for the first time in the Levant, large quantities of food-preparing equipment, made of ground stone and large structures, with stone foundations, were found. Ain Mallah in the river Hulch basin of northern Israel, Mureybit in northern Syria on the eastern bank of the Euphrates (Van Loon, 1968; Perkins, 1964) are the sites where communities of a comparatively advanced level of technology were living. Karim Shahir (Braidwood et al., 1961), Zawi Chemi Shanidar, Asiab, Ganj Dareh, Alikosh, are the other important sites in the Zagros mountain and the adjacent Khuzistan where technology was apt for the origin of agriculture. Specialised implements like grinding slabs, hand-stones, mortars and pestles for preparing food and making it more

* **The mean air temperature in western Europe had risen by 3-4°C from 16,000 to 12,000 B.P. to 7-8°C from 8,600 to 6,000 B.P. in middle latitudes; the mean air temperature was 2-5°C above the present as indicated by the palaeobotanical material.**

edible are found in these sites. Specialised harvesting tools, such as sickle blades and composite sickle were also found. Besides, various kinds of containers — stationary storage sites and basins, baskets and skin containers — have been discovered in these sites which are reasonable for the conditions set for the origin of agriculture in this region (Redman, 1977).

The third major force which is often discussed is social organisation related to subsistence pursuits and prolonged sedentarism. The communities living at Ain Mallah, Mureybit, Karim Shahir, Asiab, Ganj Dareh, Ali Kosh, Jarmo (Braidwood, 1968) and Tape Guran offer a good life-style of the period between 13,000 and 12,000 B.P. in these areas. There is evidence of domesticated animals and cereal grains, and probably farming was established in this area.

The fourth basic source of shift over to food production may have been the growth of population, which might have brought a significant change in the economic strategies. It is true that increasing pressure of population on existing resources forced man to find new resources and to find new methods of exploiting existing resources. Although there is no forceful evidence to support the rapid growth of population in the Near East around 12,000 B.P., it is quite possible that in the nuclei of population of this centre, the population had increased to a point where existing resources of the nucleus at that level of technology failed to provide sustenance. Consequently, people started domestication of animals and plants to increase and ensure their means of sustenance. This process has been explained by Reed (1977) through the pattern of a series of adaptive cultural plateau (Fig. 1.5).

1. Continued increase in population; fissioning of village populations; increase in number of villages; cultivation and incipient domestication of cereals; incipient domestication of sheep and goats; stone-based rectangular houses; incipient development of tribes.
2. Storage of grain; initial sedentarism; increase in population; stone-based round houses.
3. Worldwide warming; spread of cereal grasses and their use for food into South-western Asia; spread of use of sickles and grinding stones.
4. Hunting of game; broad-spectrum hunting, gathering; local use of grinding stones; microliths in South-western Asia; local use of sickle blades in Egypt; society organised in bands.

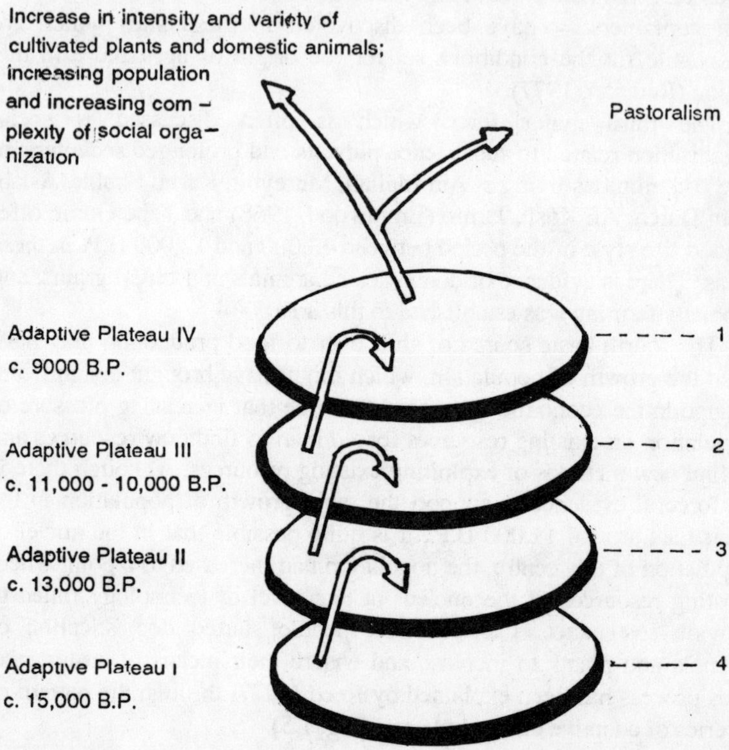

FIG. 1.5: Process of origin of agriculture (after Reed, C.A.)

However, the first archaeological evidence was for the domestication of animals and plants (cereals) from this region (Perkins, 1964). Archaeological evidences suggest that agricultural origin took place before 9,000 B.P. at Cayonu in southern Turkey and the Deh Luran plain of Khuzistan. Wild plants from which wheat and barley were domesticated are found here (Harlow and Zohary, 1966; Zohary, 1969). People who relied upon hunting and gathering were reaping wild wheat and barley with knives, grinding the grain and using storage pits (Kenyon, 1969). There is evidence of village communities growing wheat and barley and keeping sheep and goats in Greece and Crete in the West, in southern Turkey, in the Zagros mountains of Iran and Iraq, in the interior plateau of Iran and in the foothills south-east of the Caspian (Braidwood and Howe, 1962; Mellart, 1964, 1967; Rodden, 1965 and Kenyon, 1969). Gradually the number of domesticated plants grown was increased, including flax, peas, lentils and vegetables.

Although cattle and pigs had much broader distribution than wild sheep and goats, it is only in this region that the wild progenitors of all the four were found together (Fig. 1.6). Between 11,000 and 9,000 B.P. livestock-raising was very simple. Sheep are believed to have been domesticated first at Zawi Chemi Shanidar in the north-eastern end of the mountain chain. Between 9,000 and 8,500 B.P., cattle and pigs (at Cayonu) enter the husbandry which spread all over the region. Pigs could not get much popularity because of several reasons but sheep, goats and cattle became quite popular by 8,000 B.P. These three primary domesticates having their own valuable secondary products — wool from sheep, milk from goats and milk and hides from cattle — continued to flourish. Thus, a mixed herding-farming economy came into existence.

The earliest farmers of the region had reaping knives, sickles and a simple method of grinding grain. They probably had weighted digging sticks and, later primitive hoes, made at first from wood and later from stone (Butzer, 1964). Since the rainfall in the region was low, the yield of crops was also low which might have led to shifting cultivation. But continuous occupation of many excavated sites suggests that the occupants had realised the value of allowing the livestock to graze the arable stubble and possibly the value of fallowing to conserve the moisture. However, the most important change was, perhaps, the slow adoption of the ox-drawn plough which was invented by the farmers of the region around 6,000 B.P. This wooden ard could do little more than scratch the surface of the earth. Cross-ploughing is believed to have been practised on the flood plains in order to conserve the soil moisture in dry farming areas.

The lowlands of the Tigris and Euphrates have insufficient rainfall even for dry farming but the annual deposition of silt must have been a major factor in maintaining soil fertility, resulting in higher yield. But the major advance in agriculture came with the development of irrigation, although it was very difficult because floods used to come in late spring and early summer, which was neither useful for spring nor for autumn sown crops. Thus, embankments to control the floods, canals and some other form of storage would have been necessary. However, water-lifting devices, such as *shadoof*, water-wheel, etc., were present (Hamdan, 1961). New varieties of crops were added. Six-row barley replaced two-row and bread wheat joined *emmer* and *einkron*. Da'epalm and several vegetables were also grown (Clark, 1957).

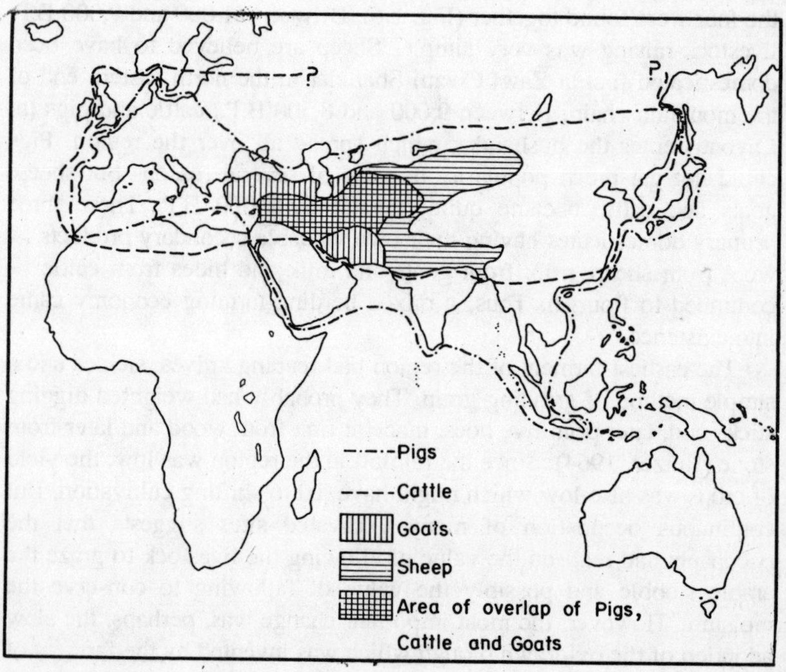

FIG. 1.6: Distribution of wild animals.

However, from here agriculture diffused (in all directions) to new areas as well as to other gene-centres of agriculture at different points in time.

The Mediterranean Gene-centre

The Mediterranean gene-centre stretches in the coastal areas all around the Mediterranean Sea. It includes the coastal areas of Spain, France, Italy, Austria, Cyprus, Syria, Lebanon, Egypt with the Nile Valley, Libya, Tunisia and Morocco (Fig. 1.4). Vavilov has listed altogether 84 major species of plants which were found here. Out of this 30 belong to vegetables, 16 to cereals and other grains, 15 to ethereal oil plants, resins and tanning plants, 11 to forage plants and the remaining 12 to other categories of plant species. In fact, in most of the studies, Syria, Lebanon, Libya and Egypt are considered to be the part of the Near East region as there had been and there still is close similarity in the physical and socio-economic conditions and the level of technological advancement in these countries. Therefore, agriculture must have originated in the period it originated in the Near East. The settlement pattern during the Natufian is a clear indicator of this fact. Agriculture originated in the Levantine countries in 8,000 B.P. From here it diffused towards the west in Egypt in 7,000 B.P. but above the flood plains of the rivers. By the middle of the sixth millennium B.P. the farmers gradually occupied the Nile flood-plains and began to use the annual floods to irrigate their crops. Due to inadequate rainfall, dry farming, possibly based on shifting cultivation with wheat and barley as the main crops, was practised in the upper areas, while in the Nile Valley irrigation farming would have been practised. Irrigation of a primitive manner was practised in the beginning but later the construction of embankments to control floods, and *saqias* (canals) to regulate water in distant fields was begun. Shadoof and water-wheel were also in use (Hamdan, 1961). However, agriculture from here diffused westward along the north African coast in Maghreb in the sixth millennium B.P.

Goats seem to have been the first widespread domestic food animal in Egypt. Sheep, dogs, pigs, donkeys, geese, ducks, etc., were domesticated later.

The excavations, especially the recent ones in northern Greece and the southern Balkans, provide an interesting and significant evidence about the origin of agriculture. According to Rodden (1965) the earliest widely known Neolithic sites at Nea Nikomedeia on the Macedonian plain in Europe provide an evidence of the remains of crops and animals dating 8,000 B.P., such as wheat, barley and lentil, and sheep, goats and pigs, respectively. From this gene-centre, agriculture spread westward into Europe and North Africa. Although domestication of plants and animals has its origin as early as that in South-West Asia, agriculture diffused from here towards the north into the Danube basin and, thence, north-westward towards the Baltic and North Sea following the fertile loess soils lying in the north of the Hercynian mountains in Central Europe. An offshoot of the Danubian culture penetrated the Ukraine steppe (Murray, 1970). It further diffused towards north-east and possibly reached the shores of the North Sea, Baltic and British Isles in the sixth millennium B.P. and the southern edge of the coniferous forest in Scandinavia and Russia around 4,500 B.P. Another wave moving towards west reached Spain and southern France by about 6,000 B.P. (Whitehouse, 1968). However, olive, fig and vine which gave the Mediterranean much of its distinctiveness had been cultivated by 4,000 B.C. Vegetables such as atrichokes, asparagus, cabbage, celery, olive, cress, beet, garlic, leek, onion, parsnip, peas, etc., have their origin in this gene-centre.

The Chinese Gene-centre

Vavilov considered the Chinese gene-centre as the largest independent centre of the world's agriculture lying in the north-eastern part of China (Fig. 1.4). He has listed 136 chief endemic plants which have been grouped into six major categories, viz., cereals and other grains, bamboos, cultivated fruits, sugarcane and plants of miscellaneous use containing 14, 22, 24, 39, 35 and 2 varieties of plant species, respectively.

The first known farmers lived in the loess uplands in the middle Hwang Ho, the Fen Ho and the Wei Ho in the eighth and seventh millennia B.P. The core of this gene-centre lies in the provinces of Hunan, Shensi and Shansi (Ho, 1969). The Pan Po of the Yang-shao culture found in Shensi represents the earliest proven phase of field agriculture, based largely on millets, of annual domestication centred mainly on pigs, of settled village communities with well-patterned grain-yards, of painted potteries and of the archetypal Chinese script and numerals (Ho, 1975). Contrary to the typical Western idea, i.e. agriculture diffused in China from the Near East and wheat, barley, sheep, goats and cattle were acquired from there, various studies related to domesticated plants — millets, soyabean and mulberry — indicate that these crops are botanically different from those of the ancient West and are indigenous plants of this region. The cultivation of millet is suggested first on the basis of several finds of jars filled with the husk of Setaria-panicum at Pan Po in Shensi. The distinctively different cropping system, the peculiarities of the natural environment of the loess highland, the freedom from the influence of great flood plains and the absence of irrigation, all testify to the indigenous origin of a Chinese field agriculture. Until the middle of the first millennium B.P., when the ox-drawn plough became more common, the chief implements were digging sticks, hoes and spades.

Unlike western agriculture where animal husbandry constitutes a vital part of agriculture, as does grain production, in China agriculture meant essentially, if not exclusively, field agriculture based on cereal grains. Animal husbandry in pre-historic and early historic China has played a subordinate role in Chinese agricultural system (Ho, 1977). The archaeological evidence from the early historic sites in Mongolia, Shansi, Shensi, Kansu suggest that pig is the first and widespread domesticated animal followed by dog (Fig. 1.7). The domestication of cattle, sheep, horse and chicken was insignificant.

The South-East Asian Gene-centre

South-East Asia is one of the largest gene-centres, covering the Indian Peninsula, Bangladesh, Assam, Burma, Thailand, Indo-China and Malaysia (Fig. 1.4). Vavilov has divided it into two sub-regions, i.e. the Indian centre excluding north-western India, but including Burma, and Indo-Malaysian centre including the rest of the countries. He has identified 117 major plant species in the former and 55 plant species in the latter. Out of this, cultivated fruits alone account for roughly 30 per cent. The other major groups of plant species are vegetables, legumes, oil plants, spices and stimulants and fibre plants, accounting together for more than one-third of the total species.

Sauer believes that this is the oldest centre of agriculture from where it has diffused in various directions as given in Fig. 1.3. Western scholars have proved South-West Asia as the oldest centre due to lack of archaeological evidences in this region. But recent excavation sites disclose interesting results. The agricultural revolution seems to have been achieved independently thousands of miles away in South-East Asia. This separate agricultural revolution involved plants and animals for the most part unknown in the Middle East and it may have begun 5,000 years earlier (Solheim, 1973). Hoabinhian sites (Gua Kechil, Sai Yok, Laang Spean, Padah Lin and Spirit Cave) suggest that an early South-Eastern Asian techno-complex first appeared during late Pleistocene and continued till 7,000 B.P. In times, the Hoabinhian complex underwent modification and gained new cultural elements,

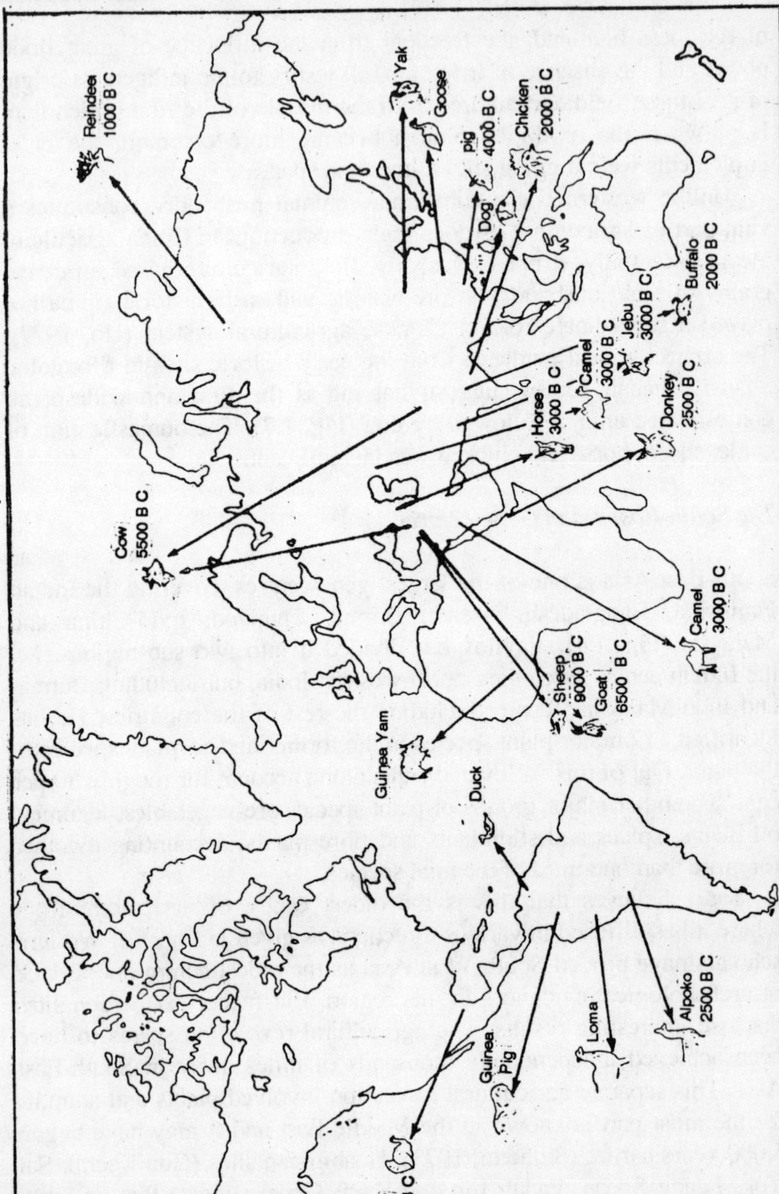
Reindeer
1000 B C
Yak
Goose
pig
4000 B C
Chicken
2000 B C
Dog
Buffalo
2000 B C
Jebu
3000 B C
Camel
3000 B C
Horse
3000 B C
Donkey
2500 B C
Cow
5500 B C
Camel
3000 B C
Sheep
9000 B C
Goat
6500 B C
Guinea Yam
Duck
Loma
Alpoko
2500 B C
Guinea
Pig
Turkey
3000 B C

including early domestication of plants, edge grinding of tools and various ceramic styles. Recent excavations of post-Hoabinhian sites (Solheim, et al., 1966; Watson and Loof, 1967) indicate a techno-complex different from Hoabinhian. It suggests piedmont and lowland settlements generally with a subsistence pattern based on rice culture. At Spirit Cave and Non Nok Tha in Thailand, latest excavation sites and the occurrence of small slate knives further support that these terminal, intrusive materials are in context with rice agriculture (Gorman, 1973). Cultivated rice is derived from two wild varieties which were found in marshy sites throughout the Indian subcontinent, South-East Asia and the Philippines. It could thus have been domesticated independently in several places within this region. Upland rice and wet rice are distinct. The former is a grain like any other cereal and relies on rainfall, while the latter is grown on lowlands and has to be partially submerged in water for most of its growing period. It is believed that upland rice was domesticated first and wet rice later as the development of the elaborate techniques associated with its cultivation were developed later. However, in addition to vegetation, the people of this region mostly relied upon hunting, gathering and fishing for much of their food supplies.

In the Indian subcontinent agricultural beginnings took place around 4,700 B.P. Until the discovery of the contrary, the practice of agriculture in India especially in north-western India, entered from outside, most probably from South-West Asia, although the humid tropics may have been part of the South-Eastern Asiatic agricultural gene-centre, and as such have had house gardens and root crops which left no archaeological evidence (Harlan, 1971). India's contribution to the agricultural origins has obviously not been much but, at present, we can say that sorghum, pearl millet, sesamum, buchnania, dolichas biflorus, etc., were cultivated or consumed first in the Indian subcontinent (Mittre, 1977).

The Central Asian Gene-centre

The central Asian gene-centre is comparatively small and consists of north-western India, Pakistan and Afghanistan, Soviet Republic of Tudjakistan, Uzbekistan and western Tian Shan. Vavilov has identified 42 major endemic plant species of which 50 per cent are cereals and other grains. In this region agriculture origin took place between the seventh and fifth millennium B.P. (later than in South-West Asia). The archaeological evidence in Turkistan suggests the existence of an agricultural community based on irrigation. They adopted mixed agriculture based on the combination of crops and livestock which characterises Mesopotamian agriculture (Lewis, 1966). Agriculture, in fact, gradually diffused toward east and south-east. The impact of chinese gene-centre on the agriculture of the region is also visible. The plant economy comprising wheat, barley, peas, lentil and flax was derived from Western Asia where these were cultivated earlier. However, peas, carrots, flax, almonds, onions, grapes, walnut, alfalfa, garlic, water-melon, radish, etc., were domesticated here.

The East Ethiopian, West African Gene-centre

Opinions differ regarding the origins of agriculture in Africa, South of Sahara. It has been claimed that there were two centres of independent plant domestication in Africa, south of Sahara: one in the western Sudan zones and the other in eastern Ethiopia. Vavilov, on the basis of his expedition to Abyssinia, Eritrea and Somaliland and subsequent comparative study of large amounts of collected material, has considered Ethiopia as an independent centre of origin of the world's cultivated plants. He has considered the origin of 19 species of cultivated grain, four species of vegetables, five species of oil plants, six species of spices and stimulants, four species of vegetables and four species of plants of miscellaneous uses in this region. Murdock (1959) supporting the West African gene-centre, argued that the Mande speaking people of middle Niger domesticated a number of indigenous plants including millets. Murdock further assumed that these Negro people first advanced from a hunting and gathering economy to agriculture and, at the same time, they made much geographical expansion and progressed innovatively at the cost of their backward neighbour. This is, perhaps, the reason for the expansion of their languages over unusually vast territories. Anderson (1960) and Porters (1950-51 and 1962) also support this hypothesis. Other scholars two are, however, of the opinion that agriculture reached Africa, south of Sahara, Egypt and north-west Africa. During the sixth millennium B.P., Sahara was moister than at present and was occupied by pastoral nomads. The inhabitants may have practised some incipient agriculture. Hence, these people might have moved towards south and started agriculture there. Pastoral people along with sheep, goats and cattle from Arabia have moved into eastern Africa. They might have grown grain crops in this region. Sheep, goats and cattle were domesticated in these regions but were not integrated with crop products (Fig. 1.7). West Africa is still one of the few areas of the world where root crops form a major part of agricultural economy (Murray, 1970). Tropical Africa is considered to be the primary generic centre of castor-beans, cotton, water-melon, rice, coffee, oil palm, cowpea and walnut and secondary generic centre for many varieties of barley and oats.

Gene-centres of The New World

Agriculture began much later in the New World, perhaps, because man traversed this area much later, but well before agriculture originated in the Old World. Thus, agriculture in this region is an independent development (Lathrap, 1977). Man reached the Americas before the invention of boats which implies that he came over here crossing the Bering Straits either between 30,000 and 25,000 B.P. or between 16,000 and 12,000 B.P. as is evident from the glaciological evidences. Because, after that the flood of melt-water at the end of the last Ice Age raised the sea level again by several hundred feet and thereby turned the key on the inhabitants of the New World (Bronowski, 1976). Thus, man came to the Americas after 30,000 B.P. and before 12,000 B.P. and would not have started agriculture immediately. The newcomers most probably entered the New World through the Mackenzie river valley taking advantage of an ice-free corridor which led them to the great plains and the big game that flourished in virgin lands till that time (Solecki, 1974). However, there were contacts between the New and Old Worlds even before Columbus (Grigg, 1977), but it did not affect the agriculture of the former. Therefore, there were major differences between their respective agricultures until the arrival of the Europeans. First, the crops, indigenous to the New World, were quite different from those cultivated in the Old World. Squash, maize, beans, manioc, potatoes and groundnuts were the major food crops. Secondly, ilama, alpaca and turkey were the only animals domesticated in the America which were not found in the Old World (Fig. 1.7). Thirdly, there were no ploughs in America but *coa* which was more akin to the digging stick than to the hoe. But there were some common techniques, such as slash-and-burn, irrigation, terracing and the use of Ilama dung for manure (Adams, 1962). However, two gene-centres have been identified, i.e., South Mexico (Central

America) and South America (Fig. 1.8).

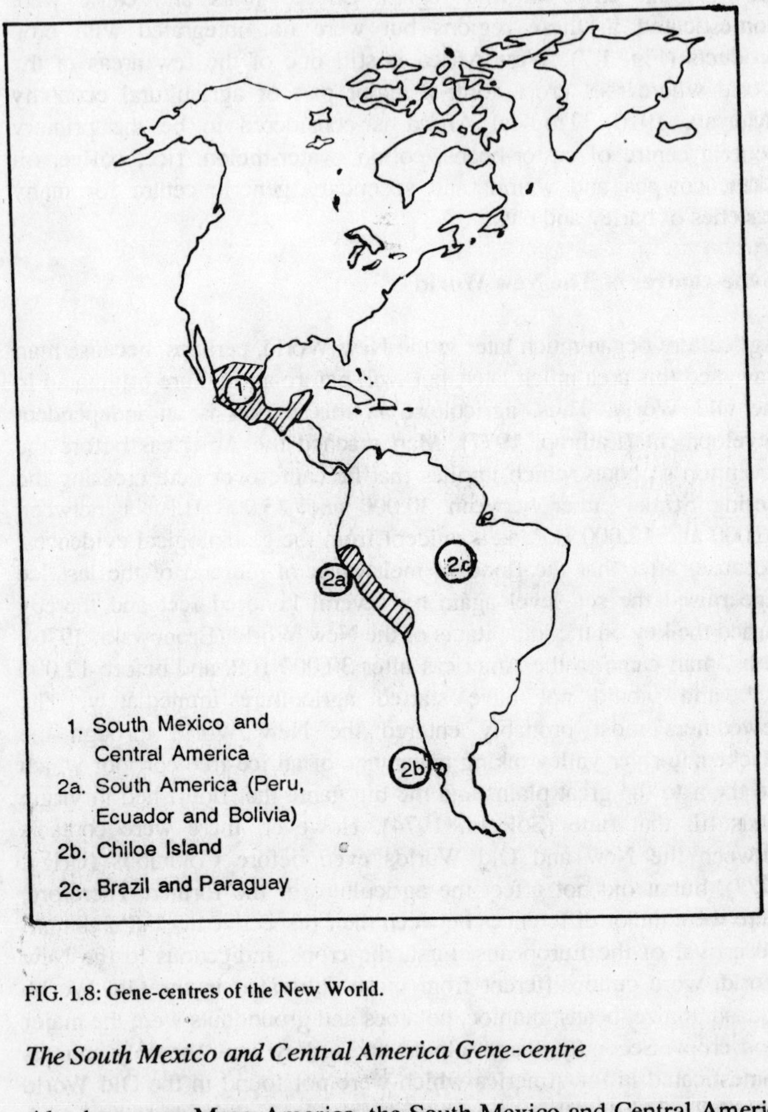

FIG. 1.8: Gene-centres of the New World.

The South Mexico and Central America Gene-centre

Also known as Meso-America, the South Mexico and Central America gene-centre includes South Mexico, Guatemala, Costa Rica, Honduras and Panama. Vavilov has listed 49 major plant species, out of which 17 are cultivated fruits, eight are cereal grains, six are melon plants and the rest are spices and stimulants, fibre plants and other miscellaneous plants. Harlan (1971) considered that Meso-America is the only north-western centre in which agriculture originated and from here it diffused to north and south. Meso-America is a region of geographical diversities resulting in great variation in agricultural adaptation. In arid highlands the practice of intensive agriculture and consequent agglomeration of dense population was the result of irrigation, drainage and terracing, while lowlands of greater rainfall resulted in more intensive agriculture, with more scattered population. Archaeological evidence suggests the existence of hunting-gathering economy with somewhat widely scattered nomadism. Although the shift from primary dependence on hunting-gathering to agriculture was very slow, it goes back to 9,000 B.P. as is evident from excavations of caves in north-eastern and south-central Mexico (McNiel, 1965). Latest archaeological, historical and zoological researches suggest that the origin of agriculture in the New World is as old as in the Old World, i.e. ninth millennium B.P. Maize appears to have been cultivated first in South Mexico around eighth millennium B.P. and gradually other plants were added. By 5,000 B.P., beans, squash, gourds, zapotoes, pumpkins, cotton and chillies were also grown. Farming villages mainly dependent upon agriculture appeared in Meso-America around 5,000 millennium B.P. However, after that plant species of this region diffused in other areas and vice versa. In the drier areas irrigation agriculture was practised, whereas in the lowlands slash-and-burn remained the basic technique (MacNeish, 1964).

The South American Gene-centre

In South america, three gene-centres are believed to have existed, viz., Equador and Bolivia; Brazil and Paraguay and Chile. Vavilov has listed altogether 62 plant species, out of which 45 are in the first, 13 in the second and only four in the third centre. Cultivated fruits account for one-third of the total species.

Harlan (1971) considers South America as a region in which many plants domesticated in different places over a very wide area but Sauer (1952) and Lathrap (1974) have identified different centres. Harlan suggests that a tropical lowland culture of farmer fishermen, growing root crops as their staple food plants, spread from an original base in northern South America about 5,000 B.P. Agriculture in the Andean area also anti-dates 5,000 B.P. Like Meso-America, it is also an independent centre of the origin of agriculture at least in actual species domesticated. Plant cultivation in this gene-centre may have begun as early as it did in Mexico, if not earlier (Kaplan et al., 1973 and Mac Neish, 1969). But some of the available evidence from the archaeological sites of Tamulipas and Tehuacan suggest that coastal Peru and south-west situated States had substantial agriculture but these are of recent origin as compared to those of Mexico. It is not surprising because cultivation on any scale on a desert coast requires development and control of irrigation system which presupposes prior development elsewhere. And it was in the highlands of Mexico. The first crops cultivated on the Peruvian coast were gourds, squash and lina beans and, possibly, cotton, although it may have arrived later. Chillies, pepper, achira, jack beans and guava were added to this list around 4,500 B.P. However, maize came around 4,000 B.P., most probably from Meso-America. By 4,000 B.P. there were farming villages based on irrigation, and a considerable number of monumental buildings (Lanning, 1965 and Collier, 1960).

From the foregoing discussion on the origin of agriculture in the Old and the New World, we may conclude that agriculture is a very complex phenomenon. Each population of plant and animal that we call domestic — each of the many kinds involved in the topic agriculture — is a subject of its own. Yet agriculture is certainly a unit — the totality of human practices involving those living secondary energy traps which man plants, breeds, makes, grows, guards, preserves, harvests and prepares for his own use (Reed, 1977).

Diffusion and Development of Agriculture

Agriculture has a history of about 12,000 years during which innumerable spatio-temporal changes have taken place. Various scholars of the concerned discipline have tried to trace the evolution and development of agriculture but they differ considerably in tracing its chronology.

Anatomically, modern man, since his emergence in this form from the *Homo-sapiens* around 250,000 years B.C., has grown through various stages of resource development, Various "stage models" have been built to explain the different stages of development (Reed, 1977; Gormann, 1977; Hasan, 1977; Bronson, 1977; Reddman, 1977; Harris, 1977) which eventually led to the practice of domestication of animals and plants from gathering and hunting. Since these models have been built for different areas they differ in their structure as well as in time sequence. But the common noticeable fact in all of them is that man in the beginning led a very hard life and had to battle his way through different climatic and non-climatic hazards. In the beginning he sustained on those things which he liked and were produced by the virgin land spontaneously. In due course of time he developed a liking for certain commodities — plants and animals — which led to the stage of gathering and hunting. He remained a gatherer and hunter for a considerably long period of time and gradually he started a sedentary life. But during the Neolithic period the discovery of fire, and subsequently man's mastery over its use, brought a momentous change in the pre-history of human life. The use of fire in some deliberate form such as setting of fire to improve pasture, clearing the forest and driving game or selection of plants — some to be encouraged and some to be destroyed by fire, undoubtedly preceded the beginnings of agriculture. Immediately after this he began in some places to domesticate some plants and animals. And this is the first and quiet explosion of agriculture.

The Radiocarbon chronology (Kabakar, 1977) and the preceding discussion clearly indicate that although there are various centres of the origin of agriculture, its evolution has taken place independently and in isolation, as wild species of the plant from which crops have been domesticated were confined to relatively very small and only a few parts of the earth surface. Gradually the improvements took place within the gene-centre and the art of domestication diffused in various directions. Based on ecological conditions, man domesticated various varieties of animals and plants. In the middle latitudes he concentrated, by and large, on cereal crops and sheep and goats, whereas in the tropical areas he domesticated root and tuber crops, and cattle and pigs. Crops and animals were exchanged and gradual development in agriculture was achieved. It changed from subsistence to specialised to commercial to market-oriented farming in due course of time. Recently technical and socio-economic changes have revolutionised agriculture so much that at the present time almost all the bounties of nature have been grabbed and almost everything is almost everywhere. Thus, the chronology of agriculture suggests that it is an evolutionary process rather than an event. Each successive step was an improvement over the preceding one. It has developed in different places and sequences over the surface of the earth, depending upon the physio-socio-economic conditions of the area.

Since agriculture has a very long history of more than 12,000 years, its gradual development would be discussed briefly during each of the following time periods in order to make it more clear and comprehensive:

(i) The Pre-historic Period,
(ii) The Early Historic Period,
(iii) The Pre-medieval Period,
(iv) The Medieval Period,
(v) The Neo-modern Period,
(vi) The Modern Period.

The Pre-historic Period

One of the major events in human history is the transition from hunting-gathering to agriculture. It is a revolutionary event in the sense that it entails radical changes in man's relationship with nature, others and himself. But despite many years of archaeological investigations culminating in an intensive period over the last two decades, the nature of the process which led to agriculture remains unclear. The hypothesis of the climatic change at the end of the Pleistocene giving the major stimulus for the origin of agriculture cannot be entertained in view of the lack of evidence for such change in South-West Asia (Braidwood, 1960; Butzer, 1971; Binford, 1968). Environmental suitability of certain regions and cultural adaptation have also been considered to be inadequate to give rise to agriculture (Meyers, 1971; Harris, 1971). Binford's hypothesis of pressure of population resulting in the origin of agriculture has little support in its favour. Whatever, might have been the primary force, according to the available archaeological evidence till this date, agriculture did originate in the Neolithic period, if not before. Leaving aside the controversy of the origin of agriculture, the early cultivators of Wadi-um-Natuf, Mureybit, Deh Huran, Cayonu, Zawi Chemi Shanidar, etc., were partially cultivating cereal grain especially wheat and barley and were partially reaping wild barley and wheat together with domesticating and hunting of animals simultaneously. The cultivated plants — emmer wheat and two-row barley — account for only one-third of the total weight of the vegetable food (Perkins and Daly, 1974) and the hunted animals accounted for the bulk of the meat supply. But gradually the sedentarism got more established and new sites came into being in these areas. The contribution of domesticated plants and animals slowly increased and by about 7000 B.C. emmer wheat and hulled two-row barley provided more than nine-tenths of the total weight of the plant food in some of the regions. Subsequently, a number of plants were added, e.g. flax, as an oil crop, peas, lentil and vetch, which was probably the result of diffusion of crops and ideas from some other areas,. The archaeological evidence regarding earliest agriculture in Hunan, Shensi and Shansi (during the Pon Po phase of the Yang-shao culture) and in Thailand (Spirit Cave, Banyan Valley Cave and Non Nok Tha) suggest rice culture along with the domestication of pigs and later cattle. During the fourth millennium B.C. farmers started cultivation of banana, yam, sugarcane, vinegrapes, etc., and domestication of sheep goats and cattle, which are believed to have come from South-Western and Central Asia. Thus, a mixed type of farming was prevalent in the earliest gene-centres. Gradually the local population densities increased and communities started so-called specialisation. Eventually, a complex pattern of socio-economic relationships among various communities developed and by the seventh millennium B.C. there was an extensive trade in obsidian and to a lesser extent in copper. Although it was a very informal trade, it however, implied a healthy trend towards switching over to more rigorous agricultural activities. Domestication of sheep and goats, followed by cattle, gained popularity in South-West Asia by the fifth millennium B.C. and they used to provide wool and milk in addition to a substantial amount of meat. But hunting of large ungulates was still a continuous process. In the South-East Asian centre, pigs, goats, followed by cattle, were domesticated and were providing a substantial amount of animal products.

Perhaps due to low amount of rainfall in South-West Asia resulting in the deterioration of soil fertility after croppings, and successive plantings in South-East Asia, farmers adopted shifting cultivation. The earliest farmers had a simple method of grinding with stones, mullers and querns for the processing of various nuts, grains and seeds. Knives, multi-toothed sickle blades of flint for reaping the grain crops were very common implements. In addition, they probably had stone axes and weighted digging sticks, and later primitive hoes, made at first from wood and later from stone (Butzer, 1964). Cultivation was mainly practised by primitive hoe-ard consisting of a curved stick to which a flint hoe was tied, later to act as a plough-share which could do little more than scratch the surface. But gradually the ox-drawn plough was adopted which brought a significant change in agriculture. Cross-ploughing was, perhaps, practised on flood-plains which would have definitely been useful for the conservation of the soil in dry-farming areas. Moreover, subsequently, various new varieties of crops, such as legumes, cotton, small seed flax, oilseed plants, onion, and various other varieties of fruits — almonds, pears, walnuts, strawberries, gooseberries, grapes, plums, apricots — were added and widely cultivated.

With the gradual improvement in agriculture and its techniques, sedentary life became more and more permanent. The growth of population increased the size of the village and slowly the tendency of urbanisation began to develop. With trial and error the early farmers realised the usefulness of fertile river plains where they slowly settled permanently, initiating the process of urbanisation. The urban civilisations of Egypt, Mesopotamia, Natuf, Mohen-jo-Daro, Harappa, Ropar, Kalibagh and China owe their origin to this very process. The river systems, besides providing a fertile cultivable land, were good sources of fish, molluscs and aquatic birds. They also acted as potential means of transport which served two purposes: (i) they encouraged the transportation facilities for agricultural and non-agricultural goods and people; and (ii) provided a network for the transmission and diffusion of farming and farming methods, inventions and inventive ideas.

However, whilst the gradual developments were taking place in these gene-centres, agriculture was diffusing in various directions around the fourth millennium B.C. From the South-West Asia and eastern Mediterranean gene-centre it diffused towards west and east. In Europe it reached the Danube Basin and moving north-eastward reached the shore towards the Baltic and the North Sea following the fertile loess soil and reached the shores by about 4000 B.C. An offshoot of this wave reached Spain and southern France during the same period. The other westward moving wave diffused along the North African coast and reached Maghreb in the fourth millennium B.C. (Grigg, 1977) (Fig. 1.9). Except cattle, which were independently domesticated there, most of the crops and animals of the gene-centre were also found in North Africa. But in Europe, in addition to domesticated plants and crops of the gene-centre, certain new crops and animals were added to agriculture. Cultivation of wheat, barley, lentil, flax, peas and vetch was extended to the north where barley became more prominent than wheat. The wild progenitors of pigs and auroch were found in Europe and they were domesticated

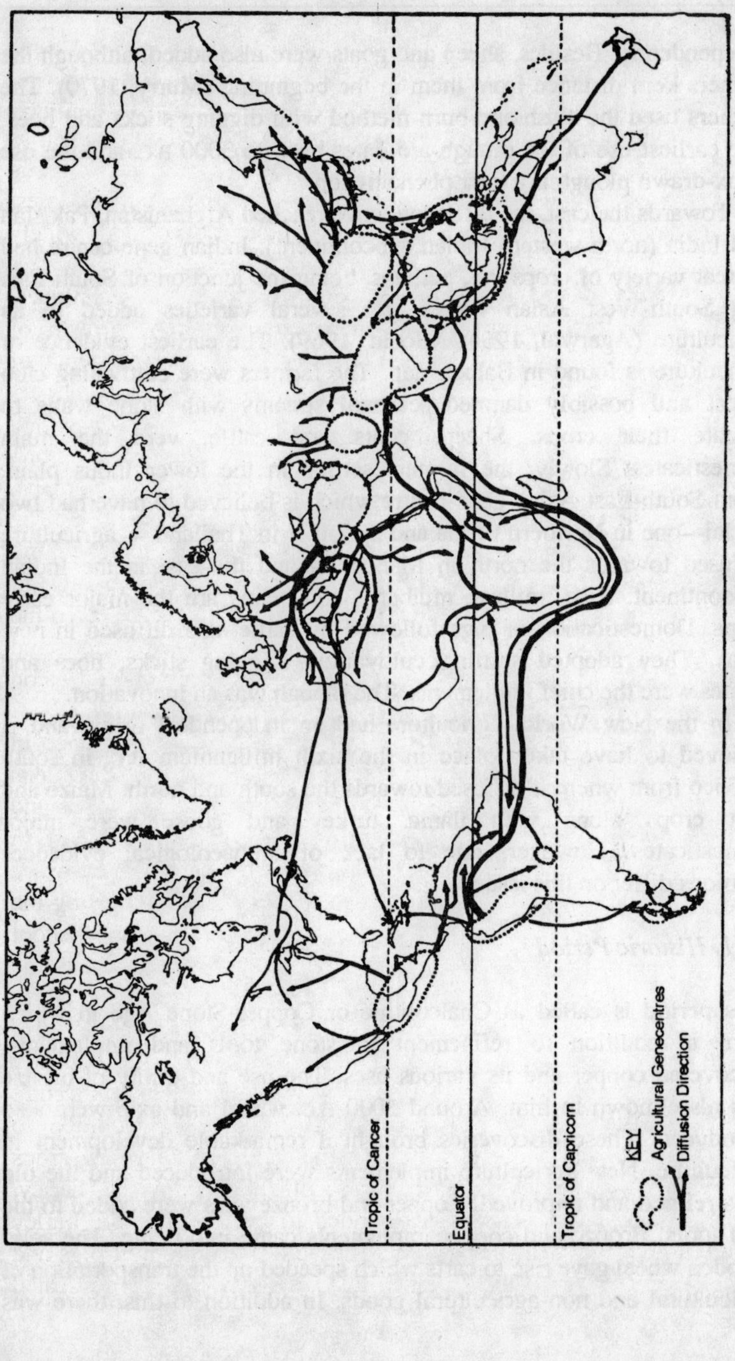

FIG. 1.9: Diffusion of agriculture.

independently. Besides, sheep and goats were also added, although the farmers kept distance from them in the beginning (Murry, 1970). The farmers used the slash-and-burn method with digging sticks and hoes. The earliest use of the plough-ard dates back to 3000 B.C. and the use of ox-drawn plough is a later phenomenon.

Towards the east, the diffusion wave reached Afghanistan, Pakistan and India (north-western Indian subcontinent). Indian gene-centre had a great variety of crops yet, perhaps, being the junction of South-East and South-West Asian waves, got several varieties added to its agriculture (Agarwal, 1964; Allchin, 1969). The earliest evidence of agriculture is found in Baluchistan. The farmers were cultivating club wheat and possibly damned seasonal streams with stone walls to irrigate their crops. Sheep, goats and cattle were the main domesticates. Slowly, the farmers settled in the lower Indus plain. From South-East Asian gene-centre which is believed to have had two nuclei—one in Northern China and the other in Thailand — agriculture diffused towards the north in Manchuria and the east in the Indian subcontinent. Rice, millets, mulberry and beans are the major early crops. Domestication of pigs followed by cattle also diffused in new areas. They adopted shifting cultivation. Digging sticks, hoes and spades were the chief implements. The plough was an innovation.

In the New World, agriculture had an independent origin and is believed to have taken place in the sixth millennium B.C. in south Mexico from where it diffused towards the south and north. Maize and root crops along with ilama, turkey and goose were major domesticates. However, due to lack of archaeological evidences opinions differ on this issue.

Early Historic Period

This period is called as Chalcolithic or Copper-Stone Age in which man, in addition to refinement in stone tools and implements, discovered copper and its various uses. The use and utility of bronze was also known to him. Around 3000 B.C. wheel and axle were also introduced. These discoveries brought a remarkable development in agriculture. New agriculture implements were introduced and the old ones refined and improved. Copper and bronze axes were added to the flint tools. Bronze and copper implements came into being. The solid wooden wheel gave rise to carts which speeded up the transportation of agricultural and non-agricultural goods. In addition to this, there was cultural change as well. The population of the early settlement sites increased resulting in an overall growth of population in the region. For example, according to Clark, the population in England and Wales had increased from 250 in the Palaeolithic period to 40,000 in the second millennium B.C. which naturally called for more food material to feed the people.

As a result of the above factors there was horizontal and vertical development in agriculture during this period. The system of nomadic and shifting cultivation gave way to cereal-fallow system. The farmers of the upper flood plain of the river Nile and Mesopotamia started descending to flood plains using flood water to irrigate their crops (Hester, 1960). In fact, the agriculture of this period may broadly be characterised as irrigation-cultural as it became the economic basis of the great valley civilisations of the world. In the Nile, Tigris, Euphrates, Indus and Hwang-Ho the irrigation systems were well developed. IN Egypt and Mesopotamia a complex irrigation system was developed. Shadoof, cerd and sagia are the examples of the then technological development.

Consequently, around 2500 B.C. two distinct types of farming emerged here i.e. dry farming and irrigation farming. The former was based upon shifting cultivation with wheat and barley as the main crop, and sheep, goats, cattle and pigs as important domesticates, while the latter was on the development of irrigation due to very low rainfall.

Farmers were growing enormous varieties of foodgrains in various parts of the world. To the earlier domesticates, zebu, water buffaloes, camels, fowl, etc., were added. Horses became very popular in Central Asia, Afghanistan and Iran. Around 2000 B.C., man learnt the art of riding the horse which gave further impetus to the development and diffusion of agriculture. Around 1000 B.C. reindeer and yak were found in the north and east of Altain Highlands.

In India, agriculture further diffused to the lower Indus and south-east from the great Himalayan foothills in the north, east to the river Jamuna, south-east to the river Narmada and south-west to the Makran coast. Cotton, sesame, and rice were added. Agriculture was intensified with the development of irrigation system. The plough reached here around 1700 B.C. after the Aryan invasion (Agarwal, 1964; Raikies, 1965). Cultivation of rice spread from the mainland into southern China and eastern India and southward into Malaysian archipelago. Agriculture in the New World also developed and expanded and, by 3000 B.C. beans, squash, zapatoes, pumpkins, avocados, gourds and chillies were added. Slash-and-burn was the major technique of cultivation.

In the third millennium B.C., farming villages mainly dependent upon agriculture appeared here. Maize reached the coastal areas of northern Peru by 2000 B.C. when farming villages based on irrigation agriculture were also established. In Africa, agriculture diffused to Savanna north of the tropical rainforest by third and second millennium B.C.

The Pre-Medieval Period

During the two millennia preceding the Christian era, considerable changes took place in human activities leading to substantial development in agriculture. The level of technology developed and people introduced bronze swords and horse-drawn bronze chariots establishing their superiority over the town-dwellers of the earlier period. In the early part of this period (Bronze Age), a light plough (ard) was introduced in northern Europe. Although it was incapable of breaking the virgin land for sowing the crops, it would have permitted more efficient tillage of land already under cultivation. The major agricultural change, in fact, took place in the succeeding Iron Age when iron and its multiple uses were discovered with improved technology. The new plough with iron ploughshare was introduced which could break heavier soil and bring it under cultivation (Evans, 1956). This is undoubtedly the most powerful invention in agriculture. Thus, the area under cultivation increased substantially all over the world.

In the beginning of the first millennium B.C., Greek cities and the Roman Empire were expanding with their developed agricultural technology. They had established their colonies around the Mediterranean coastal areas and further beyond in South-West Asia and North Africa, and had improved agricultural and non-agricultural activities with the use of iron technology. Hills were brought under cultivation by terracing and fruits — olive and vine — were grown. These fruits together with domestic animals were sent to the colonies extending over a large area. By the fourth century B.C., both Greek and Roman agriculture, together with their administration, were treated as fine arts. Consequently, it was exchanged throughout the region of their influence and beyond it through intermediate traders (Symons, 1977). Government agencies organised large estates (Latifundia) in most parts of the classical world, which were controlled by landlords engaging slave labour. But quite late, with the fall in the labour supply, the trend was reversed and subsistence farming became again more common (Heichelheim, 1956).

The development of agriculture was further speeded up with new scientific discoveries, such as Archimedes screw or the Tambour. As a result a new device for lifting water for irrigation purposes came into existence and was extensively used. This innovation brought more areas under cultivation in addition to several other improvements in agriculture. All these improvements were, perhaps, stimulated by the substantial growth of the monetary economy and at the same time abandonment of subsistence agriculture in favour of a semi-commercial one. This was also due to the growth of urbanisation leading to growing market centres for a variety of agricultural products.

In the Indian subcontinent agriculture continued to spread and develop. Around 1500 B.C. it reached Aravalli hills and further south where sheep and goats along with foodgrain crops such as wheat, flax, lentil, cotton and rice were domesticated. During the same period there was evidence of communities in the upper reaches of the Krishna river, growing distinctively Indian millets, ragi and bajra, indigenous to the continent of Africa. Contrary to the expectation, the diffusion of agriculture in the vast fertile Ganges plain, quite suitable for agricultural operation, had been much slower as compared to the undulating rugged peninsula region. According to available evidence, the earliest occupation of the Upper Ganga plain dates back to about 1100 B.C. and that of Middle Ganga plain even later between the ninth and eighth century B.C. Although improved iron-tipped ploughs, iron spades, sickles and axes were the major agricultural implements, dense forests and swamps perhaps delayed the early settlement and cultivation so much that the first evidence of rice cultivation in the Lower Ganga plain dates back only to the seventh century B.C.; and effective occupation was even later (Grigg, 1977).

In East and South-East Asia agricultural activities were further intensified and diffused further beyond their earlier limits. They reached Japan, Manchuria and towards north-eastern China. The chief implements were digging sticks, hoes and spades until the fifth century B.C. The Chinese independently developed a plough but there is little evidence in its support. Irrigation was not very common except for the use of wells in the loess uplands. The farmers were more concerned with moisture conservation than with irrigation (Chang, 1970).

From the Thailand gene-centre, agriculture diffused to Malaysia and South-East India. From the African gene-centres agriculture diffused along coastal areas and, perhaps, to the Indian Peninsula in the east and coastal areas of the New World in the west. However, in the New World also agriculture got intensified with horizontal and vertical expansion. But due to lack of archaeological evidence not much is known about the diffusion and development of agriculture in these gene-centres.

The Medieval Period

At the outset of the Christian era there had been a considerable interchange of the plants and animals between the gene-centres of the Old World (Fig. 1.9). In the second century in spite of the decline in market-oriented farming, farming techniques, tools, plants and animals diffused beyond the Roman Empire. The feudal system bound the peasants to the estate around the fourth century. Monasteries developed in Egypt and followed in the west after the fifth century by the growth of strong religious communities which kept alive scientific agriculture and land development (Symons, 1977).

However, during the early centuries of the Medieval period, the great agricultural civilisations of the Old World, South-West Asia, Southern Europe, northern India and northern China were technically on a par with each other (Arkell and Ucko, 1965; Hestner, 1968). But between the classical times and the age of discovery, innumerable agricultural innovations in all its fields have taken place and a substantial majority of them were innovated in Europe resulting in a wide gap in agricultural development in these countries. The earlier plough (ard) being light was a mere scratcher and was not much successful on the cold heavy soil of Europe. Therefore, a heavy plough consisting of a coulter to cut the turf and a mouldboard to turn it was developed in the first century. Wheels were attached later in order to make its movement easier. The heavy plough drawn by four or more oxen became popular at the end of the first millennium B.C. The Roman farmers had attained a standard in agricultural science not to be surpassed until the 19th century. The value of green manuring was known and practised. Animal manure was spread on the field; the folding of sheep was practised; extensive farming such as double cropping and inter-tillage were known; the importance of rotation of crops and fallowing to conserve soil moisture was well-known; terrace farming was practised both in the New and the Old World; irrigation was a widely used technique and dry farming was also practised (Hole, 1966; Flannery, 1965).

After the fall of the Roman Empire there came an interregnum of a dark period in the development of agriculture and dispersal of crops and animals, as the great agricultural civilisations became relatively isolated from each other. Though the tendency of *latifundia* continued in the medieval period, at the cost of insecure peasant proprietors, the prevalence of insecurity, warfare and the outbreak of major plague and Black Death were greatly responsible for the decline of agriculture. These circumstances, at places, led to the emergence of pastoralism which grew in importance.

Although before Mohammad's time Arab traders were sailing both east to India and south to East Africa and were carriers of agricultural innovations, by the eighth century A.D. the Islamic Empire extending from Spain and Morocco in the west to Indus and Jacartes in the east was well established. Thus they cut Europe's land route to India, China and Africa. The Muslims made an effective contribution in spreading citrus and other crops into the western Mediterranean. Although sweet orange, sour orange, lime, lemon, citron were indigenous to southern China and South-East Asia, the time of their arrival in the Mediterranean region is a matter of dispute. However, the diffusion of most of these crops is commonly attributed to the Arabs.

During the ninth century the developments in the techniques and administration of agriculture came at long intervals. The feudal system was established in which the vassal owed allegiance to his lord in return for a measure of security which was expressed on land by manorial organisation. The vassals supplied the lord with a portion of their produce and also had to work on the demesne lands for the lord's benefit. This system was the outcome of an economy scarce of money which probably resulted from military aggressions and overall unrest. During this period, various methods of cultivation were adopted. Open-field system with strip cultivation was identified. However, Slicher (1963) has classified the following five types of agricultural practices:

(i) Temporary cultivation system after which land is left fallow for an indefinite period to recoup fertility,

(ii) The infield-outfield system in which part of the land was cultivated continuously with heavy manuring and the other part occasionally,

(iii) The two-course rotation in which the field was tilled and left fallow in alternate years,

(iv) A three-year system in which the land was tilled for one year and left fallow for two successive years, and

(v) The three-course rotation in which winter crop (wheat or rye) was followed by spring crop (barley or Jat) and then by fallow.

In the late Medieval period, rising population pressure and circulation of money led to a great increase in cereal prices which stimulated the extension of agricultural land. However, during the last phase of this period there was considerable diffusion of crops and animals as it was the age of discovery. A number of crops began to spread in different parts of the world with the spread of Europeans. In fact, the credit goes to the navigators of Portugal, Spain, Italy, Holland, France and England who discovered sea routes to different parts of South-East Asia. The discovery of America by Columbus in 1492 is the most significant even which paved the way for the exchange of crops between the two Worlds. In this period with the efforts of such navigators regular sailing routes linked Europe, South Africa, India, China, Brazil and the East Indies. Consequently, interchange of crops and animals was accelerated in most agriculturally potential areas of the world. Colonial agriculture developed in this period. The cultivation of those crops which were either used by colonisers or were beneficial to them in any way was encouraged. Such crops were introduced in the new colonies where physico-social conditions were conducive to their growth. During this period, thus, more varied changes in the cropping pattern also emerged in most of the agricultural regions of the world.

Neo-Modern Period

In fact this is the transitional period between the classical agriculture and the modern one. During this period the pace was almost set for the revolution in agriculture. It corresponds with the later period of the Great Age of Discovery and its succeeding period. Multifaceted development in agriculture has taken place which could be characterised as inter- and intra-regional exchange of domesticated crops and animals and their diffusion on a mass scale in different parts of the world; wide expansion and intensification of colonial agriculture; horizontal expansion of agriculture; intensification of agriculture leading to more productivity; improvement and development of agricultural technology.

During this period the most significant development is the interchange of domesticates and diffusion of agriculture. After the discovery of America, exchange of domesticates started taking place. The arrival of Europeans in the New World not only led to the introduction of Old World plants, but also to the wider dispersal of indigenous American crops to other parts of the continents. Wheat, rye, oats and barley were introduced in Mexico, Peru and the Andes. Asian rice was introduced in the early 16th century which flourished after the arrival of Indian sugarcane planters. The Spanish and Portuguese were pioneers in introducing Mediterranean crops — grapes, fig, vine, olive and citrus fruits. Though viticulture had a chequered history, after trial and error, it eventually got established, by and large, in the regions of Mediterranean climate.

Sugarcane and bananas were introduced by the Portuguese by early 16th century which flourished in Cuba, Mexico, Puerto Rico and Peru, and later in the West Indies. Subsequent improvements and developments took place in their cultivation especially after the arrival of sugarcane planters from Southern Asia and Africa. The slave trades led to the exchange of crops among Brazil, Indies and West Africa. The greater yam, sorghum, *vulgare*, pearl milds; cowpeas and calcasia reached Brazil and Indies (Allchus and Allchus, 1968). Coffee plants from Indonesia also reached here via Europe which was widely spread in 18th century in Brazil, Sao Paulo, Bolivia and Rio de Janeiro, (Ho, 1969; Gorman 1969). Similarly, in return various grain and root crops diffused from the New World to the different countries of the Old World (Fig. 1.9). Indigenous crops of the continent also widely spread within the continent. Cocoa indigenous to the Andean slopes of the Amazon Basin diffused in Mexico and other areas of its present cultivation. Cotton which was independently grown in Mexico and Peru diffused to the USA and South American countries.

Among the continents of the Old World, Europe experienced maximum changes because navigators, besides diffusing innovations outside, were also feeding back new ideas, crops and plants from their area of discovery and, subsequently, colonisation. Maize and potatoes were brought back with new techniques of their cultivation. Tomato and tobacco came from the New World. Several indigenous crops were cultivated and an intra-regional exchange of crops also took place. In Africa manioc and maize were introduced from America. These food crops were adopted more quickly than other crops. Groundnut and cocoa were introduced from the New World. In Asia cash crops were quickly taken up and there was very little change in food crops. Most of the crops of the New World came via Europe or European crops diffused here. Maize, chillies, tomatoes and sweet potatoes were introduced in the 16th century. There was much intra-regional exchange of crops in Asia. Crops from South-East Asia diffused in South and South-West Asia. China was the least affected country, but groundnut, sweet potatoes and maize were introduced there in the mid 16th century.

However, in the massive phase of diffusion of crops and animals, agricultural productivity of the different countries of the New World and the Old World were almost on a par with each other in the very beginning of this period, because cultivation was done, by and large, on virgin lands and the techniques were simple and primitive. But the difference in productivity came with the introduction of technology in agriculture. Simple agricultural technology paved the way in bringing a revolutionary movement in agriculture. In the 16th century horse-drawn harrows and rollers were invented and brought in use in many European countries, while in the 17th century a proto-type seed drill, which was improved later, was in use in Italy. Most of the technological advances in farming in most parts of the world were, by and large, directed towards the ploughing of land and preparation of seed beds and, consequently, increasing the agricultural land. In the 18th century, first the all-iron plough, the Rotherham plough and, subsequently, John Deeri's plough and multi-share plough were introduced. But when more and more areas were brought under cultivation, it created the problem of harvesting and processing the crops.

In Canada and the USA, there was abundant land but few labourers. Thus, the labour shortage prompted the search for harvesting equipments but the scythe — introduced in Roman times — and the sickle remained quite common during this period.

During this period due to fluctuation in the prices of agricultural products — both crop and livestock — there has been considerable change in the cropping pattern, selection and combination of crops and animals. Before the 16th century due to a decline in foodgrain prices, emphasis was on industrial crops such as hope, flax, dye plants and oilseeds or on dairy farming, but in this century due to increase in the cereal prices indicating population expansion and increased use of horses, attention was focussed on horizontal and vertical expansion of agriculture. By the 18th century, specialisation in agriculture especially in livestock and livestock products took place. Cattle used to be domesticated for meat or milk and sheep for wool or mutton. Nevertheless, the agriculture of the neo-modern period was largely subsistence-oriented and was dominated by the production of cereals. However, at the tail-end of this period three major trends emerged, especially in Europe, which gave rise to modern mixed farming in different parts of the world in general and Europe in particular. The first was reduction in the fallow land employing the intensification of agriculture which is possible with proper selection and rotation of crops. Extinction of common rights and the consolidation of dispersed strips was the second major trend which led to the enclosure of open fields. The last major trend was a slow but steady increase in the importance of livestock (Grigg, 1977).

During this phase, colonial agriculture got further established. The colonisers were exchanging and introducing various crops in suitable areas. The main emphasis was on development and expansion of areas under those crops which were providing raw materials to the industries of their homeland. Besides, the subsistence crops were also grown to feed the people inhabiting the colony. The development of plantation farming in the tropical region is the result of this process.

The Modern Period

This is a period of profound development in almost all the economic activities of man; and agriculture is no exception to it. A number of factors have contributed to its development but the Industrial Revolution brought a tremendous change in agriculture. In fact, the agricultural development or, so to say, revolution was intimately interwoven with and dependent on the Industrial Revolution. It had set the pace for it directly and indirectly. In the beginning, on the one hand, it created a market for agricultural products, such as cotton, wool, rubber, jute, etc., to be used as raw material, on the other, it provided several basic agricultural inputs, directly and indirectly, together with creating shortage of agricultural labour, specially in the Western world, by attracting them toward industries, thus creating a necessity for mechanisation of agriculture.

However, the Industrial Revolution brought a major change in agricultural tools. Iron became cheaper for all uses. Steam power was utilised for both factory and farming. It was applied in agriculture first for threshing machines and drainage pumps in the early part of the 19th century and to ploughing in the middle of the same century. No doubt that the steam plough was capable of ploughing 10 times the area that horses could plough in a day but it was costly and complicated. Hence, it had a little impact on agriculture (Clark, 1969). The horse, therefore, remained the main source of power till the early 20th century. In 1842, the first gasoline tractor was built in the USA by John Froelich — a farmer and also a blacksmith. Subsequently, a tractor with an inter-combustion engine was developed in 1910. It was originally regarded as a versatile source of power which may be used in most farming operations such as threshing, ploughing, etc. The use of the tractor became popular during the inter-war period in the USA and after the Second World War in Europe. It is still in its initial stage of use in developing countries. By 1874, the horse-drawn reaper replaced the scythe and the sickle; and in 1930 the power harvester was introduced in Europe. In 1931, the tractor was equipped with a diesel engine which enabled the machine to work under all conditions. Drillers were introduced to sow the seeds in straight rows so as to make inter-row cultivation possible. Besides these, several other agricultural machines and tools were introduced such as weeder, leveller, sprayer, winnower, etc., which were instrumental in agricultural development.

Development in the means of transport is another factor accelerating the process of transformation of agriculture. The commercialisation of agriculture in the 19th century and the growth of new agricultural regions would have been impossible without substantial improvement in the means of transport and the consequent reduction in the freight rates. Such means link places and this does help not only in distribution of farm products or collection of agricultural inputs but also in diffusion and propagation of agricultural innovations. Although in the preceding period farmers were using manures, the use of chemical fertilisers came in the 19th century. But their popular use came after the middle of the present century. The introduction of high-yielding and improved varieties of seed also brought a major change in agriculture. Besides, there are several other factors, which speeded up the development of agriculture but it would be inappropriate to recount them here.

Thus, during this period agriculture became industrialised. The farmers began to get almost all their inputs such as agricultural implements, fertilisers and pesticides, coal and petroleum products, and oilcake from industry. Similarly, the farm products were increasingly processed in factories. But, unfortunately, this level of agricultural development was not attained all over the world. Around the middle of the 19th century the difference in the level of agricultural technology between the countries of Asia, Africa and Latin America, and Europe, North America and Australia was well marked, and subsequently the gap went on widening. Comparatively very insignificant changes occurred in the developing countries which continued with traditional methods. Consequently, the agricultural productivity in these regions is also comparatively low.

However, during the 20th century, especially in its third quarter, the mounting pressure of population on existing resources has diverted the attention towards the rapid development of agriculture in the developing countries particularly where the agricultural density of population is very high. This, along with a number of agricultural innovations, has brought horizontal and vertical development in agriculture in these regions. On the other hand, in most technologically advanced countries the trend of "capital agriculture" has emerged in which emphasis is laid on profits and re-investment.

REFERENCES

Adams, R. M., 1962 : "Agriculture and Urban Life in Early South-West Iran", *Science*, 136, pp. 109-22.

Agarwal, D. P., 1964 : "Harappa Culture: New Evidence for Shorter Chronology", *Science*, 143, pp. 950-52.

Allchin, F. R., 1969 : "Early Cultivated Plants in India and Pakistan", in Ucko, P. J. and Dimbleby, G. W. (eds.), *The Domestication and Exploitation of Plants and Animals*, pp. 323-33.

Allchin, B. and F. R. Allchin, 1968 : *The Birth of Indian Civilisation: India and Pakistan Before 500 B. C.*

Anderson, E., 1960 : "The Evolution of Domestication" in *Evolution After Darwin* (3 Vols.).

Arkell, A. J. and P. I. Ucko, 1965 : "Review of Pre-dynastic Development in the Nile Valley, *Current Anthropology*, 6, pp. 145-66.

Gorman, B. F., 1973 : "Excavations at Spirit Cave, North Thailand: Some Interesting Interpretations", *Asian Perspectives*, 13, pp. 79-107.

Binford, L., 1968 : "Post Plaestocene Adaptation" in Binford, L. R. and Binford, S. R. (ed.), *New Perspectives in Archaeology*.

Braidwood, R. J., 1952 : *The Near East and the Foundation of Civilisation*.

———, 1958 : "Near East Pre-history", *Science*, 127, pp. 1419-30.

———, 1960 : "The Agricultural Revolution", *Scientific American*, 203, pp. 130-48.

Braidwood, R. J. et al., 1961 : "The Iranian Pre-historic Project", *Science*, 133, pp. 2008-10.

Braidwood, R. J. and B. Howe, 1962 : "South-western Asia Beyond the Lands of the Mediterranean Littoral", in Braidwood, R. J. and Villy, G. (Eds.), *Courses Towards Urban Life*, pp. 134-42.

Braidwood R. J., 1967 : *Pre-historic Men*, (7th Edition).

Braidwood, R. J. et al., 1974 : "Beginnings of Village Farming Communities in South-eastern Turkey, 1972", *Proceedings of National Academy of Sciences*, USA, 71, pp. 568-72.

Bronowski, J., 1976 : *The Ascent of Man.*

Bronson, B., 1977 : "The Earliest Farmacy: Demography as Cause and Consequence", in Reed, C. A. (ed.), *Origin of Agriculture.*

Brookfield, H. C., 1969 : "The Environment as Perceived", in Board, et al. (eds), *Progress in Geography*, I.

Brown, L. R. and G. W. Finsterbusch, 1972 : *Man and His Environment: Food.*

Butzer, R. W., 1964 : *Environment and Archaeology*, pp. 416-37.

———, 1971 : *Environment and Archaeology: An Ecological Approach to Pre-history* (2nd Edition).

Caldwell, J. R., 1977 : "Cultural Evolution in the Old World and the New, Leading to the Beginnings and Spread of Agriculture", in Reed, C. A. (ed), *Origins of Agriculture.*

Condolle, A. D., 1967 : *Origin of Cultivated Plants* (Reprint of 1886).

Carter, G. F., 1977 : "A Hypothesis Suggesting a Single Origin of Agriculture", in Reed, C. A. (ed), *Origins of Agriculture*, pp. 92-3.

Change, K., 1968 : *The Archaeology of Ancient China.*

———, 1970 : "The Beginning of Agriculture in the Far East", *Antiquity*, 44, pp. 175-89.

Childe, V. G., 1929 : *The Most Ancient East.*

———, 1935 : *New Light on the Most Ancient East.*

———, 1936 : *Man Makes Himself.*

Clark, H., 1957 : "The Origin and Early History of the Cultivated Barley", *Agricultural History Review*, 15, pp. 1-18.

Clark, I. G. D., 1969 : *World Pre-history : A New Outline* (2nd edn.).

Collier, D., 1960 : "The Development of Civilisation on the Coast of Peru in Pan-American Union", *Irrigation Civilisation: A Comparative Study*, pp. 19-27.

Darlington, C. D., 1956 : *Chromosome Botany.*

De Silva, D., 1980 : "The Vanishing Seeds", in *Daily Nation*, March 26, (Quoted by Obara, D. A.).

Ekholm, G., 1955 : "The New Orientation Towards the Problems of Asiatic American Relationships", in *New Interpretation of Aboriginal American Culture Society*, 75.

Evans, E. E., 1956 : "The Ecology of Peasants: Life in Western Europe", in Thomas, W. L. (ed), *Man's Role in Changing the Face of the Earth*, pp. 217-39.

Fairbridge, R. W., 1967 : "Ice Age Theory", in Fairbridge, R. W. (ed), *The Encyclopedia of Atmospheric Sciences and Astrogeology*, pp. 462-74.

Flannery, K. V., 1965 : "The Ecology of Early Food Production in Mesopotamia", *Science*, 147, pp. 1247-56.

Freeman, T. W., 1961 : *A Hundred Years of Geography.*

Gorman, C., 1969 : "Hoabinhian: A Pebble Tool Complex with Early Plant Association in South-East Asia", *Science*, 163, pp. 671-73.

———, 1977 : "A Priori Model and Thai Pre-history: A Reconsideration of the Beginnings of Agriculture in South-easter Asia", in Reed, C. L. (ed), *Origins of Agriculture.*

Grigg, D. B., 1972 : *The Agricultural Systems of the World: An Evolutionary Approach.*

Hamdan, G., 1961 : "The Evolution of Irrigation Agriculture in Egypt", in Stamp, L. D. (ed), *A History of Land Use in Arid Regions*, Arid Zone Research, 17, pp. 119-42.

Harlan, J. R., 1971 : "Agricultural Origins: Centres and Non-Centres", *Science*, 174, pp. 468-74.

Harris D. R., 1967 : "New Light on Plant Domestication and Origins of Agriculture: A Review", *Geographical Review*, 57, pp. 90-107.

———, 1977 : "Alternative Pathway Towards Agriculture", in Reed, C. A. (ed), *Origins of Agriculture.*

Harris, J. C., 1971 : "Explanations in Pre-history", *Proceedings of the Pre-historic Society*, 37, pp. 38-55.

Hasan, A. F., 1977 : "The Dynamics of Agricultural Origins in Palestine: A Theoretical Model", in Reed, C. A. (ed), *Origins of Agriculture.*

Heichelheim, M. F., 1956 : "Effects of Classical Antiquity on Man", in Thomas, W. L. (ed), *Man's Role in Changing the Face of the Earth*, pp. 165-82.

Hester, J. J., 1968 : "Origins of African Agriculture: Comments", *Current Anthropology*, 9, pp. 497-8.

Ho, P., 1969 : "The Loess and the Origin of Chinese Agriculture", *American Historical Review*, 75, pp. 1-36.

———, 1975 : *The Cradle of the East: An Inquiry into the Indigenous Origins of Techniques and Ideas of Neolithic and Early Historic China, 5000-1000 B. C.*

———, 1977 : "The Indigenous Origins of Chinese Agriculture", in Reed, C. A. (ed), *Origins of Agriculture.*

Hole, F., 1966 : "Investigating the Origins of Mesopotamian Civilisation", *Science*, 153, pp. 605-11.

Hole, F. and K. V. Flannery, 1968 : "A Pre-history of Southern Iran: A Preliminary Report", *Proceedings of the Pre-historical Society for 1967*, pp. 147-206.

Hartshorne, R., 1939 : "The Nature of Geography", *Annals of the Association of American Geographers*, 39.

Janick, J. et al., 1974 : *Plant Science: An Introduction of World Crops.*

Kabakar, A., 1977 : "Radio Carbon Chronology Relevant to the Origins of Agriculture", in Reed, C. A. (ed), *Origins of Agriculture.*

Kaplan, L. et al., 1973 : "Early Cultivated Beans from an Intermontane Eruvian Valley", *Science*, 179, pp. 76-7.

Kenyon, K., 1969 : "The Origins of the Neolithic", *Advancement of Science*, 26, pp. 144-60.

Lanning, E. P., 1965 : "Early Man in Peru", *Scientific American*, 213, pp. 68-76.

Lathrap, D. W., 1974 : "The Moist Tropics, the Arid Lands and the Appearance of Great Art Styles in the New World", in King, M. E. and I. R. Taylor, *Art and Environment in Native America*, pp. 115-18.

Lathrap, D. W., 1977 : "Our Father the Cayman, Our Mother the Gourd: Spinden Revisited, or a Unitary Model for the Emergence of Agriculture in the New World", in Reed, C. A., (ed), *Origins of Agriculture.*

Lewis, R. A., 1966 : "Early Irrigation in West Turkistan", *Annals of the Association of the American Geographers*, 56, pp. 467-91.

Mac Neish, R. S., 1958 : "Preliminary Archaeological Investigations in the Sierra de Tamaulipar, Mexico", *Transactions of the American Philosophical Society*, 48, pp. 1-210.

———, 1964 : "Ancient Meso American Civilisation", *Science*, 143, pp. 531-45.

———, 1965 : "The Origins of American Agriculture", *Antiquity*, 39, pp. 87-94.

Mellaart, J., 1964 : "A Neolithic City in Turkey", *Scientific American*, 210, pp. 94-104.

Meyers, J. T., 1971 : "The Origins of Agriculture: An Evolution of the Three Hypotheses", in Struever, S. (ed), *Pre-historic Agriculture*, pp. 101-21.

Mittre, V., 1977 : "Changing Economy in Ancient India", in Reed, C. A. (ed), *Origins of Agriculture.*

Murdock, G. P., 1959 : *Africa: Its People and Their Cultural History.*

Murry, J., 1970 : *The First European Agriculture: A Study of the Osteological and Botanical Evidence Until 2000 B. C.*

Narr, K. J., 1956 : "Early Food Producing Population", in Thomas, W. L. (ed), *Man's Role in Changing the Face of the Earth.*

Parkins, D. and P. Daly, 1974 : "The Beginning of Food Production in the Near East", in Strigler, R. (ed), *Old World: Early Man to the Development of Agriculture*, pp. 73-92.

Perkins, D. Jr., 1964 : "Pre-historic Fauna from Shanidar, Iraq", *Science*, 144, pp. 1565-66.

———, 1969 : *Agricultural Development in China.*

Pickersgill, B. and C. B. Henser, 1977 : "Origins and Distribution of Plants Domesticated in the New World Tropics", in Reed, C. A. (ed), *Origins of Agriculture.*

Raikies, R. L., 1965 : "The Mohenjodaro Floods", *Antiquity*, 39, pp. 196-202.

Redman, C. L., 1977 : "Man, Domestication and Culture in South-eastern Asia", in Reed, C. L., (ed), *Origins of Agriculture.*

Reed, C. L., (ed), 1977 : *Origins of Agriculture.*

Rodden, R. J., 1965 : "An Early Neolithic Village in Greece", *Scientific American*, 212, pp. 89-92.

Sauer, C. O., 1952 : *Agricultural Origins and Dispersal.*

Slicher Van, B. B., 1963 : *The Agrarian History of Western Europe A. D. 500-1850.*

Solheim, W. G. et al., 1966 : *Preliminary Reports on Excavation on Ban Nadi, Ban Sao Lao, Primar.*

Solheim, W. G., 1973 : "Northern Thailand, South-east Asia, and World Pre-History", *Asian Perspectives*, 13, pp. 145-62.

Sorokin, P. A., 1928 : *Contemporary Sociological Theories.*

Symons, L., 1978 : *Agricultural Geography.*

Thomas, F., 1925 : *The Environmental Basis of Society: A Study in the History of Sociological Theory.*

Van Loon, M., 1968 : "The Oriental Institute Excavations at Mureybit", *Journal of Near Eastern Studies*, 27, pp. 265-90.

Vavilov, N. I., 1949-50 : "The Origin, Variation, Immunity and Breeding of Cultivated Plants", *Chronica Botanica*, XIII, pp. 14-54.

Watson, W. and H. E. Loof, 1967 : "The Thai-British Archaeological Expedition: A Preliminary Report on the Work of the First Season, 1965-66", *Journal of the Siam Society*. 55, pp. 237-72.

Whitehouse, R. D., 1968 : "The Early Neolithic of Southern Italy", *Antiquity*, 42, pp. 188-93.

Wilsie, C. P., 1962 : *Crop Adaptation and Distribution.*

Wissman, H. V. et al., 1956 : "On the Role of Nature and Man in Changing the Face of Dry Belt of Asia", in Tomas, W. L. (ed), *Man's Role in Changing the Face of the Earth*, pp. 278-308.

Wright, H. E., 1977 : "Environmental Changes and the Origin of Agriculture in the Old and New Worlds", in Reed, C. A. (ed), *Origins of Agriculture.*

2

ORIGIN AND EVOLUTION OF AGRICULTURE
A Geographical Appraisal

D. A. OBARA

BEFORE MAN'S earliest attempts to control or influence vegetation, plants provide him with some food directly or indirectly through animals supported by food. Even today, some people still obtain their food by hunting and collecting but over 99.9 per cent of the world's population depends on agriculture for its daily requirements (Dunlop and MacDonald, 1976). Agriculture is considered to be safer and less erratic than hunting, and provides a better return for the effort put into it, although the transition from hunting and collecting to organised agriculture took place over a very long period of time. Nevertheless, early man started to gain knowledge of the growing conditions of wild plants, and slowly learnt how to cultivate certain grasses with large edible seeds, perhaps through a process of trial and error. Similarly, primitive hunters coming into frequent contact with wild animals learnt to rear them in domestication. This paper first looks at the theories of origin of agriculture with special reference to principles governing the domestication of plants and animals. Secondly, it examines the evolution, diffusion and probable dates of origin of agriculture. Thirdly, it evaluates the significance of the theory of plant and animal introduction.

The "concept of agriculture" refers to the point in time or space at which agriculture first existed, while evolution of agriculture is used in the context of the present discussion to mean a continuous change from a simple to a more complex form of agriculture achieved by two distinct processes: domestication, the bringing into cultivation or management of wild species of plants and animals; and selection, the differential reproduction of these species. Evolution can also be used to mean that "crop plants" and animals have changed in response to environmental conditions by the natural selection of randomly occurring mutations, developing from the simplest to complex forms, which are more prolific and stronger, due to their better adaptation to their environment. On the latter basis, domesticated animals — cattle, sheep, goats, pigs, camels, asses, mules, water buffaloes, horses and dogs — have increased except the horse species which have declined in number perhaps because they are used in wars. In the present paper "crop plant" is used to mean any plant utilised by man for any purpose. They are really those useful plants that fit economically into the scheme of man's work and existence, thus being managed to some degree.

Theories of Origin of Agriculture

As has been mentioned above, for thousands of years man mainly depended on hunting and gathering wild food, living as a predator, scavenger and primitive food collector, but because his life was largely a search for food, his campsites were impermanent, and starvation was a constant threat (Brown and Finsterbusch, 1972). Consequently, surviving on wild food necessitated, for the most part, living in small groups and wandering over wide land areas, migrating with the prey and with the seasonal changes, particularly, in climate. Farming was perhaps developed to make the supply of food more regular and plentiful by selecting certain plant and animal species, and taking special care of them in order to alleviate constant movements and food shortages. The principle governing domestication of plants and animals is controlled by man of their reproduction. Agriculture is, therefore, an effort by man to go beyond the natural boundaries and modify environment to better suit man's needs. But no one knows exactly when, how and why man domesticated plants and animals. It is postulated that perhaps tamer adult animals approached humans in search of food, remaining there because they were fed and eventually reproducing. Alternatively, it is theorised that man brought lambs or kids and later calves, colts and the like, back from the hunt as pets or future decoys to be used in hunting. On the basis of these hypotheses, sheep were domesticated first followed by goats and pigs than cattle, but domestication of horses is more recent. Originally, sheep, goats and cattle were raised primarily for meat and hides, but today they satisfy manifold socio-economic demands. However, women were probably the pioneers of agriculture, because in hunting and fishing societies everywhere, women are the food gatherers, the foragers, and seekers of seeds and roots, nuts and fruits. They perhaps had better opportunities for observing the relation between seeds accidentally dropped and the subsequent germination. To date, women carried out experiments with sowing seeds, weeding and harvesting grains in their wild state. The foregoing argument is confirmed by the fact that sowing, weeding and harvesting operations are still women's job even today in many societies including those in East Africa.

Some of the theories on the origin and development of agriculture have been well-examined (Darlington, 1956; Wilsie, 1962; Janick et al., 1974), and indicate that the "concept of origin of agriculture" is not a recent phenomenon. However, agriculture is a relatively recent innovation when considered in terms of the historical development of mankind since man has merely been a collector of food for the greatest part of his existence. Nevertheless, the discovery of fire and the development of agriculture are the two innovations that form the basis of civilisation (De Silva, 1980). Agriculture coincided with the Neolithic Age, although the use of fire as an ecological tool to improve pasture and select useful plants undoubtedly preceded the origin of agriculture proper (Symons, 1978). The invention of fire as an ecological tool helped in widening the horizons of agriculture, while the advent of fire also brought innovations or technologies which until now would be unknown. Fire is often associated with the expansion of food supply, for a great number of foods are inedible, unpalatable, unhealthy unless they are cooked. It is said that only after considerable time and after considerable cultural experience could man have "learnt" to demand the satisfaction of many culturally derived "need". Before the advent of agriculture man was probably content with the satisfaction of his basic requirements: food and shelter, but only gradually developed his secondary needs including clothing, which called for more sophisticated technologies. It is also theorised that initially man depended solely on hunting, fishing and gathering and had limited time to devote to "luxury" cultural activities, for the greatest part of his day was spent searching for aliment (food) to prevent starvation. Thus, the characteristics of modern development emphasising, for example, specialisation of labour, accumulation of surplus, sophistication of religious and ritual belief and practice coincided only with permanent settlement. The foregoing theories have been verified by Ucko and Dimbleby (1969) who noted that only the material equipment available in excavation can support the claim that it would take thousands of years for cultural "development" to emerge, but this evidence is rather misleading and technically impossible to collect and analyse in a laboratory. For example, pottery manufacture did not necessarily go hand in hand with domestication since early man originally domesticated plants and animals before manufacturing pots, thus pottery manufacture would be a misleading evidence for determining the origin and evolution of agriculture. Furthermore, not all hunters and gatherers lived at a subsistence level near to starvation condition. It is believed that no group on earth has more leisure time than hunters and gatherers, who spend it primarily on games, conversation and relaxing. Although, the origin of agriculture still remains a mystery, some of the earliest evidence of the revolutionary change in man's role from hunter and food-collector to herdsman and tiller is available in Western Asia in the hills and grassy northern plains encompassing the "Fertile Crescent" (Tigris and Euphrates). Precisely, where agriculture first originated is not clear, but archaeological evidence suggests that it originated in well-watered highlands of the Indus, Nile Tigris and Euphrates. Pre-agricultural subsistence in the "Fertile Crescent" depended heavily on hunting wild sheep, goat, pig, gazelle, deer, ass, fish, avifauna and small game animals. These are often supplemented by gathered wild fruits and seeds. But Sauer (1952, 1969) provided the following six hypotheses to support the origin of agriculture:

(a) Agriculture did not originate in communities of desperately short of food, but where there was sufficient freedom from want to experiment,
(b) The hearths of domestication are to be sought in geographical areas of marked diversity of plants and animals,
(c) Primitive agriculture originated not in the large river valleys subject to heavy floods and requiring protective dams, drainage for irrigation, but in hill lands,
(d) Agriculture originated in woodlands, which have workable soils,
(e) The inventors of agriculture had previously acquired special skills, but hunters would be least inclined towards domestication of plants and animals, and
(f) The founders of agriculture were sedentary folk. Crop

cultivation requires constant attention. Thus unless constantly guarded, the crop would be destroyed.

The above hypotheses are extremely interesting, although the last thesis failed to support the concept that pastoralism developed from the modification of mixed farming as this was practised in the arid areas of Central Asia, or the idea that mixed farming originated from the merging of seed cultivation and livestock husbandry, as differing economies merged in the region of the "Fertile Crescent" (Narr, 1956). But Harris (1969) observed that agriculture probably developed with the economically more generalised gatherer-hunter fisher populations occupying the more varied natural ecosystems, particularly the forest and woodland margins, which contained abundant plant and animal species than with hunters occupying specialised ecosystems with limited species, for example, natural grasslands. Despite the various beliefs and theories advanced above, we are confronted with the evidence emanating from archaeological, historical, palaeological, philological, botanical and environmental studies confirming that man, from time to time, changed his economies, settled in one spot and lived in a radically different way from his predecessors. But increasing evidence and emergence of absolute radio-carbon dating techniques have only confirmed our doubts about the exact dates of origin of agriculture, since the climatic crisis that terminated the Pleistocene epoch and subsequent change of environmental conditions are also suggested as good reasons for changing from nomadic way of life to sedentary life. This latter reason tends to suggest that studies of the origin of agriculture cannot be based on, say, archaeological data per se. Botanists and geographers have also endeavoured to draw considerable conclusions from the present-day distribution and genetical constitutions of domesticated and wild species of plants, although corroborative facts from the past have been a real challenge. However, archaeological evidence supports the view that initial agricultural activities were often halting and differed greatly in extent, specialisation and sophistication. The latter point is confirmed by Braidwood (1952, 1958) who coined the terms "primary village farming", "settled village farming", incipient cultivation", and "intensified village farming" to describe the different forms of early agricultural economy revealed by pre-historic excavations.

According to De Candolle (1882), agriculture originated mainly from three great regions: China, South-West Asia including Egypt, and inter-tropical America, which had no communication with each other. The original sources of domesticated plants remained an impenetrable secret until the middle of the 19th century when he made the first serious study based on a botanical approach and supplemented by evidence from archaeological, palaeological, historical and philological data. He noted three necessary phases: plant availability, favourable environment and food demand between the food-gathering stage and the cultivated stage. This approach which embraced investigation of occurrence of a given "crop plant" in a locality where it also grows wild, or where wild relatives existed appeared to be a gross oversimplification of the problem of origin of agriculture. De Candolle believed that all wheats belonged to two or not more than four species and that all the cultivated races of *Brassica* were still to be found in the wild state in Europe or Siberia. Ultimately, he strongly argued that species changes occurred directly through cultivation. Darwin (1859) observed that the relationship between the cultivated plants and their ancestral wild plants was in many ways, extremely difficult to determine. Since the ancestral wild plants seem to be exceedingly rare, he suggested that primitive man was obliged to eat almost anything that was possible to chew and swallow. Despite these dubious contributions towards a solution of the problem of the origin of agriculture, it is possible to attribute the origin of "crop plants" to their special way of hybridisation and selection, for example, the close spacing of crops tends to facilitate inter-crossing among the plants themselves, and with their parent species. But Vavilov (1951) criticised De Candolle's method on the basis that many cultivated species do not occur outside of cultivation, and wild relatives or "progenitors" may represent narrow groups of forms with but a small number of varieties, which are often isolated and which cannot explain the entire diversity by which a domesticated plant represented. He established the principle that the spatial distribution of plant species is heterogeneous. Vavilov established the idea that all the major crop plants originated in the tropical to sub-tropical regions of the world, and in some 11-12 centres of origin, based on his concept that centres of diversity could be equated with centres of origin. Although he added that the origins of plant domestication were to be found in the mountain zones in these centres, modern studies show that this was probably an over-simplification (Ucko and Dimbleby, 1969). However, Vavilov was convinced that "the wild species and varieties most akin to the cultivated plants form an ecological group with the latter", for example, barley, wheat, oats, lentils, melon, carrot and hemp, all of which possess wild relatives of similar ecological requirements. To Vavilov a number of regions contained large numbers of plant varieties, while Southern China, Indochina, Malaya, India, South-Western Asia, Ethiopia, Central America, the Andean region of South America, and Southern Mexico had an unusual number of plant varieties, but Siberia, Central and Northern Europe, and North America contained very few varieties. However, a peak concentration had been reached in the Caucasus and the mountains and foothills of Afghanistan and North India. Vavilov listed eight independent centres (Fig. 2.1) which illustrate the foregoing discussion. The eight independent centres of agricultural origins are found in alpine orogenic zones, tropical and sub-tropical zones, which often experience tectonic movements and climatic fluctuations. It is in such regions of hardships where there was perhaps need to carry out agricultural experimentation. These are the original centres of the world's most cultivated plants: rice, millets, vegetables, soyabeans, sugarcane, bananas, coconuts and other fruits. Although Vavilov found Central Asia, including North-West India to be the home of some grains, legumes and cotton, South-West Asia seemed more important for wheat, rye, alfalfa, temperate fruits and nuts. Consequently, he believed that some grains, for example, sorghum originated in Ethiopia. But he credited South Mexico and Central America with the domestication of maize, cotton, sweet potatoes (Table 2.1) and the Andean region with the potato. Vavilov noted the Mediterranean area as particularly significant as a secondary source in which selection occurred and the area in which the olive, vine and fig were domesticated. To date, the olive is always an indicator of the Mediterranean type of climate.

Through a rigorous botanical analysis of the varietal composition of certain species, Vavilov proved that "the species represents a definite heterogeneous and variable morphological-physiological system, the origin of which is associated with a particular environment and geographical area". He never found mono-typic species, but all species were made up of large or small numbers of hereditary forms. The degree of success attained by Vavilov in locating the geographical centres of origin of agriculture probably can be explained in terms of the thoroughness of his botanical analysis. His identification of the principal phytogeographical regions of origin of agriculture was based on "differential phytogeographic technique" involving a number of logical steps and procedures outlined below:

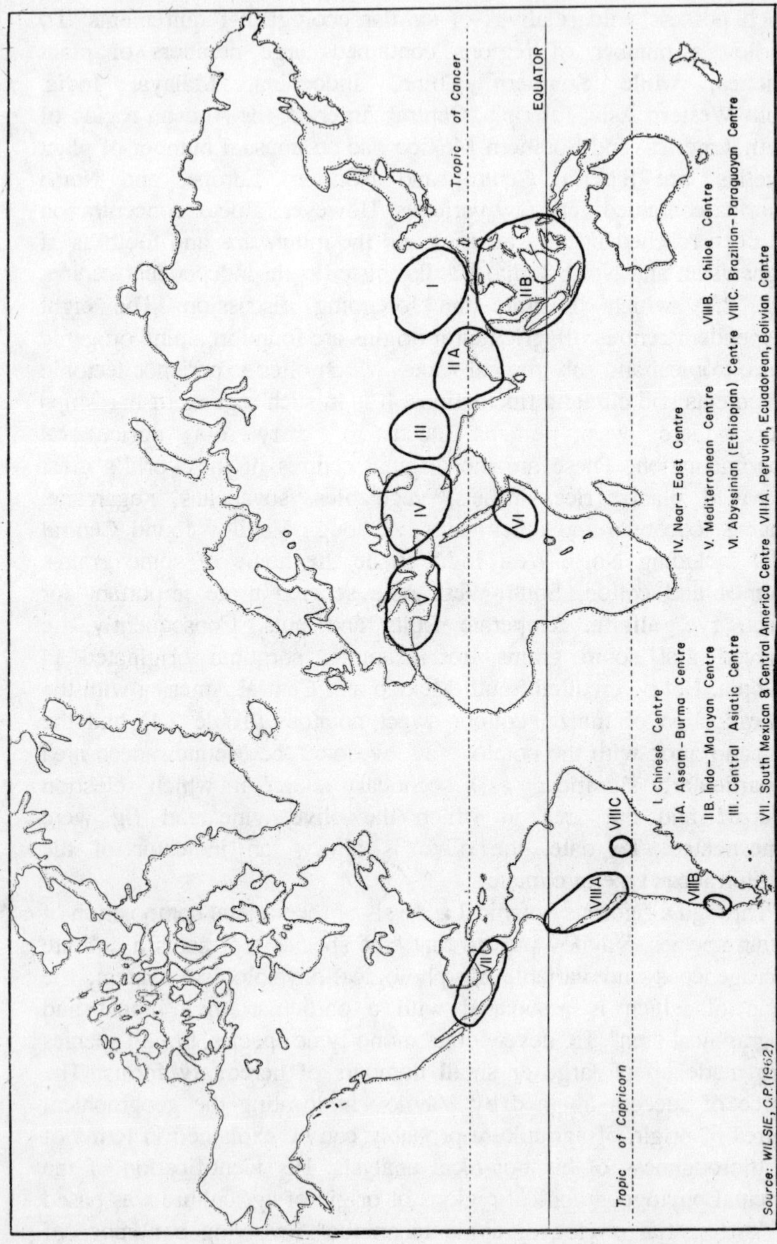

FIG. 2.1: World centres of origin of agriculture.

TABLE 2.1: Centres of origins and production of selected "crop plants"

Crop Plant	Centres of Origin	Centres of Production
Cucumber	Burma	Europe
Onions	Ethiopia	Europe
Radishes	China	Europe
Pyrethrum	China	Kenya
Cocoa	Brazil	Africa
Coffee	Ethiopia	Brazil
Maize	Mexico	United States
Monterey Pine	California	Australia
Pineapple	Brazil	Hawaii
Irish Potato	Peru	Eastern Europe
Wheat	Central Asia	North America
Cassava	South America	Africa
Sorghum	Ethiopia	Rest of Africa
Sisal	Mexico	Tanzania
Cotton (*Gossypium hirsutum*)	Southern Mexico	United States
Sugarcane (*Saccharun officinarum*)	South-East Asia	Cuba
Rice (*Oryza sativa*)	South-East Asia	Africa
Rubber (*Havea braziliensis*)	South America	Malaya
Beans (*Phaseolus vulgaris*)	Mexico	Africa
Sweet Potatoes	South America	Africa

Source: Compiled by the Author.

(a) A strict classification of plants collected using standard taxonomic methods. This was followed by arrangement into species and genetic groups based on systematics, morphological features, genetics, cytological similarities and immunology.

(b) The location of the original areas occupied by plant species in the past, when communications were more problematic than presently.

(c) A detailed determination of the nature and degree of variation in morphological and physiological characters in each species, composition of botanical varieties and races.

(d) Determining the regions and countries wherein occur the heritable forms of a given species and the geographic centres where the basic varieties are concentrated. As a "rule of thumb", in the original centres are found numerous endemic varietal characters.

(e) In order to establish precisely the centres of origins and initial formation of varietal forms, it was found necessary to determine the geographic centres of origins of varieties of closely related wild animals and crop plants.

(f) The original centres often contained a large number of genetically dominant characters, while recessive forms were considered to occur at the periphery of basic areas.

(g) Finally, a confirmation of the phytogeographic differential method may be found in archaeological, historically and linguistic data but, on the whole, these data are too general for practical purposes which require a concrete and exact knowledge of plant species and varieties.

Vavilov applying the above principles recognised eight primary geographical centres of probable origin of agriculture (Fig. 2.1). Nevertheless, it is yet to be confirmed whether these are recognised as true centres of origin or centres of great diversity. But Vavilov emphasised the need to distinguish between primary and secondary centres of origin of agriculture.

In summary, evidence available clearly shows that a majority of the domesticated plants and animals apparently originated in Asia. However, the primary centres of origins of agriculture developed independently, although there is noticeable overlapping of two centres in the Iraq-Iran region. It seemed to Vavilov that, within a centre of origin, there appeared to be small centres termed agro-ecological groups.

Evolution, Diffusion and Probable Dates of Origin of Agriculture

Domestication of plants and animals probably commenced between 12,000 and 10,000 B.C., when man began the great transition from hunter to tiller (Brown and Finsterbusch, 1972 and Symons, 1978). Available evidence shows that in Western Asia, animals were domesticated before plants, making the earliest agrarians there pastoralists. Early herdsmen frequently followed a rather pastoral, often nomadic way of life similar in many cases to those characterising fractions of populations in the Middle East, Afghanistan, and parts of Africa between 12,000 and 10,000 B.C. Further evidence suggests that the origins of domesticated plants and animals probably took place in 10,000 B.C. (Braidwood, 1958). According to Janick et al., (1974), the actual domestication of plants and animals dates back to 7,000—10,000 B.C. Bones of sheep identified as domesticated dated back to about 9,000 B.C. have been found in Northern Iraq, thus making the sheep the earliest known domesticates. But by 10,000 B.C. , varieties of wheat and barley were being sown and harvested in the Middle East, while at the same time rice was growing in Thailand. There seems to be a lot of confusion about the exact dates of domestication of plants and animals. However, the distinctiveness of the agricultural complex that developed in South-West Asia combined seed cultivation, wheat, barley, lentils with animal husbandry: sheep, goats and later followed by cattle and pigs. Here then lies the evidence of their origins of modern mixed farming, which was evolving in this geographical area between 9,000 and 5,000 B.C. (Symons, 1978). Domestication of squash, peppers, gourds, and small beans took place between 7,000 and 5,000 B.C. in Middle America while runner beans *(Phaseolus Occineus)* and common beans *(Phaseolus vulgaris)* are believed to have been cultivated by about 9,000—7,500 B.C. and 6,000—430 B.C. respectively in Mexico.

The origin and historical development of maize is even more interesting to agricultural science. Two places of possible origin have been suggested : the Mexico-Guatemala region and the Peru-Bolivia-Ecuador region. This suggestion is based on the "tripartite hypothesis" theory of the origin of agriculture which states that:

(a) Cultivated maize originated from a wild form of pop corn which was once, and perhaps still is, indigenous to the lowlands of South America.
(b) *Teosinte*, the closest relative of maize, is a recent product of the natural hybridisation of *Zea* and *Tripsacum*, which occurred after cultivation had been introduced by man into Central America.
(c) New types of maize originating directly or indirectly from this cross and exhibiting admixture with *Tripsacum* comprise the majority of Central and North American varieties.

Further evidence favours the view of Vavilov that maize probably originated in Mexico or Central America between 6,000 and 5,000 B.C.

Much later, probably about 1,000 B.C. , oats came into cultivation, while rye is considered to be of considerably later origin.

In America, the big-game hunting pattern prevalent in North America became widely disseminated through Central and South America during the period 9,000—5,000 B.C. Following the glacial retreat and the climatic shift to increased warmth and drought, modified hunting patterns developed. Population in some areas affected by drought were forced into new subsistence patterns of food-collecting, and eventually four distinct traditions of incipient farming emerged in the Americas. The first and second incipient farming known as Nuclear America included the Southern two-thirds of Mexico, all of Central America, and Andean and Coastal Colombia, Ecuador and Peru, with adjacent portions of Bolivia. A third incipient-cultivation tradition centred in the tropical forests of the Amazon or Orinoco, while a fourth and distinctly lesser tradition developed in Eastern North America in the Mississippi Valley System.

Slicher (1963) distinguished the following five types of agricultural practices carried out in Europe :

(a) Temporary cultivation, followed by an indefinite fallowing period.
(b) The infield-outfield system and similar systems in which part of the land was cultivated continuously with heavy manuring, and other parts occasionally.
(c) The two-course rotation, in which the land was tilled and left fallow in alternate years.
(d) A three-year system in which the land was tilled for one year and left fallow for two years.
(e) The three-course rotation, in which winter crops (corn, wheat or rye) were followed by spring crops (corn, barley or oats) and then followed by fallow.

The last method increased food production, but not sufficiently to sustain a rapidly expanding population. Rising population and growing circulation of money led to a great increase in cereal prices which stimulated reclamation of marshes, forests and marginal lands for agricultural purposes. New scientific devices, such as the Archimedes Screw facilitated new approaches to irrigation and other aspects of land improvement and cultivation, although the introduction of scientific agriculture was probably mainly a function of the growth of monetary economy, and abandonment of subsistence agriculture in favour of commercial farming. But the subsequent population pressure led to extensive and intensive agriculture for both food and industrial crops.

A fully developed nomadic pastoralism is said to have eventually diverged from Neolithic mixed agriculture, diffusing into the enormous grassy plains, steppes and deserts of Central Asia, stretching from Europe to China and into Northern Arabia. In time, knowledge of farming spread throughout the world moving eastward across Asia, southward into the Tigris-Euphrates, then along the Nile into Africa, and north-westward in Europe through the Danube Valley and along the Mediterranean coast (Fig. 2.2). On the other hand, cows domesticated in the Middle East dispersed in South and North America, Europe, Asia, Africa and Australia where they competed for grass with wild animals. Eventually better animal husbandry resulted in increasing supply of meat and dairy produce from the regions not suited to arable agriculture. Consequently, further adaptation to local climate, soil and transport conditions led to refinement of breeding to improve either the milking or the fattening qualities, so producing the specialised dairy breeds such as the Jersey, Friesian, Ayrshire and Dairy Shorthorn and the beef breeds, Hereford, Aberdeen-Angus and Beef or Scotch Shorthorn — all of which are now found in every part of the world including Africa. Similarly, sheep breeds were improved by selective breeding, to serve the needs of the manufacturers for different grades of wool and of butchers for different classes of mutton. Originally, sheep, goats and cattle were raised primarily to provide meat and hides. Wild sheep had hair but not wool, thus through evolution wool-bearing type emerged. Similarly, wild and early domesticated cattle gave little milk, but through selecting, favouring and breeding practices there emerged proper milk-producing strains. Consequently, man has developed chickens specialised for egg production and other types specialised for broiler production. In North America cattle ranching developed in South-east Texas by Southern Anglo-Americans diffused throughout Texas and Western America. In addition, sheep-rearing found to be an effective way of utilising the semi-arid environments of Australia later diffused to New Zealand.

The colonisation of tropical and sub-tropical areas led to the introduction of plantation system of agriculture, for example, in Africa and Mexico. At the same time the agrarian revolution was intimately interwoven with and dependent on the industrial revolution. Moreover,

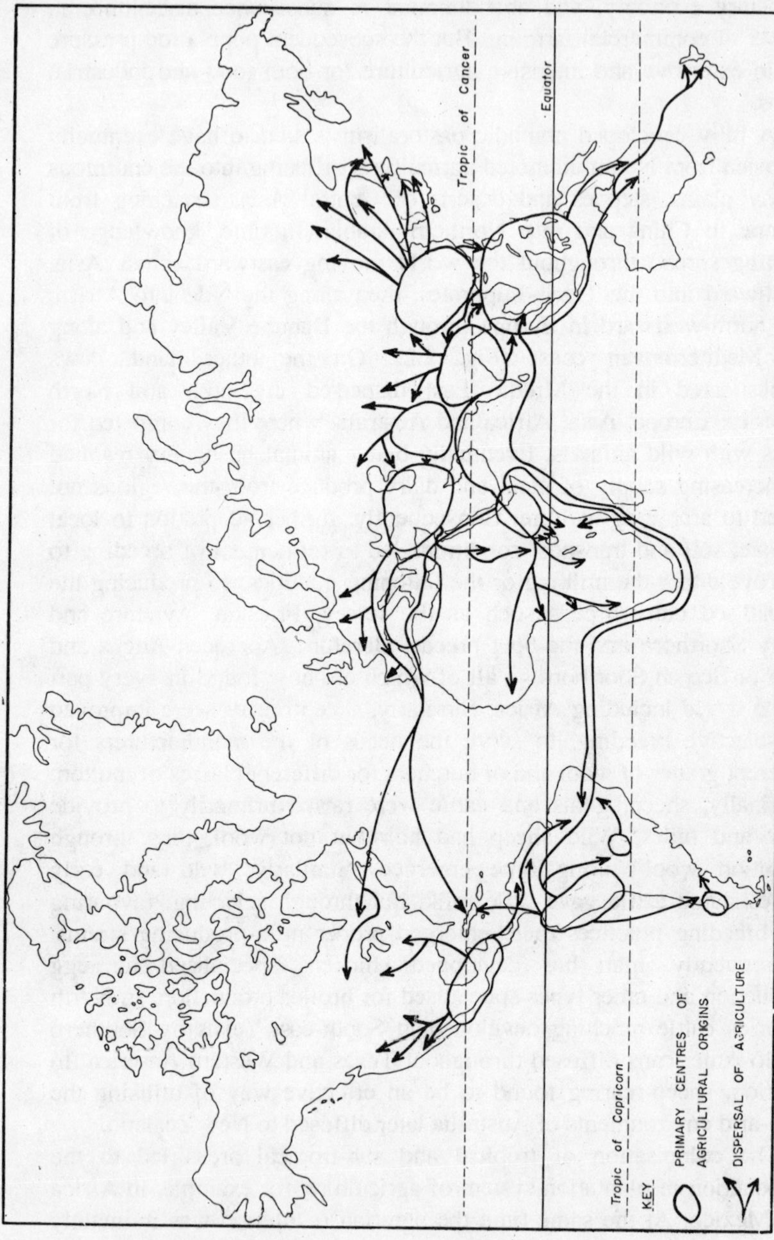

FIG. 2.2: World map showing dispersal of agriculture from primary centres of origins.

specialisation became an important feature of the industrial revolution, which stimulated attention to more efficient agricultural practices, for example, increased use of fertilisers, fungicides, weedicides, pesticides and the proliferation of manufacturing industries became an important characteristic of agricultural development during the late 19th and 20th centuries. Manufactured nitrogenous and superphosphate fertilisers minimised the dependence of the arable farmer on livestock to supply manure. But innovations in grain farming developed, for example, in America were adopted also in Australia, Africa and specialisation emerging in all new areas alleviated the market demands. After 1882, refrigeration facilitated marketing of meat and dairy produce from New Zealand, Australia and South America, especially to Britain.

Significance of the Theory of Plant and Animal Introduction

With the domestication of plants and animals, man started to change the earth's ecosystem, for example, deforestation for his well-being. Environmental consequences have multiplied with the introduction of agriculture, for example, extension of agriculture into marginal lands resulted into severe soil erosion in the mid-west of the United States often known as the "dustbowl" of the 1930s. Increased deforestation and grazing accelerated desertification not only in Africa but also elsewhere in the world. Consequently, chemicalisation of the agricultural land through application of fertilisers, insecticides, fungicides and herbicides has increased pollution and changed the equilibrium of the biosphere in many regions. In addition irrigated agriculture in the arid and semi-arid environments, for example, in Egypt poses a serious problem of secondary salinification. One of the major accusations against the "Green Revolution" is that it has brought about the massive discarding of local plant varieties, in which 10,000 years of genetic diversification is stored, in favour of a few High Yielding Varieties (HYV). But these plant materials are not indeed "High Yielding" as often stated, since their yields are realised only as a response to high inputs of fertilisers, insecticides, fungicides and herbicides, with all the environmental consequences mentioned above. Moreover, the so-called "High Yielding Varieties" are vulnerable to environmental fluctuations, for example, climatic changes and infestation of pests and diseases. Despite the foregoing criticism, through time, more varieties of crops : seeds, roots, greens and fruits, and many types of animal were adopted by farmers. The farming practices eventually diffused outwards from the primary centres of origin as one farmer copied another. The result is that modern agriculture is often a mixture of local elements together with elements introduced from abroad. The dispersal mechanism is also important for an understanding of the present features in world agricultural development. The spread of cassava and maize introduced from South America has considerably modified African diet patterns during the recent decades. Today the centres of production of nearly all agricultural crops are far separated from their primary centres of origins (Table 2.1). But out of the estimated world total of about 350,000 plant species, the number of plants that fit economically into man's activity is probably between 1,000 and 2,000 species (Janick *et al.*, 1974). On the other hand, those of major significance in world trade are relatively few, about 100 to 200 plant species. However, 15 species provide the bulk of the world's food crops : rice, wheat, corn, sorghum, barley, sugarcane, sugarbeet, potato, sweet-potato, cassava, bean, soyabean, peanut, coconut and banana.

The gradual evolution of agriculture resulted in a dependable surplus, which releases from food production those individuals with skills in other specialities. Specialities may become important features of agriculture, for example, the introduction of root crops not only enables fertility to be better maintained while yielding surplus crops, but facilitates a degree of specialisation, as between turnips for livestock and potatoes for human consumption, especially in Europe. Consequently, labour is often drawn from agriculture to work in manufacturing industries. To date, the origin of agriculture culminated in abundant food supply assured by the planting of vegetative parts or seed. Plants that grow rapidly and produce a crop within a short time season were probably the first to be cultivated because the technology involved in the cultivation of long-lived plants, for example, fruit tree, is time-consuming and requires a more sophisticated technology and thus these latter crops probably continued to be obtained from the wild plants for a considerable period of time. For example, the cassava, a major source of carbohydrate (starch) in tropical South America contains hydrocyanic acid, a deadly poison, which could be removed through the cooking process. Short-lived plants, for example, "Katumani maize" are needed for the agricultural development of semi-arid areas. Therefore, the introduction of plants and animals is effectively continuing in many parts of the world.

As has already been pointed out, crop plants were not originally evenly distributed throughout the world. Many of the current agricultural successes in the United States, Australia, Canada, Africa and the like have been achieved through the introduction and evolution of crop materials which originated in Asia, Europe, South America and the USSR. Similarly, the agricultural economy of practically every nation has been influenced considerably by the introduction of crop plant materials and animals species. Even in regions well supplied with a wide genetic diversity of a given crop plant, great benefits have been obtained through exchange of breeding materials with researchers in other countries. This has been the case in cotton development in Kenya, where breeding materials have been introduced from India, the USA, USSR, Tanzania, Uganda, West Africa, Central Africa and the like. The development of other crops, for example, sugarcane and maize have followed suit in Kenya. No doubt the breeding of new varieties of crop and livestock alike have contributed to increased agricultural productivity elsewhere in the world. But in many countries, governments have increasingly influenced farmers through price controls, marketing organisation, production subsidies, purchase, guarantees, health regulations and innumerable other ways, which may result in a positive or negative response in agricultural production. Despite government intervention in agricultural development, man's role in plant and animal dispersal has been remarkable in altering centres of diversity from time to time. In view of this latter observation, today's heaviest production of important crop species may be found in areas quite remote from the primary centres of origin of agriculture (Table 2.1).

After preliminary screening, introductions are made available to plant breeders and other interested workers for further evaluation and possible use as breeding materials or as potential new crops. The phytogeographical investigation carried out by Vavilov has particularly been invaluable to plant introduction work elsewhere in the world. Plant breeders often look to original centres of diversity for possible resistance to disease, pest, drought and other environmental constraints. It appears that even now, with all the introduction work that has taken place, we have still sampled but a small part of the total genetic diversity available in most crop plants. Moreover, climatic, edaphic and biotic conditions have received increasing attention in plant introduction work. Introduced materials should be sent to areas having similar ecological conditions, thus there is urgent need for agro-ecological approach to the study of origin and evolution of agriculture. The ecological approach is even more relevant to the world food problems due to increasing population growth, particularly in developing countries. However, often there is a planned distribution of varieties according to the agricultural needs of every country.

In summary, the introduced varieties have often been subjected to natural selection, deliberate or casual artificial selection and some local varieties have evolved that are ecologically adapted. But the introduced crops, say, from Middle Eastern U.S. and Transcaucasian, for example, wheat, barley, forage, grasses, cotton, grapes and fruit trees require great care in breeding in order to retain all the valuable properties they originally possessed.

Concluding Observations

It appears undoubtful that agriculture originated independently in various parts of the world at different times and developed through the dispersal and divergence of new plant forms in new environments. Consequently, the study of the theory of origins of domesticated plants and animals is further complicated by the varieties that man has to-date developed. Nevertheless, there is lack of data on which to base a sound theory of origin and evolution of agriculture. As agriculture diffused into more severe environments, seeding seemed to have superseded planting of vegetative parts as the dominant agricultural practice. Most of the crop plants and domestic animals, which currently provide us with food, income, employment and foreign exchange earnings are dependent on the indigenous plant varieties in Asia, Latin America and Africa. Virtually everything that we eat today can be traced back to about a dozen centres of extreme genetic diversity from which crop plants diffused into other parts of the world.

Following several years of painstaking research it has been proved that a combination of varied ecology and cultivation techniques resulted in almost all the major crops originating on less than a quarter of the earth's arable land. The principal centres of origin of agriculture are now believed to be : Mediterranean, Near East, Afghanistan, Indo-Burma, Malaysia, Java, China, Guatemala-Mexico, Peruvian Andes and Ethiopia. With the exception of a small land area around the Mediterranean, the industrialised world is excluded from the centres of genetic diversity. From these observations one would conclude that the primary geographical centres of agricultural origins are associated with : (a) the tropical and sub-tropical zones, and (b) the alpine orogenic zones, which are areas of instability often experiencing severe earth-quakes. The other areas of alpine orogenic zones are excluded from the centres of genetic diversity perhaps because they were limited by meteoro-climatic conditions. However, the genetic home of most of the world's famous grains – oats, wheat, barley, rye, lentils and flax – appears to be the Fertile Crescent and Ethiopia.

The theory of origin of the agriculture confirms that the process of cultivation has improved the heredity of plants originally found existing in a wild state. Consequently, the traditional centres of genetic diversity in the developing countries are vital to plant breeders all over the world, although most nations are facing the danger of losing this precious natural heritage because the "Green Revolution" is bringing about the erosion of these centres.

Apart from the inter-disciplinary approach to the study of plant introduction, there is urgent need for a well-coordinated international approach as one of the most effective ways of dealing with the problem for all interested countries. To date, better cooperation between the country receiving collections and the country where collections are being carried out is prerequisite. This means close collaboration of specialists in the various aspects of crop science during collection and a thorough understanding of the diversity of the species collected, related wild species, and primitive forms which may still be in existence. But the development of agriculture must be conducted with environmental consciousness in order to conserve natural resources and avoid pollution of the biosphere.

REFERENCES

Braidwood, R. J., 1952 : *The Near East and the Foundation of Civilisation*, University of Oregon Press.

———, 1958 : 'Near East Pre-history', *Sci. 127*, pp. 1419-30.

Brown, L. R. and Finsterbusch, G. W., 1972 : *Man and His Environment : Food*, Harper and Row.

Darlington, C. D., 1956 : *Chromosome Botany*. Geo. Allen and Unwin Ltd., London.

Darwin, C., 1856 : *Origin of Species by Means of Natural Selection*, Reprinted by Philosophical Library, New York.

De Candolle, A., 1882 : *Origin of Cultivated Plants*, Appleton, New York.

De Silva, D., 1980 : "The Vanishing Seeds", in *Daily Nation*, Wednesday, March 26, 1980, Nairobi.

Dunlop, S. and MacDonald, D., 1976 : *Farming and the Countryside*, Heinemann.

Harris, D. R., 1969 : *Agricultural Systems, Ecosystems and the Origins of Agriculture*, in Ucko, P. J. and Dimbleby, G.W. (eds.), *The Domestication and Exploitation of Plants Animals*, pp. 3-15.

Janick, J. et al., 1974 : *Plant Science : An Introduction to World Crops*, W. H. Freeman and Company.

Narr, K. J., 1956 : Early Food Producing Populations, in *Man's Role in Changing the Face of the Earth*, edited by W.L. Thomas, Chicago, pp. 131-151.

Sauver, C. O., 1952 : Agricultural Origins and Dispersals, *American Geographical Society*, New York.

Sauver, C. O., 1969 : *Agricultural Origins and Dispersals: The Domestication of Animals and Foodstuffs*, M.I.T. Press.

Slicher Van Bath, B., 1963 : *The Agrarian History of Western Europe, A.D. 500-1850*, London.

Symons, L., 1978 : *Agricultural Geography*, Bell and Hyman Limited.

Ucko, J. P. and Dimbleby, G.W., 1969 : *The Domestication and Exploitation of Plants and Animals*, Gerald Duckworth and Company Ltd.

Vavilov, N. I., 1951 : *The Origin, Variation, Immunity and Breeding of Cultivated Plants*, Ronald, New York.

Wilsie, C. P., 1962 : *Crop Adaptation and Distribution*, W. H. Freeman and Company.

3

ORIGIN AND DEVELOPMENT OF AGRICULTURE

S. S. DHILLON

FROM ANTIQUITY to the present times, plants, animals and men have remained in an intimately linked position by a complicated web of inter-connections, but still the problem of how, where and when agriculture originated and developed has posed an intelligent challenge to all of us. The relationship of man with the land is confirmed and is indeed inseparable, but what circumstances forced him to take the decision of domesticating plants and animals are still baffling the scholars. Different scholars hold different views about the time and place of the origin of agriculture. Highly specialised investigations into the minutiae of the problem are essential, but expertness in one science sometimes leads to myopia in other directions; it is often easier to examine the trees than to attempt to comprehend the wood (Harris, 1967). For this reason, studies made from time to time regarding the origin and development of agriculture by non-geographers, such as anthropologists, botanists, palaeobotanists, archaeologists, historians, ethnologists, geneticists etc., are reviewed here to make such an assessment.

Russian biogeographer Vavilov (1935) has noted that there were eight independent centres of origin of agriculture, where the most important economic plants of the world first domesticated. They are : China, India and the Indo-Malayan Centre, Central Asia (including North-West India), the Near East, the Mediterranean Centre, Ethiopia, South Mexico and Central America and South America (Fig. 3.1). Cultural historian Childe (1937) and anthropologist Braidwood (1952) are of the view that the river valleys and loess lands were the primary hearths of agriculture. This view is, however, rejected by geographers especially Sauer (1952) and Wissmann (1956) on the ground that such areas are always liable to flooding and drought conditions. They wrote that the favourable sites would have been the more moist mountain lands of tropical Southeast Asia as the original agricultural centres of the world. A well-known hypothesis presented by Sauer in 1952 has greatly strengthened our factual and theoretical knowledge about the domestication of plants. A decade later, Zukovskij (1962) made an attempt to comprehend a scientific investigation on the basis of available data on some 140 genera of earlier cultivated plants. His account is more detailed and dependable as it throws an abundant light on the extraordinary diversity of cultigens of man.

Recently, many specialists made detailed studies on different crops (maize, sorghum, wheat, barley, oats, rye, potatoes, forage, grasses and legumes) to highlight the processes of plant domestication. The work on crop-plant evolution edited by Hutchinson (1965) concludes on the biological principles that have governed plant domestication. He stresses the extremely rapid differentiation of major crop plants into geographical races, which is a consequence of man's vast extension of their range of cultivation into new and diverse habitats. Further, Harlan (1961), Simmonds (1962) and Stearn (1965) have explained the importance of taxonomy and genetics of plants to unmask the mystery of their domestication. They have approached the problem from different points of view, yet their common point of interest is the same. So a thorough investigation of plants' origin, both botanical and archaeological, will throw an abundant light on understanding the transition from non-agricultural to agricultural ways of life. Therefore, new techniques of investigation are amply needed to establish micro-ecological conditions under which early agriculture arose as they were not essentially the same in the past as they are today.

The taxonomist, the geneticist and the agriculturist have viewed the present problem of plant and animals domestication quite differently, but their focus of interest is exactly the same, that is, the circumstances and combinations in which it originated and developed. How agriculture began is the job of the biologist and where and when agriculture originated is the task of archaeologists? These are to be supplemented by that of palaeobotanists, paleozoologists, and paleoethnologist who identify the macroscopic and microscopic remains of cultivated plants and domesticated animals. Further, archaeological evidence duly supported by the techniques of aerial photography, carbon-14 dating, pollen analysis, etc., reveals that

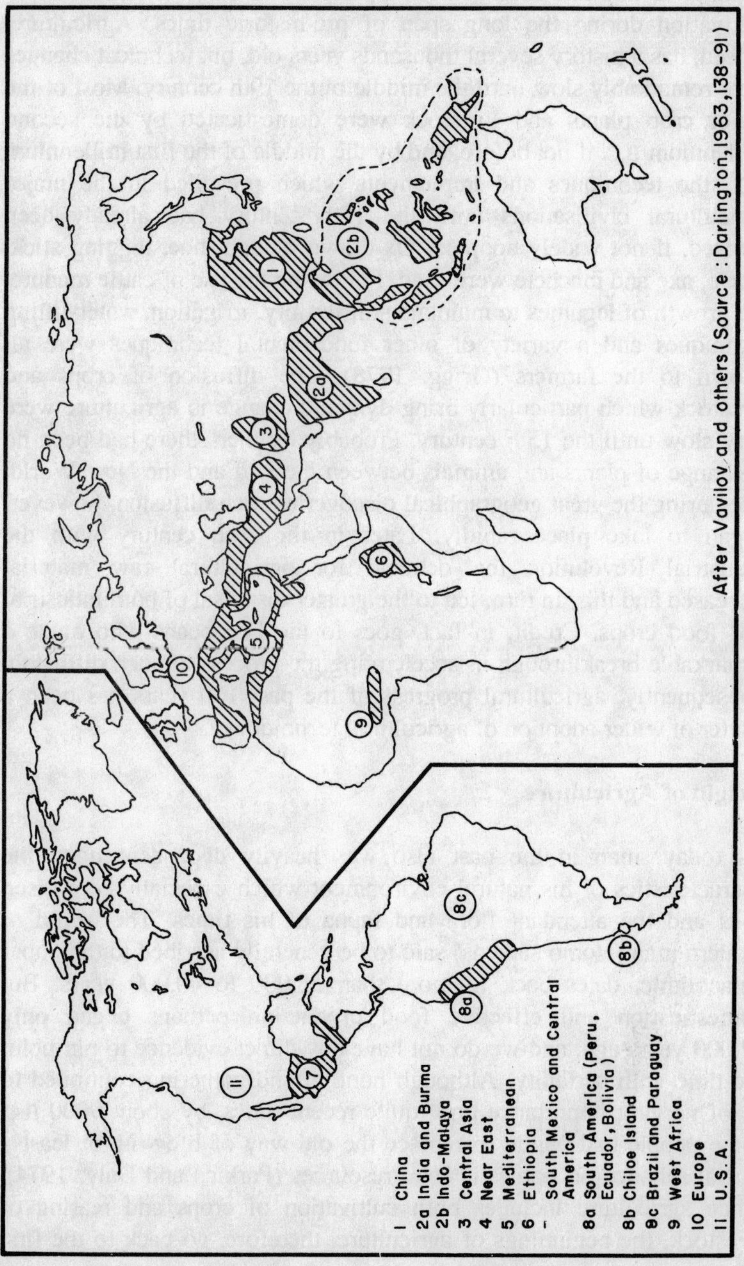

After Vavilov and others (Source: Darlington, 1963, 138-91)

FIG. 3.1: Major areas of plant domestication.

hunting, fishing and food-gathering persisted side by side with simple cultivation during the long span of pre-historic times. Agriculture, indeed, has a history several thousands years old, but technical changes were remarkably slow until the middle of the 19th century. Most of the major crop plants and livestock were domesticated by the second millennium B.C. if not before, and by the middle of the first millennium B.C., the techniques and implements which prevailed in the major agricultural civilisations until the 19th century had already been devised, if not widely adopted. Ox-drawn plough, hoe, digging stick, sickle, axe and machete were used. Following the use of cattle manure, the growth of legumes to maintain soil fertility, irrigation, water-lifting techniques and a variety of other fundamental techniques were all known to the farmers (Grigg, 1978). The diffusion of crops and livestock which particularly bring dynamic change in agriculture were very slow until the 15th century. Probably by then, there had been no exchange of plants and animals between the Old and the New World. But during the great geographical discoveries such diffusion, however, began to take place rapidly. Later in the 19th century with the Industrial Revolution the demand for agricultural raw material increased and this, in turn, led to the greater dispersal of both industrial and food crops. Credit, in fact, goes to the Europeans who made a remarkable breakthrough in accelerating the process of such diffusion. Subsequently, agricultural progress of the past 100 years has been a matter of wider adoption of agricultural techniques.

Origin of Agriculture

As today, man in the past also was heavily dependent upon the characteristics of his natural environment which especially comprised soils and the attendant flora and fauna of his times. The origin of modern man (Homo sapiens) said to be generally ascribed to the upper Palaeolithic, dates back to more than 35,000 to 40,000 years. But domestication and effective food production perhaps began only 12,000 years ago, and we do not have any direct evidence to pin-point the time with certainty. Although hunting and gathering continued to be of varying importance until quite recent times, by about 9000 B.C. some people had begun to replace the old way of life with at least a partial reliance on domestic food resources (Parkins and Daly, 1974). Since agriculture includes both cultivation of crops and rearing of livestock, the beginnings of agriculture, therefore, go back to the first domestication of plants and animals at different times in different parts of the world. At the initial stage plant domestication gave rise to "seed" agriculture and "vegeculture" which are to be differentiated. The former includes the cereals which are now staple food crops, in almost all the parts of the earth. In seed agriculture seeds are sown en masse and crops are harvested usually in very short period. They have marked growing seasons. This necessitated arrangements for storage facilities during off-seasons. It has formed the basis of the major agricultural civilisations and was associated with an early plough and drought animals. The latter refers to plants reproduced by vegetative propagation. They include tropical roots e.g., taro, manioc, yams, sweet potatoes, arrowroot.

It seems likely that primitive agriculture developed in the tropics, on the boundary between forest and grassland in both the Old and New Worlds. It is almost certain that vegeculture preceded seed culture. It has since been reduced by the expansion of seed culture. Little is known about implements and methods practised in farming. The digging stick and the stone axe were the main implements, and shifting cultivation was practised by clearing natural vegetation with fire, though hunting, gathering and fishing were contributing much of the food supply.

In the Old World

Archaeologists have long pin-pointed the land masses of the Old World as the most likely scene of man's maiden experiment with agriculture. The variegated ecology of hills and valleys of South-West Asia is considered an ideal by many to be the birth place of agriculture. The fact is that such a topography is rich in species, richer than plains, as the myriad micro-climates of mountains and valleys offered all chances of survival and multiplication of innumerable plants. Consequently, this tract has probably yielded the richest and the earliest evidence of plant cultivation and animal husbandry as well. Excavations of the sites of Neolithic villages pushed back the earliest date of agriculture by more than 7000 B.C. Current excavations also continue to claim the importance of South-West Asia as a primary hearth of domestication of flora and fauna, the first urbanisation, the invention of the wheel and the first use of iron weapons. These interesting changes in fact took place in a relatively short span of time, and were of such fundamental importance that the question promptly arises : Why did they occur ? There are no definite answers at the moment but there are various explanations having varied degrees of probability. The climatic change is one which perhaps occurred near the close of the last phase of continental glaciation. The late Pleistocene/Holocene is one of the periods about which there are sufficiently precise carbon-14 datings and palaeo-temperature indicators quantifying the climatic phenomenon. One of the most interesting points of the period is the rise of air and sea-water temperature. The mean air temperature in Western Europe, according to Fairbridge (1967), had risen 3-4°C from 16,000 to 12,000 B.P. (14,020 to 10,020 B.C.), 7-8°C (that is 1-2°C higher than today) by 8,600 B.P. (6620 B.C.). By 6000 B.P. (4020 B.C.) in middle latitudes, the mean air temperature was 2-5°C above the present as indicated by Palaeobotanical material. Such anomalies over a sufficiently long period might have led to climatic changes which seems obvious and the same is being involved as the crucial factor in causing the change from hunting and gathering to food production. The retreat of glaciers seemed to have shifted the rain belts, with which desiccation increased all over South-West Asia. Consequently, crowding men, animals and plants happened to concentrate into a few dwindling oases where increasing close association might have led to domestication of animals and plants. Butzer (1964) too took up the environmental conditions and discussed the geographical circumstances under which South-West Asian agriculture originated. He believes that there may have been a minor cold phase during the eighth millennium B.C. and possibly a moister period locally in Palestine and Egypt between 9000 and 7000 B.C. might have occurred. Such hypotheses suffer from numerous drawbacks. It seems more and more doubtful that any of the agreed upon climatic crisis, i.e., the retreat of the last glaciers (*c.* 11,000 B.C.) or the latest altithermal (*c.* 4000 B.C.) or other conditions, according to Parkins and Daly (1974) bear a useful relationship to the evidence for early food production in the area. The fact is that human behaviour furnishes an even stronger argument that retreat of glaciation might have made men and animals as close neighbours in a closed environment and the most probably results would be that men would hunt animals to extinction rather than domesticate them.

Now there is another hypothesis which explains that the prevailing conditions at the last glacial period caused either extinction or the retreat of the bigger game animals from this part of the world. This perhaps necessitated man to switch over to smaller and more amenable animals. It is believed that most of the animals hunted by man as early as the Middle paleolithic persisted until historical times. Some of the faunal records are being revealed by Shanidar Cave and Mount Carmel in the area. The principal food animals remain the same for 50,000 years : sheep, goats, cattle, pigs, red deer, roe deer, fallow deer, gazelle and various equids. Elephants, lions, leopards and bears are equally durable in the record, although obviously never an important dietary source. Sheep and goats were important prey for millennia before they were brought under culture control, but there is nothing in the record to relate their change in status to any major climatic event. Neither palaeontological nor palynological data show any significant climatic change for many thousands of years on either side of the period which apparently saw the first attempts at food production (Parkins and Daly, 1974). Such notions, however, still require more detailed investigation in the historical ecology which are expected to unravel the problem of origin of agriculture.

Rejecting the climatic change as a fundamental cause of shiftover to food production, another equally interesting explanation is that of population increase which is most likely to have brought about a marked change in economic strategies. It is true that increased pressure of population forces man to find new resources and discover new methods of exploiting existing resources. But there seems to be no powerful evidence which may reveal what factors could have caused increase in population under the almost automatic biological controls on population and an abundance of both animals and plant foods.

In the ecosystem the human hunters are in fact predators and their role as predator differs in its zoo-geographic effect from that of carnivores. Wolves preying on herds of wild sheep or deer commonly take the newborn, the very old or the diseased thus, in effect they cut the herd of its weakest members. The most promising result is that the breeding capacity of the herd is in no way diminished, and the predation may even be beneficial in many a way. Human hunters, on the other hand, act in a quite different and more destructive way since all evidence seems to show that they hunt a quite high percentage of healthy adult animals which are seldom preyed by carnivores. This sort of behavioural practice inflicts a heavy loss upon flocks of herbivores and if it persists over a long period, it would undoubtedly deplete the supply of hunted animals. This in turn would rightly have its effect as the game diminished the carnivores would be affected much to find food, whilst omnivorous man could turn to other food sources — wild cereals, fruits, seeds and nuts. From his differing role as a predator, man could also learn, perhaps, to move into such ecological niches as might not support hunters but would rather be more favourable for those people who had initiated to utilise a broader spectrum of resources. The hunters remained hunters until this time but the others expanded the economic possibilities slowly but steadily to achieve better and better levels of living.

Mellart (1958-61) made the most impressive contribution after having studied two sites of South-Western Turkey, namely, Hacilar and Catal Huyuk which exhibit the presence of substantial farming communities in the sixth and seventh millennia B.C. The development of agriculture was based on the cultivation of three most commonly domesticated grains : einkorn wheat (*Triticum monococcum*), emmer wheat (*Triticum dicoccum*), and two-row barley (*Hardeum vulgare* ssp. *distichum*). The other domesticated plants were peas (both the field pea and the purple pea), lentils, bitter vetch, etc. Archaeologists have assumed that the domestication of these cereals occurred roughly contemporaneously with the domestication of some important food animals, namely, sheep, goats and cattle.

In the western flank of the Zagros Mountains and the Jordan Rift Valley numerous earliest sites occur which yield conclusive proof of agriculture. A great many grinding stones, mullers, and querns presumably for the processing of various seeds, grains and nuts are available about 9000 B.C. It seems that Mesopotamian lowlands did not have sufficient rainfall for non-irrigated agriculture, and possibly, plant and animals domestication first occurred in the upland areas of greater rainfall where Neolithic farming communities found the best agroclimatic conditions. This piedmont intermediate zone between the Mesopotamian lowlands and the Zagros Mountains of Kuridstan and Luristan had sufficient rainfall in winter for dry farming and large rivers (the Tigris and the Euphrates) for irrigation in dry season or in areas of insufficient rainfall.

Before 10,000 B.C. nomad peoples used to follow the natural migration of wild herbs. But sheep and goats have no natural migrations. They were first domesticated about 10,000 years ago, and the nomad had to lead the helpless herd. It is often assumed that the first domesticated animal was the dog. But there is no indication of this animal anywhere to the effect that it was ever important in the Neolithic economy as a food animal. It, however, might have been an important animal as a camp follower as is today. So the most commonly raised stock comprised sheep and goats. During the period spanning 2000 years from 9000 to 7000 B.C. stock raising was the most simple. It is said that sheep were first domesticated at the north-western end of the Zagros Mountains; goats were first domesticated at the south-eastern end of that mountain chain. But during the period 7000 and 6500 B.C. cattle and pigs entered the husbandry which spread throughout South-West Asia. Pigs particularly, at no time were an important economic factor, perhaps because, unlike other potential domesticates, they do not normally eat rough fodder but actually compete with man for food. By 6,000 B.C. animal husbandry became more diversified. A single domesticate is usually sufficient. If it is kept solely for meat; but each of three primary domesticates has its own valuable secondary products — wool from sheep, milk from goats, and milk and hides from cattle. Therefore, quite simply, all these domesticates continued to flourish as they are economically more productive than the others. By about the same time, a mixed herding farming economy came into existence. The donkey and ox were used for labour. The horse was late and not much used, while cows and calves were obviously important for milk and meat, and sheep, goats and pigs were also plentiful. It seems that sentary agriculture villages first characteristically developed in the uplands and highlands. The agricultural development in the alluvial lowlands of the Tigris-Euphrates, however, took place much later as these lands were not made habitable, until irrigation became fully available, which had to make man free from a partial dependence on hunting and gathering wild food.

Until recently South-West Asia has been the main focus of interest for agriculture origins; but the new excavations in other parts of the Old World have also yielded fresh evidence in recent years. Many details are also available of how and when agriculture spread from South-West Asia to the North-West temperate areas. Cultural diffusion is, however, relatively a well-known and dependable story. The most interesting and significant new evidences have been noted as derived from the excavations in Northern Greece and the Southern Balkans. Rodden (1965) explains that the earliest widely known Neolithic site at Nea Nikomedeia on the Macedonian Plain in Europe provided an evidence of the remains of 6000 B.C. of crops and animals such as wheat, barley and lentils; and sheep, goats, cattle and pigs respectively. Further, northward spread of Neolithic cultures across Europe has long been accepted. The greater details may be had from the carbon-14 dating and pollen analysis which have now made it more feasible to trace the spread of agriculture with much greater precision.

Towards the south in the continent of Africa, the antiquity of man as a toolmaker, according to Clark (1964), goes back to nearly two million years; but man as an agriculturist hardly reached back to the fifth millennium B.C., and restricted to the north-east corner of the continent. By about 4000 B.C., sedentary villages practising farming were scattered throughout the Nile's flood plains. The Neolithic cultures of the Faiyum Depression and the lower Nile Valley exhibit some proof of the domestication of flora and fauna. Cereals, legumes and vegetables were sown on the land ploughed by oxen. During the early period goats were most common but not sheep. Dogs, donkeys, pigs, geese, ducks, etc., were also domesticated. But neither horse nor camel was used. The Nile Valley lent itself rightly to intensive irrigation agriculture and to massive riverine transport. It is an exceptionally well-casted and highly productive agricultural area, but bounded by areas of very low productivity. On both sides, the valley is bound by deserts — the Libyan on the west and the Arabian to the east. The great delta to the north was, of course, open to conquest only from the sea, which would have been difficult since it had no natural harbours. To the south were the Nubian Desert and the enormous cataracts of the Nile Valley (Service, 1975). Egypt was thus remarkably sheltered from the intrusion of rival alien ethnic groups (such as the Sumerians, the Semites, etc.). It seems that the Egyptian agriculture was much more of a self-contained entity having the autochthonous character.

The aforementioned type of agriculture is also believed to have been introduced from South-West Asia. Subsequently seed agriculture and animal husbandry further diffused in other parts of Africa, i.e. westward along the North African littoral and southward along the Nile Valley during the fifth and fourth millennia B.C. The Nile archaeology is particularly beset with problems. The annual Nile floods and the deposition of silts have undoubtedly played havoc with a number of early archaeological sites, either by washing them away or by burying them too deep to recover.

Some tracts of the Sahara also demonstrate the domestication of crops and livestock as is evidenced from plant remains, fossil soils and rock drawings. This shows that most of the mountains and valley sites of the desert may have experienced favourable moist climatic conditions about between 6500 and 2500 B.C. It may also be ascertained that there had been the initial transmission of agricultural knowledge from the Nile Valley to further south via the Sahara.

Murdock (1959) proposed West Africa as an early centre of plant domestication. He advocates that ancestors of the Mande-speaking Negro people of the middle Niger independently invented their agriculture about 5000 B.C. and domesticated a number of indigenous plants including a variety of millets. He further assumed that they were the people who first advanced from a hunting and gathering economy to an agricultural one, and at the same time, they made much geographical expansion and progressed innovatively at the cost of their backward neighbours. Perhaps because of this reason the group of their languages spread over unusually vast territories. The renowned botanists Anderson (1960) and Porterers (1950-51 and 1962) too have suggested that West Africa may have been a primary centre of plant domestication of the continent (though Vavilov (1935) thought Ethiopia to be a centre of plant domestication). According to Murdock, there were about 30 cultivated plants; but still nothing can be said whether they were locally domesticated or introduced from the East or carried in the opposite direction. Conclusive statements, however, about the age of agriculture in this part of the continent must await the accumulation of archaeological evidence. At present little is known of Neolithic cultures in either the humid or the dry tropical regions of West Africa, and positive evidence of agriculture in the form of plant remains is lacking (Harris, 1967). It may perhaps be the result of decomposition of organic material and absence of proper archaeological research.

On present evidence, it may, however, be derived that agriculture in several parts of Africa is of considerable antiquity. In this context, the Ethiopian and East African highlands together with the Sudanic zone (between the Sahara and the Congo forests) may particularly be noted. Some evidence of Neolithic culture practising sedentary cultivation has also been found in Ethiopia, Kenya and Northern Tanzania. It is believed that the main crops raised in Ethiopia were probably wheat and barley which were expectedly introduced from South-West Asia, presumably via Egypt, where sorghum, finger millet, pearl millet, teff, etc., were of African origin. The age of agriculture of Neolithic cultures that flourished over here is not definitely known, but Kenya had such an opportunity by 2000 B.C. Porterers (1950-51 and 1962) is of the view that the agriculture that evolved in the Ethiopian and East African highlands is the impact of external influence. Most probably, it is the result of diffusion of agricultural knowledge from the Nile Valley via the Ethiopian highlands near about 3000 B.C. Later, agricultural practices happened to spread gradually from the primary foci in Western and Eastern Africa to other parts of the continent. European penetration into the parts of the tropical rain-forests suggests that the crops like banana or plantain, tara, etc., of South-West Asian cultigens have been the staple food of the natives. Now, a very controversial problem cropped up as to when and by what routes such crops were introduced. Murdock (1959) postulated that maritime people might have first carried them to the littoral of East Africa. Subsequently they spread to West Africa along the Sudanic zone, perhaps, at the beginning of the Christian era, and later on to other parts of Africa. Another hypothesis is that of Sauer (1952) originally proposed by him, but a decade later modified by McMaster (1962), that banana and many other vegetatively reproduced crops of economic importance which were the part of ancient agriculture of South-West Asian tradition happened to reach Africa along the coastal fringe of Southern Arabia. Thereafter onward transmission of such crops was made by way of the Ethiopian highlands.

The exploration of agricultural origins toward east from the South-West Asian hearth is much more limited. Some recent investigations carried out in Afghanistan and Baluchistan have revealed that a rudimentary form of agriculture was being practised by 5000 B.C. But for about 1000 years, beginning about 2500 B.C. a full-fledged literate Bronze Age civilisation of classic type arose in the valley of the Indus river and its tributaries. Geographically, this was perhaps the largest of the four "primal" civilisations of the Old World — Egypt, Mesopotamia, the Indus and China — stretching some thousand miles along the great alluvial river valley. Unlike the Nile Valley and the Indus Valley is not closely confined but rather bordered by the foot-hills of Baluchistan at varying distances to the west and by arid desert to the east. However, the extensive floodplains might have had proved an ideal factor for any type of agriculture in the remote past. Further east and south beyond the low-lying plain of the river Indus a well-known affluent Harappan Civilisation had flourished by the third millennium B.C. Neolithic sites have also been identified in India right from Kashmir in the north to Mysore in the extreme south, but explorations provided scanty information about the domestication of plants and animals, by the earliest Indian yeomen. In central India, at a single site of Navdatoli on the river Narbada, some remains of wheat peas, broad beans, lentils and rice together with bones of cattle, sheep, goats and pigs have thoroughly been explored and identified. This central tract of India seems to be the confluence of agricultural traditions from both south-west and south-east. Further the detailed investigations also reveal that the distribution of early agricultural sites is confined to the major river valleys from the Ganges in the north to the Krishna and the Tungabhadra in the south.

Beyond India towards east, the evidence of the beginning of early agriculture is even more meagre. In Northern China, the Huang Ho (the Yellow River) basin is the well-known area usually called the Great Bend where once developed the Neolithic farming economy. Within the middle Huang Ho basin, the valleys of the Fen Shue and the Weiho are particularly noted as they have apparently yielded remains of both plants and animals such as rice, millet, Kaoliang and soyabean; and domestic dog, pig, chicken, cattle, sheep and goat respectively. This interesting feature of agriculture is due to the loess soil which is easily worked with primitive hand implements. This deep wind-blown soil is free from stones. The porosity of the loess causes it to absorb water rapidly without run-off and erosion. The water is drawn again to the surface to nourish the crops. The rainfall is sufficient but is irregular. Probably, all this necessitated man to develop a simple form of agriculture with small irrigation schemes. Since only a few sites have been excavated and much more is yet to be exploited, so, as yet, nothing can be established as to how and when agriculture originated in China. Despite this fact, some scholars seem to agree that the total pattern of agricultural development in China was strongly autochthonous as there was no agricultural model nearby for the Yellow River people.

Sauer (1952) proposed South-East Asia as the cradle of the earliest agriculture. It is most likely that it was the progressive fishing people who initiated agriculture in mild climate along fresh waters. He further emphasises that no area is so equally well suited for development of a fishing-farming culture. In contrast to the heard animals of South-West Asia the most common and earliest domesticated animals are dog, pig, fowl, duck and goose. This of the Old World also proved as the major centre of planting techniques and vegetative reproduction. This process perhaps man adopted before he learnt the growing of crops from seeds. It is believed by some pre-historians that rice was the first crop domesticated on the mainland by the fourth millennium B.C., the farmers who had hitherto been confined to the upland colonised the river valley. Subsequently, wet-rice cultivation began to spread to other parts and refinements in its farming methods such as transplanting, terracing the levelling of fields, the construction of bunds, water control and irrigation were sufficiently achieved. Despite this fact, still not much is known about the beginning of agriculture because of non-availability of archaeological evidence to establish the antiquity of the domestication of plants and animals. It is perhaps due to both the high rate of decomposition of organic materials in the humid tropical areas, and lack of intensive and accelerative breakthrough in research in archaeological fieldwork. Harrison (1963-64) in his works presented the largest cultural sequence preserved in the great Niah cave of northern Sarwak, Borneo. He made an attempt to elaborate an unbroken succession of deposits dating from the Middle Palaeolithic to early Iron Age times (i.e. about 50000 B.C. to A.D. 1300). Of this long period nearly four millennia (4000-250 B.C.) are tentatively taken as the Neolithic. No proof of plant and animal domestication has yet been traceable (i.e. *c.* 2000 B.C.). However, some remains of mats, nets, boats, etc., suggest that the early man had been liberally using the plant materials.

In the New World

Unlike the Old World agriculture, the New World agriculture began much later. Why did it so happen? Evidently because man came to the New World much later, but well before plant and animal domestication had appeared in the Old World. He probably came before boats were invented. This implies that he came over here across the Bering Straits when they formed land bridge during the last Ice Age. The glaciological evidence points to two possible times when men might have wandered from the easternmost promontories of the Old World beyond Siberia to the rocky wastes of Western Alaska in the New. One period was between 28000 B.C. and 23000 B.C., and the other between 14000 B.C. and 10000 B.C. After that the flood of melt-water at the end of the last Ice Age raised the sea level again by several hundred feet and thereby turned the key on the inhabitants of the New World (Bronowski, 1976). This explains that man came from Asia to America not later than 10,000 years ago, and not earlier than about 30,000 years ago. Moreover, he did not necessarily make this attempt all at once as archaeological evidence of early sites reveals that two separate streams of culture came to the Americas in two small and successive migrations. They carried with them a hunting tradition. Massive choppers and cleavers made by percussion techniques and resembling Siberian counterparts have been found in the Arctic of North America. Burins and long, well-made bifacial points, shaped by the pressure technique similarly occur in ancient contexts. These newcomers appear to have entered the New World by the way of (now known as) the valley of the Mackenzie River, taking advantage of an ice-free corridor which led them to the heart of the North American continental plains area and the big game that flourished in a world empty of man until that time (Solecki, 1974). In the New World agriculture was, therefore, certainly invented independently. But, according to Grigg (1978) there were also contacts between the New and Old Worlds before Columbus. Such contacts, however, did not significantly affect agriculture of the former. With the result there were major differences between their respective agricultures until the arrival of Europeans. First, the crops indigenous to the New World were quite different from those cultivated in the Old World. Maize, squash, beans, manioc, potato and groundnuts were the major food crops. Second, the domesticated animals confined to the New World were the llama, alpaca and turkey. Third, there was no plough but the only available agriculture tool was the *Coa* which was similar to the digging stick. Thus, the agricultural civilisations of the two worlds were quite different. Recent archaeological discoveries in the New World have also completely revised the old concepts on the time and place of agricultural origins of the Old World. Particularly in Messoamerica[1] the geographical differences are enormous and the consequent variation in agricultural adaptation is correspondingly so. On the one hand, in arid highlands the practice of intensive agriculture and consequent agglomeration of dense population were the result of irrigation drainage and terracing. On the other hand, lowlands of greater rainfall resulted in more extensive agriculture, with more scattered populations. The archaeological evidence also reveals that in Mexico there existed a hunting-gathering economy with somewhat little nomadism widely scattered over the land-scape for a very long period. The beginnings of cultivation in Mexico were small in scale, and the shift from the primary dependence on hunting-gathering to agriculture was also very slow. Mac Neish (1965), however, advocates that American agriculture goes back to antiquity, that is, the seventh millennium B.C. as series of Mexican caves have yielded wonderful evidence of cultivated plant remains. As a result of these remarkable discoveries the earliest proven date of plant domestication and cultivation in the New World is now seen to be approximately the same as in the Old World: in both, the beginnings of agriculture go back at least to 7000 B.C. (Harris, 1967). There are still many more sites which are yet to be explored in this part of the world.

Long cultural sequences are found in the caves of north-eastern and south-central Mexico. The oldest remains found in Southern Tamaulipas reveal that from the period of 7000 to 5500 B.C., the plants like annual pepper, pumpkin, the bottle gourd were cultivated. By 1500 B.C., many more new additions were made to the local stock of cultigens, namely, maize, amaranth, sunflower, squash, common bean. But the Tehuacan caves have yielded a strikingly parallel and longer cultural sequences. During the period of 7200 and 5200 B.C., the most common cultigens were annual pepper, amaranth, avocado, squash and cotton. In the subsequent period, i.e. between 5200 to 3400 B.C., crops such as maize, common bean, bottle gourd, squash, black sapote and white sapote were commonly cultivated. This type of agriculture flourished until about 2500 B.C., before agricultural products constituted nearly 70 per cent of the people's diet. By this time many of the important Mexican crops were widely cultivated. The fact is that there were densely-settled enclaves where conditions were favourable for continuous or intensive agriculture with the indigenous crops and technology, separated from one another by extremely rugged areas. Differences in rainfall, in soils, and, above all, in elevation greatly affected the complex of cultivable crops. But at some later times with dispersion and diffusion a further selection and introduction of new crops had to be made as of tepary bean and jack bean by 2300 B.C.; pumpkin by 900 B.C.; runner bean by 200 B.C.; peanut, tomato and guava by A.D. 700; and finally lima bean came too late, i.e. by A.D. 1540.

The origin of the ancient domestication of a few plants is unknown whether they were domesticated independently in the Old and New Worlds, or whether they could have been transmitted accidentally to America by ocean currents after they had been domesticated in Asia and Africa or whether they were introduced deliberately by man in the remote past. However, the time and place of American agricultural origins largely rest on the available evidence from Tamulipas and Tehuacan, as elsewhere the archaeological records of early agriculture are seen to be scanty. Both the coastal Peru and the South-Western United States do yield substantial agricultures but these are of recent origin as compared to those of Mexico. The coastal sites of central Peru date back to a period of 4000 to 2500 B.C. But from about 2500 B.C. other new cultigens have been added. Many of these cultigens are believed to have been of Mexican origin and their diffusion and cultivation techniques steered southward to other parts of South America. On the whole, South America was the scene of early vegeculture, and cereal domestication took place quite late. But the Peruvians had only the llama for its wool and as a beast of burden in the highlands. Gradually, slash-and-burn, irrigation, terracing, and the use of llama dung for manure gained significant importance among agricultural techniques.

Development of Agriculture

Agriculture, indeed, has a history of at least 12,000 years, but it is of great interest how a variety of changes in agriculture took place. In supporting this issue, various non-geographical contributions have been called for to make such studies more fruitful — the reason being that opinions advanced on the specific chronology of agricultural forms are most polemical.

Man in fact had faced incredible hardships over the past million years and battled through different climatic and other hazards and finally moved into a different kind of life. Varro (1912) accepts and explains that in the remotest stage, perhaps 50,000 years ago, man sustained on those things which he liked and the virgin earth produced spontaneously. Gradually he became pastoral and began to pluck fruits from wild woodland trees and shrubs and also began to store them for subsequent use. Twenty. thousand years ago man in all parts of the world that he had reached was a forager and a hunter, whose most advanced technique was to attach himself to a moving herd as the Lapps still do (Bronowski, 1976). In this way and for this reason, he captured wild animals and began to domesticate them. Perhaps sheep was the first animal to be domesticated because of her meek gentle nature, meagre feed requirement, and enough supplying capacity in return in the form of meat. This led man to a *wandering pastoral life.* The stage sequence of collecting, hunting and pastoralism runs as an undercurrent. From pastoral mode of life he attained the fourth stage of the agriculture which perhaps retained in it much of the characteristics of the two former stages. This stage, which probably emerged as nomad's life, is too narrow to have given him time or skill to develop some specialisation. The fact is that there was no room for innovation, as there was no time to develop a new device or a new thought as he remained on the move all the time. In a way, after this trichotomy man actually took quite a long time before he could attain sedentarisation. About 10,000 to 12,000 years ago, he began to cultivate in some places some plants and to domesticate some animals. It was the first and quiet explosion of agriculture.

It is a known fact that evolution of agriculture had taken place independently and in isolation as the wild varieties of plants from which crops have been domesticated were confined to relatively very small and a few parts (i.e. ecological niches) of the earth's surface. Later, from time to time dispersal and diffusion of crops, livestock and farming methods have brought changes in agriculture. Both crops and livestock are now widely distributed. Likewise farming methods have also been carried out from their areas of origin to those of new environments. Recently, technical and socio-economic changes have greatly revolutionalised agriculture and with which at the present time *almost* everything is *almost* everywhere. Since the inception of the first agriculture and right up to the agriculture of present civilisation, each successive step was an improvement over the preceding one. In short, the "food-producing revolution" was a *process* and not an *event* in that it developed both gradually and steadily or by small steps over a considerably long period and at the same time followed different emphases and sequences according to differing local conditions in different parts of the world.

In the Neolithic Period

From an ecological standpoint it is difficult to make the distinction between terminal food-collecting and the onset of food production, but the archaeological evidence of living sites and the artifacts (pottery, polished stones, tools, etc.) are our unwritten history of the origin and development of agriculture. Having this in view the first agriculture began in the Neolithic period, and in which the domestic plants and animals preceded such artifacts. All that implied that an economic base of that agriculture and first food-producing culture evolved.

By the tenth millennium B.C. in most of the parts of South-West Asia and the Eastern Mediterranean people who relied much upon hunting, fishing and food-gathering during the long spell of pre-historic time began to reap wild barley and wild wheat with sickles and curved knives. They used to keep some surplus grains in storage pits or bins, perhaps, for unforeseen deficit periods. Somewhat similar characteristics of the period of 10,000 — 9000 B.C. are also available in the Western Zagros. The earliest farmers had simple methods of grinding with stones, mullers and querns, presumably for the processing of various seeds, grains and nuts.

Some scholars believe that new vegetation blossomed at the end of the Ice Age and a hybrid wheat appeared in the Near East. It might have happened in many places but a more typical one is the ancient oasis of Jericho (Israel) which is said to be older than agriculture. The people who came first to settle over here during the spring must have found the wild wheat. They harvested it with their tools, though they had yet to know how to plant it. They were Natufian[2] and by the period of 9000 — 8000 B.C., they became more or less sedentary collectors of wild grains. Earlier to this period wheat was not the luxuriant plant as it is today, rather it was merely one of many wild grasses that spread throughout the Middle East. But by some genetic accident, the wild wheat crossed with a natural goat grass and formed a fertile and luxuriant hybrid. That accident might have occurred and repeated many times in the springing vegetation of the last Ice Age with the result the Natufian began to collect wild grains in large quantities. It is likely that the early cultivators practised some from of shifting cultivation. Some excavated sites suggest that they also learnt the value of livestock grazing. But the continuous occupation of some places further developed sedentarisation among them. Nearly all the early sites in South-West Asia and the Eastern Mediterranean, for example, suggest that crops were grown and livestock kept, consisting of a primitive form of mixed farming. Gradually, they made permanent village settlements of relatively large populations. That was Natufian agricultural civilisation which emerged from a happy and promising conjunction of natural and human events. The available evidence of settled villages as well as the kill-sites and seasonal camps of hunters reveals that local population densities increased. Eventually, a fairly complex pattern of socio-economic relationships among various communities resulted, and by 7500 B.C. there was an extensive, though probably informal, trade in obsidian, and to a lesser extent in copper. This, however, implies basically a healthy trend towards switching over to more vigorous agricultural activities. They were using mortars and grinding stones apparently for grinding grains and seeds. Implements such as multi-toothed sickle-blades of flint, perhaps for reaping the wild grain crops, were most common. The wild emmer wheat was in the area but it was particularly important in the Jordan Valley. The agricultural surplus probably had made sedentism possible. With the result, in the following period, other cereals, legumes and seasonal fruits added to the foodstock. In south-eastern Iran (particularly Tape Alikash located on the Deh Luran plain) during *c.* 7000—6750 B.C. the subsistence was characterised by high dependence on the collecting of seeds from plants native to the Deh Luran plain. Emmer wheat and hulled two-row barley were cultivated but comprised only a third of the total weight of vegetal food (Parkins and Daly, 1974). Though people herded domestic goats, yet the hunting of wild ungulates supplied much of the bulk of meat supply. By *c.* 6750—6000 B.C., there was a great change in the economic base. Emmer wheat and hulled two-row barley now began to provide more than 90 per cent of the total weight of the plant food. Despite this fact, seeds of native wild plants continued to be collected for food. Goats were only domestic animals though hunting of large ungulates was still a continuous process. But by 6000 B.C. multiple domestication of plants and animals was not uncommon. For example, in South-West Asia and Eastern Mediterranean the village communities began to grow wheat and barley and to keep sheep and goats. Subsequently, a number of plants were added, e.g., flex for oilseeds rather than fibre, peas, lentils and vetch. This is perhaps the result of increased contacts with other areas or invading people might have brought such a diffusion.

In the Proto-Historic Period

Diffusion of agricultural activities rightly began as the Neolithic culture spread in a series of waves to the Aegean, the Levant and Egypt in the sixth millennium B.C.; South Russia, the Balkans and the Danube Valley in the early fifth millennium B.C.; and Italy, France and Spain in the late fifth millennium B.C.; China too experienced the diffusion, for example, farming started in the loess uplands of the middle Hwang Ho of Northern China near about 6000 B.C. Farmers perhaps initially adopted shifting cultivation. Later these farmers spread to other parts of the country. South China received banana, rice, yam, sugarcane and tea from South-East Asia, other parts acquired grapevine, sheep, goats and cattle from South-West and Central Asia. But such a diffusion reached India (particularly Sind, Punjab, Rajasthan and Gujarat) too late, probably in the fourth millennium B.C. Such a diffusion, however, initiated a food-producing mixed-farming economy which was established well before 5000 B.C. in other countries. Cultivation was mainly practised by primitive hoes consisting merely of a curved stick to which a flint hoe was tied to act as ploughshare. It was probably of shifting type because of the little knowledge of manuring the fields. But by the fourth millennium B.C. primitive plough and ploughing techniques were almost given up in Egypt and Mesopotamia. Cross-ploughing was practised on flood plains which were then considered as the most useful for dry-farming which without conservation of soil moisture is impossible. In the late fourth millennium B.C., particularly in the temperate parts, the crops like legumes, cotton, small seed flax, many oilseed plants, onions and most of the fruits such as plums, apricots, peaches, almonds, cherries, apples, pears, grapes, walnuts, strawberries, gooseberries, blackberries, etc., were widely cultivated. The agroclimatic conditions of Abyssinia made it the home of hard wheats, barley, millets, sorghums, sesame and coffee.

Stock-raising seems to have been limited. Cattle and pigs are thought to have been domesticated after sheep and goats. Both the cultivation of plants and the raising of animals were simultaneously developed. Until the late fourth millennium B.C., no evidence is available to indicate their use as milch animals in the South-West Asian tract of the world. Nevertheless, cattle played an important part because they were used as the beasts of burden, draught force and for procuring meat.

Settlements became more permanent. Relatively bigger villages of large populations and even tendency of urbanisation began to develop. Near about 3500—3000 B.C., the peasantry class came into being and the farming communities comfortably began to settle in the fertile alluvial plains, where the sedentary village pattern stabilised. Consequently, establishment of effective food production led these settlements to urbanism. However, from this process, for example, grew the urban civilisation of Harappa, Mohen-jo-Daro, Kot Diji, Ropar, Kalibanga, Sumer, Egypt, etc. The river systems of these civilisations provided not only fish, molluscs and aquatic birds in abundance but also acted as potential means for transportation. Easy and efficient transport of that period had two important aspects: first, it encouraged the transportation facilities to agricultural and non-agricultural goods and people, and second, it stimulated a wide transmission and diffusion of farming and farming methods, inventions, discoveries and innovative ideas. Before that there were only primitive groups of tribal people. But with the passage of time the peasant societies became more and more complex, and stable as well.

In the Early Historic Period

The wheel was invented and found for the first time near about 3000 B.C. in what is now called Southern USSR. The solid wooden wheels gave rise to carts which speeded up transportation of various agricultural and non-agricultural goods. From then on, the wheel and axle became the double root from which new inventions grew. The result was that along with polished stone implements, brass and copper implements came into use. Such developments may be the result of increased prosperity and technology. Perhaps because of this reason, this period is called Chalcolithic or Copper-Stone Age. The system of nomadic and shifting cultivation gave way to cereal-fallow system. But a major advance in agriculture came with the development of irrigation. Irrigated farming began to be practised and to this effect inundation and perennial canals began to be dug and flood waters managed to be stored in reservoirs. New schemes were also made and implemented to irrigate cultivated lands by annual floods in rivers. The irrigation agriculture became the economic basis of the great valley civilisations of the world. During the period 3000—2500 B.C., in Egypt, for example, the large irrigation systems developed upon which its culture was based. The artesian well, the nilometer for gauging streams and the development of great canal systems (for both irrigation and navigation) were the great achievements of the time. The more complicated devices such as the *shaduf* and the *sakia* were the result of the then technical developments. The mechanical engineers of Sumer and Assyria turned the wheel into a pulley to draw water for irrigation. It was really a hydraulic culture.

The copper and bronze axes were added to flint tools. At the same time, metal works in copper, bronze and gold, and the making of glass and glass goods had been achieved. All this, in fact, implied an excellent progressive economic base of agriculture. Farmers commonly grew wheat, barley and cotton (coarse type) and raised shot-horn zebu cattle, water buffaloes, camels, sheep, goats, pigs, fowl, etc. But somewhere around 2000 B.C., man discovered how to ride a horse, since it was a serious use of the horse, it probably began first in the nomad tribes who bred horses. They were men from Central Asia,

Persia, Afghanistan and beyond. In the occidental world they were called Scythians. Almost in the same period, the buffalo and chicken were bought under domestication particularly in South Asia. But the reindeer and yalk began to be tamed in the north and east of the Altain highlands in about 1000 B.C.

It is believed that in this period there was somewhat rapid growth of population which stimulated further development of agriculture. For example, the population of England and Wales, according to Clark, was estimated at the most at 250 humans in the Palaeolithic period, 4,500 in the Mesolithic, 25,000 in the Neolithic and about 40,000 in the second millennium B.C. So it can safely be deduced that many countries then might have shown a continuous substantial growth of their respective populations. With this trend of population growth the agricultural activities should, therefore, be much more activated, as is evident. According to Kosambi, that the most efficient hunting and food gathering can hardly support one person per square kilometre, pastoral life can support three, and agriculture about 100 (as cited in Randhawa, 1974). Obviously agricultural activities gained much importance as a sustenance for the increasing population.

In the Pre-Middle Ages

During the period of 1500—500 B.C., a good many changes in almost all human activities occurred in the world. The Bronze Age people with their bronze swords and horse-drawn bronze chariots established their hegemony over the earlier town-dwellers of different areas. Though a majority of the people were living in the countryside, yet continuous growth of population and improved economic conditions emerged from better agricultural farming and turned towns into cities.

The aforementioned conditions largely accelerated the process of dispersal and diffusion of various crops. For example, *Ragi* and *Bajra* indigenous to the continent of Africa were cultivated in South India near about 1500 B.C. These crops probably reached India via South-West Asia. In the first millennium B.C., rice, cotton and possibly sugarcane were commonly grown along with other crops in Mesopotamia and may have also been in the Levant by Romans. At the same time, all these crops further dispersed to other parts of the world. For arable farming though a light plough, the *ard*, was designed and first used long ago in northern Europe. But it would not have been capable of breaking the virgin land as it was a little more than scratching the land, despite giving an efficient tillage of land already in cultivation. The agricultural changes, in fact, took place in the succeeding Iron Age after the discovery of iron near about 1000 B.C. Soon thereafter iron technology was mastered. With the result iron plough-share came into existence and large areas of heavier soils were brought under cultivation in most parts of the world. This is, of course, the most powerful mechanical invention in all agriculture. It is a lever which lifts the soil. Since then one ploughman of the Middle East had been saying: "Give me a lever and I will feed the earth".

The spread of farming in the Ganges Plain seems to have been much slower than in Peninsular India. For example, the earliest settlements of peasants in the Upper Ganges plain came into being near about 1100 B.C.; in the Middle Ganges plain further penetration was made in the period between 900 and 800 B.C. By then settlers were probably equipped with better iron-tipped ploughs and other improved agricultural implements such as iron-spades, axes, sickles, etc. There may have been dense forests and swamps which probably delayed the settlement and the cereal cultivation. So much so, the first evidence of rice being grown in the Lower Ganges plain and the delta margins dates back only to 700 B.C.

Both the Phoenician and the Greek, according to Plato, had established colonies around the coastland of the Mediterranean Sea and developed agricultural and non-agricultural activities "like ants and frogs around a pool". During the period of fifth and fourth centuries B.C., both Greek and Roman agriculture and their administration were treated as a fine art — with the result agricultural products and farming techniques were exchanged throughout the great region of their influence because of having contrasted climate, vegetation and civilisation with the geographically favoured and precociously civilised Mediterranean world. Beyond this region intermediate traders also explored and exploited numerous activities pertaining to agriculture. New scientific devices such as Archimedes Screw or tembour, a lifting device invented by Archimedes of Syracuse near about 200 B.C. came to be extensively used. Such irrigation devices brought many more improvements in agricultural landuse. All this development was, perhaps, stimulated by the substantial growth of a monetary economy and, at the same time, shifting over to semi-commercial agriculture. With the growth of urbanisation subsistence agriculture had to be discarded particularly in favour of ever-growing urban markets for a variety of agricultural products.

Large estates usually organised by government agencies came into existence and gradually they became dominant in most parts of the classical world. These large estates (the *latifundia*) were controlled by big landlords who were engaging slave labour. But quite late, with fall of supplies in the slave labour the farmers began to take the place of peasants in most of the Roman world as there was a reverse in organising those estates, and again, peasant tenancy and subsistence farming became more common.

In Middle Ages

In the millennium that preceded the agrarian revolution of the 18th century and the development of commercial farming as we know it today, developments in the techniques and administration of agriculture came at long intervals. The feudal system, in which the vassal owed allegiance to his lord in return for a measure of security, was expressed on the land by manorial organisation. The vassals supplied the lord with a portion of their produce and also had to work on the demesne lands farmed for the lord's benefit (Symons, 1967). This system was, perhaps, the outcome of an economy in which money was scarce, which probably resulted due to military aggressions and overall unrest.

In early Middle Ages, the great agricultural civilizations of Northern India, Northern China, Southern Europe and South-West Asia were, indeed, on a par technically. But the period between classical times and the Ages of Discovery is said to be the greatest innovations. The earlier ploughs were scratchers, e.g., the *ard* which was not very much successful on the heavy soils. A heavy plough was, therefore, developed having a coulter to cut the turf and a mouldboard to turn it. Wheels to the plough did come but some time late. By the first century of the Christian ear, Roman farming had reached its apogee. Agricultural knowledge had attained a standard not to be surpassed until the 19th century. The value of land fallowing and green manuring was fully known to the farmers. A wide range of agricultural implements was available. Ploughshares and blades of sickles were made of iron. At the same time, many more agricultural implements like scraper (or trowel), pitch-fork, reed-winnowing baskets, etc., were developed. Dry farming was widely practised and irrigation was supplemented with lifting devices. By early Christian times there had also been a considerable interchange of plants between South-West

Asia, Europe, North Africa, Turkistan, North-West India. The South-East Asian complex of roots and rice had expanded into India, China and Malaysia while there had been diffusion between Northern China, Turkistan and South-West Asia (Grigg, 1978). But during the last half of the first millennium A.D., particularly in European countries, there were many changes in the methods of cultivation. The most common were : (i) temporary cultivation system, in which land was cultivated temporarily and was left as a wasteland for an indefinite period, (ii) the infield-outfield system in which a part of the land was cultivated continuously with heavy manuring and rest of the land was brought under plough occasionally, (iii) the two-course seasonal rotation in which one winter crop was sown in alternation with a bare fallow (e.g. in the Mediterranean region a summer crop is impossible because of lack of summer rainfall), (iv) the two-course annual rotation in which the arable land was cultivated and left fallow in alternate years, (v) a three-year system in which the arable land was cultivated for one year and left fallow for two consecutive years, and (vi) the three-course rotation in which winter crops (corn, wheat or rye) were followed by spring crops (corn, barley or oats), and then by fallow.

There was a set-back to agriculture with the fall of Rome because the great civilizations became relatively isolated from each other. Consequently, the process of diffusion of plants was considerably slowed down. Even in the later period of Romans, the *latifundia* increased at the cost of the peasant proprietors who were not secure enough. Though this tendency continued into the Middle Ages, yet the prevalence of insecurity, warfare and major plague outbreaks of A.D. 600 is greatly responsible for the decline in agriculture. At places, this led to the emergence of pastoralism which grew in importance and reached its height in the late Middle Ages. Circumstances, however, forced man to grow more food with improved farm implements. During the eighth and ninth centuries the horse collar and shoe allowed ploughs that were invented to be pulled by horses. Horses were faster and brisker than oxen though expensive to feed. They did replace oxen but very slowly. In the 18th century, they outnumbered the latter in most of the countries of the occidental world.

By A.D. 800, the Islamic empire extended its limits from Spain and Morocco in the west to the Indus and the Jacartes in the east and, therefore, became an important agent of plant migration by cutting Europe's land route to India, China and Africa. The Arabs, however, made an effective contribution in spreading citrus and other crops into the countries of the western Mediterranean. The sweet orange, the sour orange, the lime, lemon and the citron were all indigenous to Southern China and South-East Asia. The time of their arrival in the Mediterranean is a matter of dispute (Grigg, 1978). Though some citrus fruits had earlier been taken into that part of the world, yet these are more commonly attributed to the Arabs. Subsequently orange was introduced into Spain in the 10th century, the lemon, in the 12th century and the lime in the 13th century. Until the 15th century, the diffusion of crops was slow and steady. Crops grown in Europe, for example, either derived from South-West or South-East Asia. But thereafter a number of crops began to spread to different parts of the world with the spread of Europeans. The credit, in fact, goes to the European navigators of Portugal, Spain, Italy, Holland, France and England who discovered sea routes to different parts of South-East Asia. In the same period the American continents also came to be added to the world map. Vasco-Da-Gama happened to reach India via the Cape of Good Hope in 1498 and Magellan managed to circumnavigate the world in 1500. With the efforts of these navigators regular sailing routes eventually linked Europe, Brazil, South Africa, India, the East Indies and China. Consequently the process of diffusion of different crops began to accelerate in most agriculturally potential areas of the world. It was during this period that more varied changes in cropping patterns also emerged in most of the agricultural regions of the world.

In the Post-Middle Ages

By A.D. 1500 there were as yet no great differences in agricultural productivity between Europe, India and East Asia. In East Asia crop production per unit area was higher than that in Europe though output per man was lower. Indian agriculture, in any way, did not lag behind the European one. The major difference was in the use of the heavy plough which was either not known or adopted outside Europe, Further, there was a major difference between the plough civilizations with their rudimentary mixed farming, and Africa, the Americas and Polynesia where the plough had yet to penetrate (Grigg, 1978). But in the post-Middle Ages in many parts of the world some remarkable changes in agriculture have been noted. Introduction of simple agricultural technology paved the way in bringing a revolutionary movement in the field of agriculture. By the 16th century horse-drawn harrows and rollers were invented and brought in use in many parts of Europe. In 1602 a prototype seed drill was in use in Italy, and later, the modern seed drill was constructed out of it. But its adoption was very slow as most crops even in the United States were still commonly broadcast in the 1840s.

Most of the technical advances in farming in most parts of the world were much concerned with bringing more cultivable land under plough and improving the methods of preparing seed beds. Changes in the methods of harvesting came more slowly. The Scythe was probably introduced into Europe by the Romans but none could replace the sickle or scythe until the periods of labour shortages of the 14th century. Even at the end of the 18th and towards the middle of the 19th centuries at the time of labour shortages sickle and scythe were the most common and important implements and were widely used to reap grain crops. During the 16th and 17th centuries, there were quite notable changes in agriculture methods or institutions, and the beginnings of modern agriculture can be traced back to this period. There were major trends which gave rise to modern mixed farming. They were: (i) the reduction of fallow, (ii) the extinction of common rights and the consolidation of the dispersed fields, and (iii) slow and steady increase in the importance of livestock.

At the time of frequent agricultural depressions when cereal prices fell sharply, livestock prices either stagnated or fell much less — with the result there was a tendency to shift toward either the industrial crops or the raising of livestock. In most cases, cereal production was abandoned for dairying. In the mild agricultural depression between 1650 and 1750, the corn fields were switched for the grass or fodder fields. Viticulture also emerged and widely spread at this time. But it was in the 18th century that specialisation in livestock and livestock products took place as cattle began to be bred for meat or milk, and sheep for wool or mutton. Animals were not treated for general purposes. But by the beginning of the 17th century only the land masses of the southern Pacific remained isolated. The fact is that Tasman circumnavigated Australia in 1642 and also sighted New Zealand in the same year. Introduction and development of agriculture obviously took place much later.

After 1500th year there were also significant development in the dispersal and diffusion of crops in many parts of the world. Potatoes and maize, for example, adopted from America, gave much higher yields than the grain crops grown in Europe and became instrumental in sustaining the phenomenal increase in population in the forthcoming periods. According to the climatic conditions, both crops were complementary as potatoes are mainly grown in Northern Europe whilst maize is restricted to the Balkans and the Mediterranean. On the whole, Africa adopted food crops more rapidly. Two new food crops, manioc and maize were introduced from America nearabout 1525–1535, and by the middle of the 19th century they were grown throughout most of the continent. The sweet potatoes were introduced into the continent in the 16th century. Groundnuts were adapted from Brazil. Since 1500 plantation crops, particularly tea, coffee, rubber, etc., transformed much of the agricultural economy of South-East Asia. Maize, chillies, tomatoes and sweet potatoes reached in the early 16th century and groundnuts in the late 17th century. But the arrival of Europeans in the Americas in 1492 and the subsequent colonisation not only stimulated the process of introduction of Old World Plants into the New World, but also dispersal of American crops to other continents. The first settlers of both Australia and New Zealand found no indigenous plants of any value. They carried with them their modern crops which flourished well in their new environments.

In brief dispersal and diffusion of food, beverage and industrial crops had brought changes in the agricultural landscape in many areas of the world and their production in large quantities was the result of the great economic and technological changes. But it must not be overlooked that in most densely populated parts of the world, agriculture was still remarkably backward. Oxen were still yoked to a plough and the most common farm land transport was by oxen or horse-drawn cart which began to be replaced only in the late 18th century. Even at the beginning of the 19th century the sickle was still used for harvesting. Most of the crops were sown by broadcasting as the drill was a rarity. Agricultural production was low though crop yields were rising gradually. It was, probably, due to meagre responding of farming to the new changes. On the whole, the period of the post-Middle Age is, in brief, said to be the period of substantial agricultural change in which the seed of broader and intensive agricultural revolution germinated.

In the Present Times

Earlier agriculture in most parts of the world satisfied the requirements of the cultivators for food, clothing and shelter. Yields were low, and recuperation of the soils under plough was inadequate — the reason being that the agriculture was carried on by peasants who were occupying an intermediate place between tribal and urbanised societies. At the same time, at places the bullock-based technology was hardly modified since its invention. But it was the industrialisation with which transformation in agriculture began to gather pace. In brief, the Industrial Revolution paved the way for Agricultural Revolution. Revolutions do not always come in totally unexpected forms; rather they are complexes of slowly evolving technological improvements which gradually disseminate far and wide. More explicitly, it is a relatively sudden set of changes that yield a state of affairs from which a return to the situation just preceding the revolution is virtually impossible (Hockett and Ascher, 1964). Having this in view, the agricultural revolution was not only interconnected but also intimately interwoven with the Industrial Revolution. Therefore, this phase has been the period of real agricultural revolution, particularly in the Western world.

With the Industrial Revolution iron became cheaper for all uses. Better and newer agricultural implements began to be invented. Steam power became generally available for both farmyard and factory. Further, the use of steam power and the introduction of the factory system led to urbanisation. With the development of factories and ancillary services the demand for labour became insatiable. At the same time, greater employment opportunities and relatively higher wages in industrial centres led to migration of the rural population. Obviously, labour shortage was experienced, as such opportunities relieved some of the population pressure on the land. Moreover, population growth slowed down with the fall in birth rate, and simultaneous emigration of some rural population to the Americas further aggravated the problem in the countryside. The agricultural population was in effect residual. Therefore, necessity arose to adopt more sufficient agricultural methods by employing labour-saving devices. Simultaneously, higher incomes of industrial workers also led to the development of substantial urban markets demanding foodgrains, vegetables and livestock products. At the same time, industries based on agriculture generated an equally high demand for cotton, wool, jute, rubber, etc. All such activities further compelled the farmers to use the labour-saving machinery.

Steam power was first used for threshing machines in the early part of the 19th century and for ploughing in the 1850s. The steam plough, indeed, was able to plough 10 times the area that the horse could plough in a day. But it was a cumbersome and costly affair and its impact on farming was limited. During the period when rural populations were more or less stagnant, the great problem of harvesting the crops emerged. The labour shortage, in other words, prompted search for reaping or harvesting machinery. By 1874, the horse-drawn reaper replaced the scythe and the sickle. But in 1910, tractors with inter-combustion engines were developed; originally, the tractor was regarded as a versatile source of power which may be used in most farming operations, e.g., ploughing and threshing. In 1931, tractors were equipped with diesel engines which enabled the machines to work under all conditions. They began to replace horses and mules rapidly. Less labour had to be hired and less land was needed to produce feed for draught animals. They made the farmers more-mobile enabling them to cover a large hectareage in a given time or a given hectareage in a short time. Arable fields were cultivated continuously to keep down the weed growth. Seeds were sown in straight rows by drills so as to make inter-row cultivation possible. But the adoption of the combine-harvester has come only since, 1945. In the United States, for example, the number of man-hours per hectare fell from 148 in 1830 to 37 in the 1890s, but with combine-harvesters to only 6.25 hours. Apart from tractors and combine-harvesters, fundamental technological changes are clearly indicated in the adoption of HYV seed, chemical fertilizers and plant-protection chemicals which increase yields and lower labour costs.

The transformation in agriculture was rapid especially in Europe and later brought into the overseas countries. But no change occurred in Asia, Africa and much of the Latin American countries which continued with traditional agriculture. Gradually, a little change occurred in farming methods. But the unprecedented increase in population during the last half a century with no parallel increase in urban employment to lessen the increasing rural congestion led to an increase in the sub-division and fragmentation of holdings, underemployment and malnutrition. Farmers in these countries were unable to find new inputs which, in fact, have brought transformation in Western agriculture. By and by the farmers made an attempt to reduce the fallow land and introduce multiple cropping to intensify landuse. They further intensified labour input in preparing the seed-bed, weeding and harvesting. Subsequently they brought new land under cultivation. At present, mechanisation has swayed over many parts of these continents.

Mechanisation, in fact, improves the timeliness of farm operations and, therefore, selective breeding and improvement of new varieties of crops has helped to extend production and maintain yields. The new scientific techniques have brought refinement in animal breeding to improve both the milking and fattening qualities. A drastic change made from the 1880s onwards was towards specialisation in animal husbandry. A series of cooperative agencies were set up to make possible more efficient and profitable farming. Large tracts not suitable for arable farming were devoted to improved pastures and meadows to yield hay sufficient to keep livestock in increased numbers and improved health so as to permit an increasing supply of meat and dairy produce.

The improvement in the means of transportation further accelerated the process of transformation of agriculture. Such means linked the different places to facilitate the transportation of farm inputs at much reduced rates — with the result that farmers received an incentive to boost all types of agricultural production. Thus, commercialisation of agriculture and the growth of new agricultural regions would not have been possible without substantial improvements in transport and the consequent reduction of freight rates. Probably because this reason there was a significant shift in new investment which went to the temperate, European settled areas in North America, Chile, Argentina, Uruguay, Southern Brazil, South Africa, Australia and New Zealand. But a little proportion of the investment also went to tropical areas.

With the development of agricultural science and technology the farming is now becoming a business rather than a way of life as spectacular changes in agriculture have occurred. In other words, there is a reorientation of agricultural economy from subsistence to market. For example, in 1886 in the United States, 95 per cent of the population lived on farms, fed themselves and fed the remaining five per cent, but now hardly nine per cent live on farms and feed the remaining 91 per cent. Moreover, their produce is so much surplus that it is even supplied to the world market. In most technically advanced countries a newly emerged trend is that of "capital agriculture" in which emphasis is laid on profit and reinvestment. Farmers are entrepreneurs who take up agriculture as a business for handsome profits. They plan their agricultural production according to the resources at their disposal and the demand in the market. They are socially, politically and economically integrated with the wider society.

Conclusion

In the past three-decade period, there have been spectacular advances in our knowledge about the origin and development of agriculture. Many scholars have painstakingly probed the beginnings of agriculture in some regions of the world. The present discoveries are, indeed, impressive, but they are not to be overestimated as viewing them in the world context, they contribute a meagre knowledge of the origin of agriculture. Right from the earliest phases of cultivation numerous specialisations entered into agricultural systems. The fact is very clear as minor cultigens have gone out of cultivation with the spread of progressive commercial farming.

In most parts of the world, man has changed the old landscape altogether along with his agricultural plants and animals to such a degree that at present it is hard to believe that hunting and gathering would ever have been a possible way of life. Man the Hunter came to be Man the Pastoralist, the Farmer, the Villager, the Townsman and the most sophisticated Urbanite. Agriculture had its beginning some 10,000 or 12,000 years ago; whereas man took more than half a million years to be a sedentary grower. During his history, viewing the whole world, man has used more than 3,000 species of plants for his food. Of these about 150 have ever been exclusively cultivated and at present only a few dozens are economically significant. But the domestication and raising of livestock were carried on by people who led a nomadic life. Some evidence shows that there was some association or trading of products between herdsmen and those who raised crops. By the time agriculture had gained importance and spread to different parts of the then known world, the distinction between herdsmen and cultivators was disappearing and the two forms of agriculture were practised together.

The diffusion of knowledge of the plough to ease cultivation and the manure to restore soil fertility gave a great fillip to agriculture. In the course of the growth and movement of human population, new food crops began to develop. With the result a very wide range of combinations of cultivated plants had been established in the farming patterns of different regions. The primitive forms of cultivation with simple hand tools disappeared and the plough cultivation with draught animals gained much importance. Despite the fact, until about A.D. 1600 there were very little differences in agricultural productivity between different countries of the world. Thereafter, the elimination of medieval institutions and at the same time introduction of relatively better crops brought some increase in agricultural production.

On the whole until 1850, there was no agricultural revolution but everywhere there was evolution. Change was there but it was very slow as the peasant way of life is several millennia old and still continues to survive. This explains that the descent of modern farming system is clearly traceable to the Neolithic Age in many parts of the world. It was the Western world that first experienced industrialisation which provided new inputs which are always and everywhere necessary to make a break with the past. The past is still found to be imprinted firmly on the agricultural landscape, but less now upon the minds of the farming community than it was some three decades ago. There are tow important factors which explain this fact, namely, the slow change in agricultural technology, until the 19th century, and the very limited areas which adopted the new techniques during the 19th and 20th centuries. Therefore, much that is old and traditional in farming practices persists, much too is now achieved by methods rooted in scientific knowledge and by the use of machines, good management and organised transport facilities, and much is yet to be achieved as the common aim is clear, that is, to produce more food for an ever-growing population and, at the same time, fodder and a range of raw materials for livestock and modern industry respectively.

Agriculture, at present, seems in a state of crisis. Particularly in the Western world it faces the problem of competing for resources with manufacturing industry. In some countries, there is also a serious problem of surpluses of some staples and industrial crops. On the other hand, in the non-Western world the major problems of agriculture crop up from a very low productivity. There is always a race to overcome a deficit between supply and national subsistence requirements of foodgrains. Successful solutions to these problems, however, lie in the more efficient use of the present unused, underused and misused resources by farming communities all the world over. The higher agricultural productivity in the Western world is the result of industry-type organisation particularly through an increased scale of activities, high degree of mechanisation and higher levels of management.

The non-agricultural sector, in the case of oriental countries, may also be called upon to provide increasing support for the agricultural sector as it is common in the Western World. In many instances, State agencies involve themselves directly and indirectly to boost agricultural activities. So, with the adoption of new system of cultivation and agricultural organisation it seems a rapid and conspicuous change would come in the next half a century compared with the previous rates of change that occurred in the period of several earlier millennia.

NOTES

1. Messoamerica refers to the complex geographic region including the highlands and lowlands of Central and Southern Mexico and Guatemala and the lowlands of Salvador, British Honduras and a part of Western Hunduras.
2. A series of sites in the Levant are designated Natufian, after the type-find in the Wadi-en-Natuf in Israel.

REFERENCES

Anderson, Edgar, 1960 : The Evolution of Domestication, in *Evolution after Darwin* (3 Vol. Chicago), Vol. 2, *The Evolution of Man* (edited by Sol Tax), pp. 67-84.

Braidwood, R. J., 1952 : *The Near East and the Foundation for Civilisation* (Eugene : Oregon State System for Higher Education).

Bronowski, J., 1976 : *The Ascent of Man*. London : Sir Joseph Causton and Sons Ltd., pp. 59, 62-77, 92.

Butzer, Karl W., 1964 : *Environment and Archaeology* (Chicago), pp. 416-437.

Childe, V. G., 1936 : *Man Makes Himself*, London : Watts.

Clark, J. G. D., 1965 : Radiocarbon Dating and the Spread of Farming Economy, *Antiquity*, Vol. 39, pp. 45-48.

Darlington, C. D., 1963 : *Chromosome Botany and the Origin of Cultivated Plants*, pp. 138-139.

Fairbridge, R. W., 1967 : "Ice-Age Theory" in *The Encyclopaedia of Atmospheric Sciences and Astrogeology* (edited by Rhodes W. Fairbridge. R.W.), New York : Reinhold Publishing Corporation, pp. 462-474.

Grigg, D. B., 1977 : *The Agricultural Systems of the World : An Evolutionary Approach*. Cambridge : University Press, pp. 6-7, 20, 26, 53, 225-226.

Harris, David R., 1967 : "New Light on Plant Domestication and the Origins of Agriculture : A Review". *Geographical Review*, Vol. 57, No. I, pp. 90-107.

Harrison, Tom, 1963-1964 : "100,000 Years of Stone Age Culture in Borneo", *Journal of Royal Society of Arts*, Vol. 112, pp. 174-191.

———, 1964 : 'Tom Harrison : Inside Borneo', *Geographical Journal*, Vol. 130, pp. 329-336.

Harlan, J. R., 1961 : "Geographic Origin of Plants Useful to Agriculture in Germ Plasm Resources" (edited by Ralph E. Hodgson). *American Association for the Advancement of Science Publications*, No. 66, Washington, D.C.

Hocket, C. F. and Ascher, Robert, 1964 : The Human Revolution. *Current Anthropology*, Vol. 5 (3), pp. 135-168.

Hutchinson, J. B. (Editor), 1965 : *Essays on Crop Plant Evolution*, London.

Mac Neish, R. S., 1965 : The Origin in American Agriculture. *Antiquity*, Vol. 39, pp. 87-94.

McMaster, D. N., 1962 : "Speculations on the Coming of the Banana to Uganda", *Journal of Tropical Geographers*, Vol. 16, pp. 57-69.

Mellart, James, 1958-61 : Excavations at Hacillar : First, Second, Third and Fourth Preliminary Reports. *Anatolian Studies*, Vols. 8, 9, 10, and 11; pp. 127-156, 51-65, 83-104 and 39-75.

———, 1962-64 : Excavations at Catal Huyuk; First, Second and Third Preliminary Reports, *Anatolian Studies*, Vols. 12, 13 and 14, pp. 41-65, 43-103 and 39-119.

Murdock, George Peter, 1959 : *Africa : Its Peoples and Their Cultural History* (London), pp. 64-77, 207-11, 22-225, 233-238, 245-252, 261-274.

Parkins, Dexter, Jr. and Daly, Patricia, 1974 : "The Beginning of Food Production in the Near East". In *Old World : Early Man to the Development of Agriculture* (edited by Robert Stigler). New York : St. Martin's Press, pp. 73, 75-76 and 92.

Porterers, Roland, 1950 : *Viulles Agriculture de l'Afrique Intertropicale.*

———, 1951 : 'Geographic alimentaire, bereaux agricoles et migration des plants cultives en Afrique interropicale', *Compte Rendu des Seances Soc. de Biogeogr.*, No. 239, pp. 16-21.

———, 1962 : "Bereaux agricoles primaires sur le continent africian", *Journal of African History*, Vol. 3, pp. 195-210.

Randhawa, M. S., 1974 : *Green Revolution : A Case Study of Punjab*, Delhi : Vikas Publishing House, p. 5.

Rodden, Robert O., 1965 : "An Early Neolithic Village in Greece", *Scientific American*, Vol. 212, No. 4, pp. 82-92.

Sauer, C. O., 1952 : "Agricultural Origins and Dispersals", New York : *Amer. Geogr, So. Brown in Memorial Lectures*, Ser. 2.

Service, Elman R., 1975 : *Origin of the State and Civilization : Process of Cultural Evolution*. New York : W.W. Norton & Company, pp. 166-167, 203-207, 225-227, 238-239, and 247.

Simmonds, N. W., 1962 : "Variability in Crop Plants, Its Use and Conservation". *Cambridge Philos. Soc. Biol. Review*. Vol. 37, pp. 422-465.

Solecki, Ralph, 1974 : "The Old World Paleolithic", in *The Old World : Early Man to the Development of Agriculture* (edited by Robert Stigler). New York : St. Martin's Press, p. 69.

Stearn, W. T., 1965 : "The Origin and Later Development of Cultivated Plants. *Journ. Royal Horticultural Soc.*, Vol. 90, pp. 279-291 and 322-340.

Symons, L., 1967 : *Agricultural Geography*, London, p. 15.

Varro, Marcus Terentius : *On Farming* (Translated by Lloyed Storr-Best; London, 1912), Book II, Chapter 1, paragraphs 3-5.

Vavilov, N. I., 1935 : Teoreticheski osnovi selektsi rasteni. Moscoe : Selected Writings translated in "The origin variation immunity and breeding of cultivated plants", *Chronica Botanica*, Vol. 13, 1949-50. Waltham Mass, 1951. p. 366.

Wissman, H. V. *et al.*, 1956 : "On the Role of Nature and Man in Changing the Force of the Dry Belt of Asia" in *Man's Role, in Changing the Face of the Earth*, (edited by W.L. Thomas, Jr.) Chicago : University of Chicago, pp. 278-303.

Zukovskij, P. M., 1962 : *Cultivated Plants and Their Wild Relatives* (Abridged Translation by P. S. Hudson : Commonwealth Agricultural Bureaux, Farham Royala, Bucks.

4

INDIAN AGRICULTURE DURING THE RĀMĀYAṆIC AGE

R. K. SHUKLA

AGRICULTURE FORMS the principal source of human food the foremost primary necessity. From the primitive stage of human civilisation, agriculture has been claiming the greatest attention of man. The development of agriculture in India dates back to the *Vedic* times when majority of people subsisted on it. Likewise, the *Rāmāyaṇa of Vālmīki* (600 B.C.) is replete with allusions to this economy and such reference may form the basis of substantial discussion, tracing the type, form and spatial organisation of agriculture. It is, therefore, apposite to study the salient features of agriculture of the *Rāmāyanic* age from a geographical view point.

Methodology

The *Rāmāyaṇa* is versatile in the field of knowledge and marks out the most splendid information relating to its various social, economical and political institutions. The subject matter belonging to geographic facts appears incidentally. These facts are garbed within the religious and social expressions and performances. The facts have been traced out of the ritual procedures, discussions and mythologies by way of logical interpretation from geographical view point, and therefore, the qualitative analysis has been adopted. Incorrect and legendary information is frequently omitted, unless its discussion and analysis have become inevitably expedient for marshalling and tracing the link. If a fact is simply repeated at different places, only one or two references to it have been cited and deemed adequate. The references cited from the epic are given in the text in brackets.

General Facts

The Sanskrit word *Kriṣi* derived from the root *Kriṣa* is used in the *Vedas* and *Saṁhitās* (Saxena, 1976) to indicate the agriculture. In the *Rāmāyaṇa* too this world recurs. The epic mentions that the agriculture as the science and art of raising plants for the human use was known as *Anrita* (VII, 74.17). So, it is evident that the people of that age knew well the art of farming and they were well-versed in the mixed farming.

In a predominantly agricultural country like India where the cultivators are poor and backward in all respects, the responsibility for the development of agriculture lies mostly on the State. Traditionally in India, the kings have been the guardiáns of agriculture since the remote historical times and as such the *Rāmāyaṇic* period monarchs protected the fields and contributed largely to the development of agriculture by adopting sound planning policies (Shukla, 1984).

The kings being the master of the land (IV. 18.6) was supposed to have a complete knowledge of agriculture — a subject included among the eight fold administrative items of the state (II. 100.68) and he used to take the sixth part (III. 6.11; VII. 74.32) of the agricultural produce as agricultural tax for the facilities provided by the state to the cultivators. It was the duty of the king to keep agriculturists free from six calamities, i.e., fire, flood, epidemic or endemic diseases, famine, pestilence and human agencies (II. 100.44). The State maintained its own granary called *dhānyakośa* (II. 36.7). Agriculture was the major industry during the *Rāmāyaṇic* age. Along with trade and cattle rearing, agriculture was one of the items to be included in the science of *Vārtā* – a special branch of learning (II. 100.47).

Factors Affecting the Agriculture

Agricultural peculiarity of a country is inter-related by various factors, therefore, conformity and disconformity are evinced in agricultural conditions of the different regions. The people of the epic age were well-acquainted with the knowledge of physical factors affecting the agriculture e.g. relief features, soil structure and its fertility and various climatic conditions. The *Rāmāyaṇa* (II. 43.16) refers to the possibilities of better crops due to timely tilling and the availability of adequate rainfall. Similarly, visualising the different aspects of cultural traits, it is evident that the kings were responsible to the welfare of the farmers and advancement of agricultural activities by the administrative policies (II. 100.47-48).

General Landuse

A characteristic feature of ancient land policy in India was the division of all available land of a village for different uses. As far as the utilisation of land is concerned, it was categorised into the following four types (Shukla, 1984) :

(i) Habitation lands which included houses, villages and towns,
(ii) Agricultural lands,
(iii) Pasture lands, and
(iv) Forests and uncultivable wastelands.

The forests in the *Rāmāyaṇic* age must have been more extensive than they are at present. The pasture must have included large spaces needed for cattle ranches. The *Rāmāyaṇa* mentions (II. 20.52; VII. 86.14) the land which was not under plough as *Ūṣara* (Wasteland). The *Aryans* knew well that the seed sown in a barren soil is fruitless (III. 40.3). The king was supposed to have a detailed knowledge of land, its fertility and its barrenness and thus to determine the productive areas more likely to be attacked by the enemies (II. 100.68). There are references of white, pale (II. 91.42) and ochre soils (VII. 14.27).

The *Rāmāyaṇa* refers to the agricultural villages as *grāmas* (II. 57.4), the pastoral villages as *ghoṣas* (II. 83.15), the larger provincial towns and the big markets for agricultural produce as *nagaras* (II. 1.46) and the smaller moffussil towns dealing as trade centres for the neighbouring villages as *pattanas* (IV. 40.25). The villages had their fields on the outskirts. The inhabitants of Ayodhya had their fields just in the suburbs (II. 33.17) and the town is mentioned as full of cultivators (I.6.7). The vicinity of the town must have intensively encouraged the agriculture as the town provided marketing facilities to the farmers to sell the produce of their land.

Agricultural Implements

The *Vedic* literature gives us the picture of well organised rural economy in which agriculture carried on with the plough and bullocks played an important part. The ploughman or cultivator of the soil was called *Kīnāśa* or *Krisāna* who is today popularly known as *kisāna* (Agnihotri, 1963). There are several references in *Vedic* literature to various types of ploughs which were drawn by six, eight or even twelve bullocks. Sometimes, even twenty-four oxen were yoked (Saxena, 1972). Such ploughs, most probably, must have been so effective as a modern tractor for the purpose of deep ploughing. The *Rāmāyaṇa* refers to various implements used in agricultural processes such as *Kuṭhāra* (I. 4.23; II. 80.7), *Kshurā* (III. 47.41), *Hala* (VII. 17.39), *Piṭaka* (II. 36.25), *Ṭaṅka* (II. 80.7; VII. 5.25), *Lāṅgala* (I. 66. 13), *Khanitra* (I. 40.27; III. 4.27), *Fāla, Kuddāla* (II. 32.29, 31), *Dātra* (II. 80.7) *Kalasa* (II. 63.36); *Kaṭhinakāja* (II. 55.17) and *Śula* (I. 39.19).

Irrigation

Water is the basic requirement for life. At a very early period, the ancient Indians recognised the paramount importance of irrigation lest the vagaries of nature and uncertainty of rainfall lead to ruining of crops resulting in famine conditions. During the *Rāmāyaṇic* age, the farmers depended mainly on rain for the cultivation (II. 112.12) but to overcome the difficulties created by the uncertainty and uneven distribution of rainfall, drought and famine, artificial irrigational facilities were provided to the cultivators by the kings. The canals, wells and tanks were constructed by *Yantrakas* (Engineers–II. 80.1). Canals termed as *kulyās* (I. 53.3) were taken out from the rivers and tanks. The epic refers to the perennial streams on which the dams were constructed (VI. 58.58). The *Āryans* made no efforts to dam the rivers that were not perennial (II. 9.54; II. 18.23). The reference in the *Rāmāyaṇa* that *Kārtaviryārjuna* the ruler of Māhismati, blocked the Narmada river like a dam with his thousand arms (VII. 32.4-15), confirms the idea of building a dam across the river. The epic refers to the dams made of sand (III. 69.50) which were subjected to damage during the rains (II. 105.5; VI. 128.4). The *Rāmāyaṇa* also mentions the need of wells and tanks in waterless tracts for agricultural purposes (II.80. 11-13).

During the *Rāmāyaṇic* age, there were two kinds of crops i.e. the dry crops depending on rain (IV. 14.16) and the wet crops depending on wide-spread irrigational channels (VII.36.4). The fields of *Kosala* which were irrigated and did not depend only on rain-water termed as

Adevamātrika (II.100.45) and wet crops grew well all over the land with the help of irrigational facilities and that it was *sphita* (prosperous), presumably on that account (I. 5.5; II. 100.45). The description of the charming fields adorned with rows of crops along both sides of the *Sumāgadhî* (modern Son) river (I.32.10) proves that these fields were irrigated by the water-lifting machines.

Agricultural Crops

There were two types of crops i.e. *Kristapacya* and *Akristapacya* (VII. 84.8). *Kristapacya* were those crops which were cultivated and *Akristapacya* included self-growing and ripening crops without tilling the land e.g. *Nîvāra, Sānvān* etc. *Nîvāra* (wild rice) appears to be inferior grain and was specially eaten by the ascetics living in the forests. *Anna* or *dhānya* was a common term to denote the grains (I.5.5; II. 91.20). The cultivated cereals and wild ones were both considered as confering vital strength to the sacrificer. Even today barley and rice are regarded as auspicious on ceremonial occasions and have an important place in *Hindū* rituals.

Paddy (Śāli)

Paddy (variously known as *tandula, śāli* and kalama) become the most important crop and it was used in various forms. The unhusked grains of rice (*akshata*) were used in various religious performances (I. 73.23, II. 20.17). Besides, various forms of rice e.g. *tandula* (the unhusked rice– I.5.17), *Pāyasa* (boiled rice with milk and sugar– I.16.15), laja (parched rice — I.53.2), *odana* or *bhakta* (boiled rice– II.69.10). *Krisara* (a mixed preparation of rice, beans and Sesamum — II.20.18), etc. are also referred to in the *Rāmāyaṇa*. These references support the view that rice was the staple foodgrain of the people.

Sāli was comparatively a larger variety of paddy and was the principal stable crop of Kharif season. It was sown in the beginning of the rainy season and harvested in autumn. The autumn rice was called *śāli* (III. 16.17) because it was transplanted in the waterlogged fields. The poet in the description of autumn season refers to ripe sprouts in the paddy fields (IV, 30.47, 53). A paddy field was known as *Kedāra* (VII.7.3) or *Kalamakshetra* (IV. 14.16; VI. 4.91). The epic clearly mentions that the lands near the banks of the rivers were the main paddy producing areas (I. 32.10). *Ayodhyā* (I. 5.17) *Kiskindhā* (IV.

26.27) and *Bharadvāja āśrama* (II. 91.35) are mentioned as having abundant supply of paddy and rice. This clearly shows the sufficient production of paddy in the fertile lands of the northern India as well as in the southern India.

Barley (Yava)

It is believed that it was first produced by the gods on the fertile banks of the *Saraswatī* river (Saxena, 1972). It was spring crop sown in the autumn season and harvested in the spring season. It did not require much rain for its cultivation. According to the epic (II. 50.47; III. 16.16) the *Gañgā-Yamunā doab* and the *Pañchavati* region were the principal barley producing areas. New barley grains were offered in the 'First fruit sacrifice'(III. 16.6). The principal rite of this ceremony was offering of the new produce to the gods and the manes, presumably as a mark of gratitude towards them.

Wheat (Godhūma)

It is noteworthy that the word *godhūma* is not mentioned in the *Rigveda.* Probably, it was neither known nor cultivated during the *Rigvedic* times and must have been introduced from elsewhere in a later age, for we find it mentioned in the *Yajurveda Saṁhitas* and the *Brāhmaṇas* (Das, 1979). Wheat did not become an important crop during the *Rāmāyaṇic* age. It was *ravi* crop sown in autumn season and cultivated mainly in the fertile areas where water supply from irrigation was available (III. 16.16; VII. 36.4).

Other Cereals

Among other cereals *bhadrakas* (a kind of bean), *mudgā* (kidney beans), *maṣa* (a bean-urda), *Caṇaka* (gram), *kulattha* (a kind of pulse) and *tila* (Sesamum) (II. 32.20; II. 91.68; VII. 91. 19-20) were produced during the *Rāmāyaṇic* age. Sesamum — the important oilseed was sown with other cereals and was also used in offerings.

Cash Crops

Sugarcane and spices (I. 53.20; II. 91.67) were the most important among the cash crops. Sugarcane and its products — *guda* (molasses —

I.53.4), *sarkarā* (sugar-II. 91.73) and *khāṇḍava* (sugarcandy — VII. 92.12) were extremely popular. The main areas of its cultivation were the fertile alluvial plains of the northern India (I.53.2; II.91.56).

Food was dressed with various spices and condiments to increase its flavour and tastes. Generally, seasoning of food (II. 91.67, vide Vyas, 1958, 75) by clarified butter, oil and spices was common in the society. We find that the cultivation of spices flourished in the coastal region of India (III. 35.23). Dry ginger, cumin and ptychotis ajowan (II. 91.67, 72-73) were main spices of those times. Besides, aloes, pepper (III. 35.22-23), and nutmeg (V. 10.23) were also known.

Cotton, the most important fibre crop (II. 30.12; V. 53.6; V. 58.152) was cultivated in *Satluja-Ganga* basin and black loamy soils of the peninsular India, where the climatic conditions were favourable. Besides cotton, silk (I.74.7; II. 4.30; III.47.27) and hemp (V.48.46; V. 58.152) were also important fibre crops during the epic age.

Natural Calamities

Climate played an important role in affecting the agricultural activities in ancient India. The *Rāmāyaṇa* mentions that the people during the reign of Rama, were free from famines (I.1.90). The fact that the king was considered to have a clear conception regarding six calamities of agriculture and the methods to counteract them in those days, indicates that famines and floods were not unknown phenomena (II. 100.69).

There are many references to famines having occurred before the reign of Rama. These famines were produced by drought damaging the agricultural crops, as a consequence of which the world became unproductive, the forests, tanks, lakes and rivers dried up and all living being faded and decayed (I.9.1 ff. II.117.9). The performance of an *Aśvamedha* restored the prosperity of the earth (VII. 86.4-6). From these references it may be pointed out here that the *Aryans* sincerely believed that it was *Indra* worship that gave them regular rains in abundance and made their country immune from the visitations of such dire calamities as famines and droughts. Besides, the *Rāmāyana* also refers to crop destruction by hailstorms (III.34.9) and insects (VII.7.3).

Social Bases of Agriculture

The economic basis of the *Rāmāyaṇic* India was principally formed by two component organs i.e. agriculture and stock-raising. The agriculture was regarded as quite honourable pursuit and it was not the incompatible occupation of the *Vaiśya* caste only (II. 100.47) but it was open to all other castes (II. 32.29). For *Kshatriyas,* agricultural activities were not regarded as disgraceful. Janaka, the king of *Mithila* is described as ploughing the field and clearing the weeds and grass of the field (I. 66. 3-14; VII. 98.7) This indicates the importance of agriculture among the ruling classes.

Agricultural crops were destroyed by the climatic vagaries and insects, therefore, the *Āryans* had developed divices to avert such agricultural misfortunes by performing sacrifices calling for the help of agricultural gods like *Indra,* (I. 9.9 ff.; VII. 86.4 ff), and *Parjanya* (II, 31.12; II.112.12). There are numerous allusions in the epic for *Indradhvaja* festival. It was a public festival celebrated on the full moon day of *Āṣvina* (IV. 16.37) when the crops were ripe and ready for being reaped. A week earlier than the actual festival a lofty pole decorated with flags and festoons was raised in honour of *Indra;* the guardian of rain and of agriculture. On the final day, the ropes or the other contrivances supporting the pole were removed (II. 74.36; II.77.9, 25). Besides, *Agrāyaṇa* sacrifice (III. 16.6) was a religious ceremony for partaking in the first fruits of the harvest, in different seasons such as in spring by offering rice and barley, in autumn by rice and millets. The aim of this ceremony was to please dieties for averting any harm in eating the new grains or fruits.

The farmers of the *Rāmāyaṇic* age considered that the deep ploughing with timely and adequate rainfall is a work of religious importance and shall result in the greater possibilities of good harvests (*cf.* I.9.9; II.43.16; II.67.9). The king had the prerogative to collect *bali* on the agricultural produce, but the prerogative arose from his sovereignty and not necessarily from his ownership of land. These religious performances brought homogeneity not only micro-regional but on macro-regional basis as well. These systems penetrated so deeply in the society, that even today they can be seen traditionally among the Indians.

Agricultural Prosperity

The *Kosaladeśa* was provided with efficient system of irrigation so that the farmers had not to depend upon rain-water and on that account it was *sphīta* (prosperous) state full of foodgrains (I.5.5, II.100.45-46; VII.41.20). *Magadha* (I.32.10), *Vatsa* (II. 52.101), *Kiskiṅdhā*

(II.33.13) *Kerala, Pāṇdya* and *Cola* (III. 35.25-26; IV. 41.12-19) *Janapadas* are said to be full of abundant foodgrains which proves that these *janapadas* were more productive due to better climatic conditions and fertility of the soil. During the reign of *Rāma*, the agriculturists kept their fields well ploughed (II. 100.44), because clouds rained at proper time (VII.41.20; VII.70.10), therefore, the earth was covered with copius crops (VII.128.72). The kingdoms were prosperous and happy and the country furnished a plentiful supply of food-stuffs (VII. 39.7; VII. 59A.8). While considering the agricultural prosperity during *Vrita's* reign, the earth is termed as *Sarvakāmadughā* (VII. 84.7), thereby, acknowledging the prosperity of an all-round the country through the development of agriculture.

Conclusion

Concludingly, it may be said that during the *Rāmāyaṇic* age, agriculture was naturally the primary mainstay industry of the country. Needs of agriculture were classified and the efforts were made by the administrative machinery to fulfil them. It may be also pointed out that the land, the source of all raw materials for industry and still more important for food stuffs, had been carefully considered in the lay-out utilisation and exploitation.

N.B. The present study is based on authentic English version of the *Rāmāyaṇa* of *Vālmiki* published by Gita Press. Gorakhpur in 1969. The references cited from the *Rāmāyaṇa* are given in the text in brackets.

REFERENCES

Agnihotri, P. D., 1963 : *Patanjali Kalina Bharata*, Bihar Rastrabhasa Parishad, Patna.

Das, A. C., 1979 : *Rigvedic Culture*, Bhartiya Publishing House, Varanasi.

Saxena, D. P., 1972 : *Indian Agriculture During the Vedic Period*, Proceedings of Symposium on Landuse in Developing Countries, A.M.U., Aligarh.

Saxena, D. P., 1976 : *Regional Geography of Vedic India*, Grantham, Rambag, Kanpur.

Shukla, R. K., 1984 : *Rāmāyana : A Study in Ancient Indian Geography*, (Unpublished Ph.D. Thesis) Bundelkhand University, Jhansi.

Vyas, S. N., 1958 : *Rāmāyana Kālīna Saṁskriti*. Sasta Sahitya Mandala, New Delhi.

5

CONCEPT OF AGRICULTURAL REGION

A Review of the Ideas and a Mathematical Interpretation

JYOTIRMOY MUKHOPADHYAY

FOR VISUALISATION and comprehension of the intractably complex surface features of the earth, geographers have for a long time been using the concept of region that is meant for territorial classification of landscapes at various scales. Its role in a geographic study is therefore fundamental, though the concept itself is yet to have a logically sound and generally agreeable basis. The commonly held notions about a region as a territory characterised by a uniformity or homogeneity as to the phenomenon/phenomena concerned, or as a "unitary entity", or a "coherent whole", etc., have been placed before persisting questions (Hartshorne, 1959; Kimble, 1951) and the geographers have now been divided into two contending schools on the point of a philosophic debate as to the reality or very existence of a region (Mitchell, 1973). The geographers of one school view regions from subjective point and consider them as nothing but "mental constructs" (Kolars & Nystuen, 1974) or merely as "descriptive tools" and for them the concept of a region as a single real object has "passed into history" (Hartshorne, 1959). According to them, the traditional region could be simply dismissed by the argument that wherever there is vagueness as to the significance of a regional boundary, there is no proper region. In the words of D. Grigg (1967), 'if a region cannot be delineated accurately can it really exist?' And since 'there are more intermediate areas than there are "regions",' it is theoretically arguable that 'complete partitioning of space implies mis-classification in areas where features are not uniquely separable'. There are, however, strong arguments from the objective school in favour of the real existence of regions

(Alampiev, 1961) and some geographers have now come to an accord that the reality of region can hardly be empirically denied (Carter & Thomas, 1969; Dickinson, 1964). Enrid Alayev (1986), in connection with the conceptual-terminological systematisation of social and economic geography, has defined a region as a territory which in the sum-total of objects filling it differs from other territories and is characterised by the unity and interconnection of the objects constituting it, i.e., by its integrity which serves as an objective condition and natural result of a given territory's development.

The concept of an agricultural region must necessarily stem from the concept of a region in general. It is, however, pertinent to point out that at present the geographers have come to deal with not one, but several types of region. In common geographic parlance, regions are often designated as 'uniform' or 'homogeneous region' (formal region), 'nodal region' (functional region) and 'planning region' or 'programming region'. Geogrescu-Roegen (1972) has however pointed out to a clear-cut distinction between 'arithmomorphic' and 'non-arithmomorphic' regions; the former ones are quantitative models or "theoretical constructs with geometrical . . . connotations", whereas the latter are "empirically derived from observation, . . . and indeterminate, . . .". All these types of region, now common in the geographic literature, appear to be different, but it is imperative on the part of the practitioners of geography to probe into their base that holds them over the single general term 'region'.

At this point an agricultural geographer could simply ask the question: What kind of a region an agricultural region is? Though traditionally one is used to think that an agricultural region could not be anything but a special kind of a uniform or homogeneous region, it would not be unwise to investigate the properties of a nodal region in an *a priori* defined agricultural region. Secondly, from the viewpoint of applied agricultural geography, an agricultural region should not be regarded as an inflexible, static and permanent partition on an agricultural landscape, but it should be considered as a purposive, flexible and dynamic territorial planning unit that could better be designated as an agricultural planning region. It is therefore necessary to consider an agricultural region against the commonly held notions of uniform, nodal and planning region, and the resolution of its basic and general properties is an important step in the development of the proper concept of an agricultural region.

For the agricultural geographers in particular, the concept of an agricultural region lies in a tangle of problems. Apart from the controversy between the adherents of realism and those of nominalism or idealism about the nature of a region in general, there exists considerable difference in the views as to the formalisation of the agricultural regions by the different authors of agricultural geography itself; the latter fact was pointed out as early as 1936 by Derwent Whittlesey, but it still characterises the situation (Kostrowicki, 1968), if not more conspicuously. In fact, it is now very difficult to have an *entente cordiale* between the different ideas and thereby to arrive at a well-shaped, acceptable concept of agricultural region.

The Nature of the Variables in Agricultural Geography

The major obstacle on the way of formulating 'arithmomorphic' agricultural regions is that the agricultural variables are not all quite susceptible to quantification and a good many of them are described and studied only at nominal scale. The types of farming or the major crops and livestock that are often portrayed on the maps as labels of agricultural region, are an example of employing the simple and straight-forward nominal scale data. The common types of farming, e.g. Shifting cultivation, Intensive subsistence farming, Extensive commercial farming, Mixed farming, Plantation agriculture, Mediterranean agriculture, Kollektivnoe khozyaistvo (Kolkhoz), Sovetskoe khozyaistvo (Sovkhoz), Wet farming, Dry farming, Irrigation farming, etc., are nothing but mere empirically derived nominal classes (cross-classes) and their employment as terms in agricultural regionalisation without a formal scrutiny and taxonomic structuring would obviously result in the formulation of agricultural regions (of the world or of a country) without an agreeable logic. It is so, not because that the nominal scale data are hardly acceptable in a modern geographic study, but simply because the nominal scale variables are yet to be defined properly and to be structured (Cole & King, 1968) in a scheme of agricultural regionalisation, so that no logical error, e.g., a cross-classification, results.

One of the important characteristics of the use of the nominal scale data in agricultural regionalisation is that they are in most cases virtually composite in their meaning; for example, the labels such as Migratory agricultural practice, Intensive subsistence agriculture of Mixed farming actually mean much more than that expressed explicitly by the one or two qualifying terms and they signify all the inherent features of the type of agriculture concerned. They are, therefore, synoptic and certainly of more advantage than the other scale data in this respect. But the boundary of an agricultural region delineated on the basis of nominal scale data has seldom been precise.

The use of other scale data in the context of agricultural regionalisation has been relatively infrequent. Ordinal scale values are found in the ranking of the crops, though such rank data have seldom been used for delineation of an agricultural region. Similarly, interval and ratio scale data are yet to have frequent use in agricultural regionalisation. This has been so, perhaps due to the fact that the agricultural geographers have been a little late to shift to the quantitative methods. Nevertheless, it is difficult at this point to comment that such a complex matter of study as agricultural geography could ever be completely 'arithmomorphic'.

There are, in fact, some difficult problems as to the development of quantitative indices, which account for the persistence of the conventional qualitative and descriptive attitude in agricultural geography. As for example, the practitioners of the subject are yet to formulate satisfactory definition of the terms of the agricultural typology. Is there any definite quantitative limit that distinguishes intensive farming from the extensive one? Should intensive agriculture be only defined in terms of man-land-ratio? Or, for the case of capital-intensive farming the ratio of the input of capital per unit of land or man be considered for determining the intensity of farming? And, to make the case still worse, the question may be raised whether there is any farming in the present-day world that could be called purely labour-intensive or capital-intensive? These are only few amongst the host of perplexing questions that reflect the intrinsic difficulties of definition of some of the terms in agricultural geography, and thereby the basic conceptual—terminological problems that are inherent in agricultural regionalisation.

The Typical Ideas about Agricultural Region : A Review

The variation of agriculture has both spatial and temporal dimensions, and the historians have recognised different stages of evolution of the practice of farming. What interests the geographers is that the different forms of agriculture that mark the particular phases of temporal development still co-exist simultaneously on the earth's surface with distinct location and characteristic ecology. These forms of agriculture, which are commonly designated as Shifting cultivation, Intensive subsistence farming, Extensive commercial farming, Mixed farming, etc., are conventionally depicted in several text-books as the regions of agriculture or agricultural landuse (Alexander, 1977; Jones & Darkenwald, 1965; Thomas, Conkling & Yeates, 1962, to mention but a few).

Whether the distinct territorial expressions of these forms of agriculture could be considered as agricultural regions is undoubtedly a problem to decide upon. These forms of agriculture have been recognised in the context of agricultural regionalisation of the world as the main types of agriculture (Freeman, 1971) by Whittlesey (1936).* But, as Kostrowicki (1968) has pointed out, type and region are two different sorts of notion. Type, according to him, being based on similarities between the objects with respect so some attributes, may recur over the earth's surface given similar environment. On the other hand, a region is a unique area, "characterised by a particular set of characteristics different from all others," and unlike type, is spatially continuous.

Whittlesey, in dividing the world into the agricultural regions employed five criteria, which are : crop and livestock combinations, intensity of landuse, level of technological development, processing and trading of agricultural output, and types and association of buildings and other structures associated with the farming. But his scheme suffers from the malaise of cross-classification. It is evident, for example, where the Primitive subsistence farming has been distinguished from the Intensive subsistence agriculture on the basis of intensity of farming (or landuse), the latter differs from the Mediterranean agriculture mainly in respect of crop association, capital input and degree of commercialisation.

Though the scheme of Whittlesey is found to be logically unsound, its popularity has nevertheless been impaired. It has been so, perhaps, because the conventional type-areas of farming roughly correspond to the natural regions of the world. That the idea of an agricultural region was sometimes imbued in the idea of a natural region, has been pointed out by Freeman (1971). But "most agricultural geographers today would reject the notion using 'natural regions' as agricultural regions . . .", because, ". . . agricultural region, if it is to be defined at all, should be defined in agricultural terms". (Morgan & Munton,

* The text-books represent the modified versions of the regions of Whittlesey.

1971). Kostrowicki (1968) has also expressed similar view. Whittlesey, however, is explicit as to the bases for regionalisation, which are no doubt agricultural terms, but the manner of their employment in his scheme is questionable.

As a sample of the early ideas about agricultural region, the agricultural regions of Scotland as delineated by Wood (1931) may be considered. Stamp and Beaver (1971) have commented on the work of Wood that the basis of his regions was the environmental (physical) factors and it does not seem necessary to alter the regional division of the agricultural landscape of Scotland as proposed by Wood. The Highlands and the Western Isles, delineated by Wood as an agricultural region, merely reflects a particular physical landscape where a particular type of agriculture is expected, but no particular criteria have been explicitly employed to determine specifically the uniformity of the agro-geographic-phenomena of this area. The significant intra-regional variations in agriculture, e.g., the predominance of crofting in the western part have been ignored by Wood.

Other cases might be cited as examples to prove that the agricultural regions according to Wood appear more as merely territorial compartments with their respective physical environment or simply location than well-defined, unique agricultural landscapes. It is, in fact, difficult to understand on what basis/bases Wood has differentiated the Southern Upland from the Highlands of Scotland when both have been described as areas dominated by hill-sheep farming; or, the South-Western region from the West-Central region — both known as areas dominated by dairy farming (Types of Farming in Scotland, 'H.M.S.O.' Edin, 1952, in Stamp & Beaver, 1971). As Kostrowicki and others have now decidedly expressed the dictum that agricultural regionalisation must necessarily be based on the attributes of agriculture, the regional scheme of Wood, being based on environmental factors, therefore, becomes unsound.

Though Wood's concept of agricultural region has been denoted as one of the early ideas, this sort of concept did not disappear from the geographic literature quite early. In the recent edition (1971) of *The British Isles*, the idea of Stamp and Beaver in dividing England and Wales into three broad agricultural zones, nineteen agricultural regions and a number of sub-regions does not seem to differ much from that of Wood, in the context of agricultural regions of Scotland. In distinguishing the Highland Zone from the Lowland one, the variation in the relative importance of type of farming has been pointed out, but the agricultural regions do not clearly expose the basis of regionalisation, particularly in agricultural terms. As for instance, it is not easy to understand how the Lake District being an area of predominant hill-sheep farming with bordering strips of cattle rearing could differ from the region of Wales and the Welsh borderlands. The authors have mentioned that the region of Lake District describes considerable intra-regional variation from agricultural point of view; but because of the movement of sheep from one area to another according to season, there is a certain unity in the entire tract. Had this been the basis, it could be stated that the authors have not been explicit enough as to how this can isolate the Lake District from the Welsh plateau, or the latter from most of the northern and western pennines and seemingly they have ignored the other aspects of variation.

The notion that an agricultural region is not necessarily co-extensive with a natural region or a physical division of a country has been put forward by Baker (1926-1932) in the scheme of agricultural regionalisation of the United States. Unlike the regions of Wood, and Stamp and Beaver, the regions of Baker have been specified in purely agricultural terms. Baker is found to have considered a number of bases, viz., predominant crop/livestock, association of crops and/or stock animals, type of farming, etc. As for instance "in the case of Corn belt" the basis has been "the intensity of the production of some characteristic crop"; whereas, "in the case of Cotton belt, the relative importance in terms of value, of one characteristic form of production" (Jones & Bryan, 1942). However, the difference between the Corn belt and the Cotton belt visibly lies in the variation of the predominant crop, and that between the Grain and Fruit belt, and Corn and Winter Wheat belt in the particular association of major crops, and in the north-western part of the country 'Mixed farming' labels the region as an area of a particular type of farming. Moreover, the region named 'Wheat' can only be distinguished from that which has been called 'Spring Wheat' only in terms of period of cropping. Consideration of several bases in agricultural regionalisation is not methodologically unsound, but Baker's concept of agricultural region falls back when it is found that the bases are not equally considered for all parts of the country.

In the work of Vince (1951) on Belgium, one finds a more developed form of the concept of agricultural region. Here, the author is found to be quite explicit not only as to the bases of regionalisation, which are truly agricultural variables, but also he has used the technique of intersecting isopleths for the variables selected. Two of his variables, viz., density of crops (wheat, sugar-beet and industrial crops) or the percentage of exploited land under the particular crops in each canton and the size of holding are quantifiable and yield easily to method adopted by the author. But it is difficult to understand how the author has manipulated effectively the other variable, i.e., the type of farming (which belongs to the nominal scale) with the aforesaid method of study. However, in the work of Vince, it is evident that the concept of agricultural region has evolved from mere empiricism and qualitative judgment to the stage of quantification and structuring on the basis of sophisticated method of analysis. From mathematical point of view, it would not be unjust to comment that in the work of Vince, the concept of agricultural region has emerged as an intersection set of several variables.

The impact of Vince's work on the later workers has not been so much as it should have been. Of course, it cannot be stated that the idea of Vince on agricultural region is comprehensive and quite perfect, yet his approach and method could have been the right stimulus for the subsequent authors.

The development of the concept of agricultural region since the publication of Vince's work has not been in a particular direction. Much later, Harrison Church, Clarke, Clarke and Henderson (1964) in the context of agricultural regionalisation of Egypt have in the main considered the territorial variation of the dominance of crops, crop association, and such aspects as condition of land-surface, soil, climate, irrigation, land reclamation and/or density of population. The regions according to Church and his colleagues appear to be loosely defined as they are founded on indistinct bases that have been varyingly applied.

Similar thought behind agricultural regionalisation is not uncommon in other cases that have come out in about the same period. As for instance, Randhawa (I.C.A.R., 1958) has classified the agricultural tracts of India into five major regions on the basis of consideration of maximum similarity in respect of latitude, altitude of land, air temperature, average annual rainfall, natural vegetation, soil, crops and stock animals (Mamoria, 1975). Within the regions, the tracts showing variation from the common pattern have been considered as to make the sub-regions, though it has not been always specified in what respects the variation occurs. The sub-regions, in fact, have not been formed by considering uniformly the attributes concerned. Moreover, the designation of the Temperate Himalayan

Region as Fruit region, or the Eastern region as Rice region only indicates the fact that the predominance of one crop or a set of few crops (crop association) has kept its mark on the concept of agricultural region.

Lately, the I.C.A.R. (1972) has proposed a new scheme of agricultural regionalisation for India, where the concept of agricultural region appears to be different from its earlier version, i.e., that of Randhawa (1958). According to this scheme, the country has been divided into four zones for each of the most common crops on the basis of consideration of the relative yield and relative spread of the crops. The four zones have been formed not according to any subjective choice of the relevant attributes, but simply as a consequence of combination taking any two sets at a time from the four possible sets of yield and spread defined quantitatively, viz., High yield, High spread, Low yield, Low spread. The actual number of derived sets of combination is two less than (4_2), because the two of the derived sets, viz., Low yield-High yield and Low spread-High spread are evidently empty. The zones, however, like the regions of Vince represent the unique intersection of the sets of the particular attributes.

That crop pattern is a significant variable in differentiating agricultural landscape into a set of regions is a dominant idea amongst the authors. According to Dikshit (1973), the crop pattern is the most sensitive indicator of the condition of agricultural landscape and activity, and therefore, "agricultural regionalisation can be attempted on the basis of the similarity of crops". His selection of crop pattern as an attribute in the context of the agricultural regionalisation of Maharashtra does not differ fundamentally from the most commonly selected variable, i.e., predominance of one crop, as Dikshit himself points out that in real situations "more-cultures are seldom found and quite often the crop cover of an area in a combination with variable share of several crops". He has however applied the method of Spearman's correlation between the crop ranks so that "a high correlation between the crop ranks of the two areas would mean a similarity of crop pattern enabling them to be grouped in one region". He has established the agricultural regions of Maharashtra on the basis of linkage analysis of the rank correlation coefficients. Thus, the concept of agricultural region may be considered as to have reached a stage of sophistication based on the modern statistical techniques. Nevertheless, the question may arise whether the agricultural regions of Maharashtra according to Dikshit are based on partial consideration of the facts or not, as he has omitted a number of facets of agricultural landscape and activity.

In defining an agricultural region, consideration of only one agricultural variable, however important, is not the idea which is quite prevalent in the recent time. Spencer and Horvath (1963) have defined an agricultural region as a tract "with a high degree of homogeneity of one or more criteria". Grigg (1969) has considered a number of criteria, that are : degree of commercialisation, intensity of farming, type of tenure and scale of enterprise, method of farming and combination of crops and livestock. Here, as in the other cases previously mentioned, the problem of manipulation of the several variables to find out their syn-spatial uniformity or what is called territorial homogeneity arises and the effective application of the criteria to define and delineate an agricultural region becomes difficult due to the lack of proper definition of some of the criteria (Knowles & Wareing, 1976).

From the preceding discussion on the ideas about agricultural region, at least four trends can be discerned. Some, particularly the earlier authors, consider that an agricultural region would not be very different from a natural region or a physical (or physiographic) division of a land. The second trend of ideas tries to define an agricultural region as an area in the agricultural landscape where a particular crop/livestock is predominant, or where a particular assemblage of crops/stock animals is found, or a particular type of farming practice is the most prominent feature. The third one attempts to visualise an agricultural region as a unique intersection set of the different agricultural attributes (or variables). The fourth one tries to classify the different agricultural landscapes on the basis of the quantitative measures of similarity (correlation coefficients) of a number of agricultural variables, though the authors of the rigorous quantitative analyses have not usually considered all the important agricultural variables. However, it is very difficult to arrange these trends of ideas in a chronological order. Though the first trend can be considered as older than the others and less revered at present, the other trends still persist with their own impact and control on the current thought.

The Basic Problems

The existence of so many ideas about agricultural region is undoubtedly a stumbling block for any attempt to formulate a precise definition of an agricultural region. Though the importance of the notion of applying the concept of a 'natural region' to delineate an agricultural region has been on the wane, no definite, generally agreed resolution has yet been made as to the question of acceptance or rejection of the different ideas about the issue.

Secondly, the various ideas about agricultural region already outlined, draw attention to the fact that they are partial, and therefore incomplete, as regards the consideration of the relevant and significant attributes. Often, in the case of a particular scheme of regionalisation the application of the criteria or the bases have been unjudicious which has resulted in the cropping up of a number of cross-classes or cross-divisions. In fact, the authors of most of the classification (regionalisation) schemes are found to have ignored the structural or formal aspect of the classification (regionalisation) model. The attributes or the variables concerned often lack a proper definition (e.g., intensity of farming, predominant crop/livestock, crop pattern, to name but a few). Moreover, there is as yet no standardised, generally accepted norm in the format of a list of attributes or variables that are to be considered together in delineating agricultural regions.

Thirdly, if an agricultural region be considered as an area that is uniform or homogeneous with respect to agriculture, the term 'uniformity' or 'homogeneity' should properly be defined in the context of regionalisation. It is, however, very difficult to apply the concept of uniformity or homogeneity for solution of the problem of the classification of agricultural landscapes into distinct units. One of the fundamental characteristics of the spatial phenomena of agriculture is their variability not only in respect of their several aspects, but also with respect to their scale of occurrence. For instance, in a micro-level study one often finds a significant variation in crops with the micro-geomorphic and hydrogeologic change of the terrain, and, two adjacent plots can seldom have an equal quantity of average yield. These variations are detectable only in the field and representable on a map of sufficiently large-scale. But the agricultural landscape appears rather differently on an aerial photograph taken at a certain height. Micro-territorial variations progressively disappear as the cartographic scale becomes gradually smaller. In the agricultural regionalisation of India, Randhawa (1958) identified the Eastern zone as the Rice region, but when the Eastern zone alone be considered, a number of agricultural regions (uniform crop areas) could be identified. Therefore, scale of study is one of the most important aspects for definition of an agricultural region — the one whose role has so far been quite overlooked in many of the geographic investigations. It must be noted that an agricultural region should not necessarily have a dimensional limitation as to its territorial extent. The concept of uniformity or homogeneity of an agricultural landscape could better be understood from the point of view of the geographers of the perception school and be defined statistically for the positivist geographers. A uniform or homogeneous agricultural landscape is a phenomenon where the agricultural attributes may describe variation which is not significant but random.

Fourthly, if it is agreed that an agricultural region is to be considered as a composite entity that is unique in respect of the association of the variables, or that could be described as the intersection of the sets defined in terms of the several attributes of agriculture, a problem often arises due to the common pattern of arrangement of the sets in relation to each other. The following Venn diagram (Fig. 5.1), which has been prepared in accordance to the nature of the territorial variation of some of the attributes of agriculture in northern and north-central India would illustrate the nature of the problem.

The part of the Fig. 5.1 named A could be considered as a unique assemblage of the attributes, viz., wheat, rice and the tracts under canal irrigation. The part labelled B differs from A only in respect of the presence or absence of a particular crop (presumably a predominant one). C varies from B with respect to the type of farming and irrigation, though the variation of these attributes does not describe a covariation of the crop cultivated. On the other hand, the predominant crop may vary under the same type of farming. D differs from B only as to the crop, but F differs from B in terms of the crop and irrigation together. Again, it should be noted that, the presence or absence of irrigation does not induce a corresponding territorial variation in crop type in the area dominated by Jowar *(Sorghum vulgare)* and Bajra *(Pennisetum typhoideum)*.

The Venn diagram, which is a model of the agricultural regions of the Ganga plains and its adjacent tracts, shows that the different parts of the landscape that have been labelled by the capital letters do not stand on the same base. Though an agricultural region must be based on a number of different criteria, the uniformity of distribution as to the several criteria does not in reality describe a definite spatial correspondence. As a result of this disparity in spatial correspondence, a number of overlap — and underlap areas occur. For example, the set under rice is a definite crop region, but there are intra-regional variations in respect of irrigation on one side and the type of farming

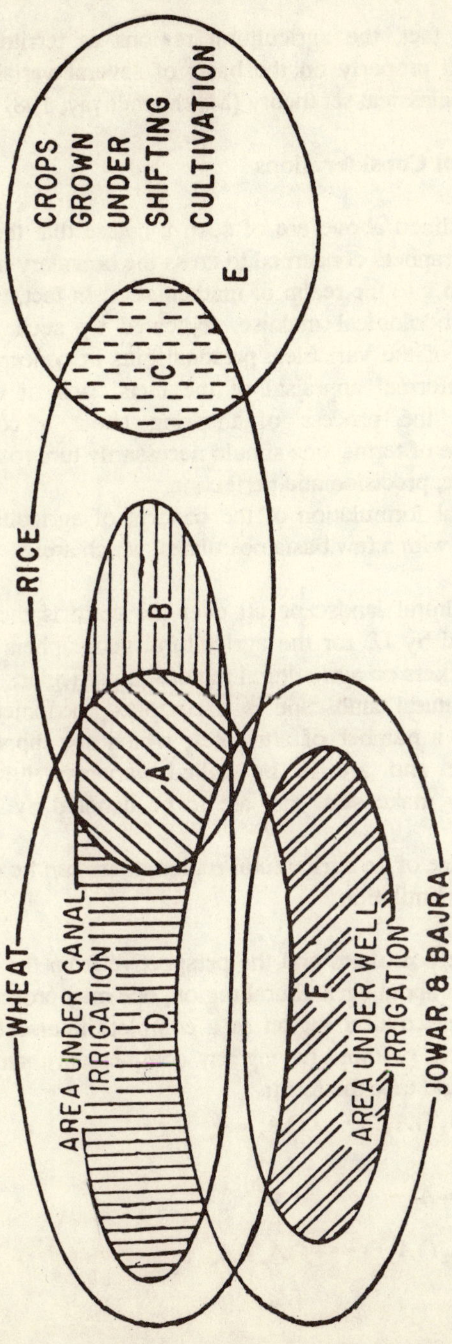

FIG. 5.1: Venn diagram showing areas under different types of cultivation.

on the other. In fact, the agricultural regions as territorial classes cannot be defined properly on the basis of several variables by the application of the classical set theory (Mukhopadhyay, 1981).

The Mathematical Considerations

The problems outlined above are of such a nature that their solution demands the geographers concerned to cross the boundary of their own discipline and step into the realm of mathematics. In fact, to get rid of the chronic methodological malaise, indicated by such features as subjective choice of the variables, personal bias in conceptualisation, qualitative and informal appraisal of the facts, lack of a structural uniformity under the process of thinking about a concept and inconscientious use of terms, one should necessarily turn to a discipline that stands on logic, precision and perfection.

A mathematical formulation of the concept of agricultural region may be developed with a few basic postulates, which are:

(i) The agricultural landscape all over the earth is the Universal set, denoted by U, for the agricultural geographers in general and the workers on agricultural regions in particular.

(ii) The agricultural landscape is a complex phenomenon which consists of a number of attributes, which are supposed to be quantifiable and are to be called variates; these variates individually make sets, that are to be denoted by A_1, A_2, A_3, ..., A_n.

(iii) As to the size of an agricultural region, there can be no specific quantitative limitation.

With the above postulates and the perspective from the discussion on the typical ideas about agricultural region, one may proceed to form a concept of an agricultural region as a complete intersection of the sets of the n variates (without having any complement); symbolically, this proposition could take the form:

$$A_1 \cap A_2 \cap A_3 \cap \cdots \cap A_n = A_n$$

or $$\bigcap_{i=1}^{n} A_i = A_i$$

$$A_1 \cap A_2 \cap A_3 \cap \cdots \cap A_n = A_n$$

But this proposition seems to be untenable as a real definition of an agricultural region. In the Venn diagram it has been found (as in the other cases) that there are considerable underlaps in the intersection of the different sets; these underlaps or the complements of the intersection set being unique on account of the absence of the elements belonging to the different sets, do not have the same status with the intersection or the common set.

Such a feature of the real-world has led the geographers to take recourse to the method of principal component analysis and its statistical extension — factor analysis. It is possibly due to the nature of the data and the nature of the development of the subject, the agricultural geographers have been comparatively slow to appreciate these methods in the context of their problems, particularly of that related to regionalisation, and thereby to conceive of an agricultural region in terms of a few mutually uncorrelated factors, *p*, that underlie the *n* variates, *p* being less than *n*.

Factor analysis is directed to interpret quantitatively the variation in the characteristics among the areas (or objects). Common sources of variance are investigated amongst the variates concerned and the analysis thus seeks to isolate the characteristics that the variates have in common and that account for their intercorrelation. The problem might be considered as one of finding out the intersection of sets (Cole & King, 1968). One of the assumptions of factor analysis is that a set of intercorrelated variates has common factors running through it (Fruchter, 1954) and it is these factors that are extracted by this particular analytical procedure. In the analysis of the principal components or of the factors, each observation (the units of agricultural landscape in the present case) is represented in a multivariate space (the *n* dimensional space) as a vector. The principal components are nothing but the eigenvectors of a variance- covariance matrix. The original values are subjected to a linear transformation, whereby they are converted to a score by projection onto the principal axes. The variance along the major axis is greater than the variance along any other possible straight line through the scatter of points. In fact, the eigenvectors represent the total variance of a set of data, and the purpose of projection or linear transformation is to sort the variances of the individual variates along the eigenvectors of a hyperellipsoid, by which the factor analysts visualise the domain of the data. The regions may be defined on the basis of the amalgamated variates whose variances describe a communality (common variances).

On consideration as points in a *n*-dimensional cartesian space, the units of observation, i.e., the agricultural landscapes can be regarded as to have a 'situation' or a 'location' in relation to the others. Thus, the agricultural landscapes may be said to have a fundamental relational property — the *n*-space Euclidean distance — that can be used to determine their relative proximity or similarity, where multivariate distance coefficient is :

$$d = 1 - \left\{ \frac{\sum_{i=1}^{n} (x_{i1} - x_{i2})^2}{n} \right\}^{\frac{1}{2}}$$

It should be noted here that geographers have now-a-days become interested to explore the characteristics of distances with other metrics, as for example, the Taxicab or the Manhattan metric, where the exponent for the difference value is 1 instead of 2, from a pool of *n* metrics. The general expression of such multivariate distance is :

$$D = \left\{ \sum_{i=1}^{n} (x_{i1} - x_{i2})^n \right\}^{\frac{1}{n}}$$

But how could an agricultural region be identified in such a space? Intuitively of course, it is considered as a cluster of points in a *n*-dimensional space, which should have a specific multivariate distance coefficient (or other similarity coefficients). The clusters or the *a priori* regions could be distinguished by discriminant function analysis, which is meant to find a transform that gives the least ratio of the difference between a pair of cluster-multivariate means to the multivariate variance within the two clusters (Davis, 1973). Geometrically, the analysis searches for one orientation along which the two neighbouring clusters have the greatest separation while simultaneously each cluster has the least inflation. The co-ordinates of the axis of such orientation are the linear discriminant function. The projection of the distance between the multivariate means of the clusters on the axis of separation (i.e., discrimination), which is known as the Mahalanobis' distance or the generalised distance, D^2, has significance in defining or delimiting regions in general (Davis, 1973). It measures the amount of separation between the multivariate means expressed in units of the pooled variance. The significance of a discriminant function can be established by a suitable statistical test.

In reality, however, clusters do not frequently occur as discrete units, and the problem of regionalisation, therefore, cannot so easily be solved by setting a discriminant function that would stand as a mathematical/statistical support of an already established empirical, non-quantitative or often a perceptual finding. As a matter of fact, agricultural regionalisation necessitates the consideration of all the relevant agricultural variables, and the experience all over the earth for a long time has firmly established the dictum that the variables do not describe a perfect territorial covariation (Knowles & Wareing, 1976). It is, of course, not true that the 'unit' agricultural landscapes cannot be clustered on the basis of their similarity coefficients, but the clusters are less likely to be defined in terms of all the relevant variables. The problem of the complements of the intersection sets, i.e., the underlap areas is yet to have a satisfactory solution (other than statistical) in the mathematical formulation of a region.

To overcome the methodological impasse in developing a logically sound and quite a satisfactory concept of an agricultural region, we should now turn toward a recent development in mathematics that has come to be known as the theory of fuzzy subsets. This theory may be considered as an extension of the classical set theory, where, in the latter, a set is clearly defined on the basis of the membership of its elements. An element is considered either to belong to a set or not (disjunctively) and its characteristic function $\mu_A(\chi)$, where χ is a member of the set A, is either 1 or 0. But in the theory of fuzzy subsets, the characteristic function may have any value from the domain of real numbers within the interval (0.1). Thus, an elements χ_i of the superset E may not be a member of A (μ_A = 0, where $A \subset E$) could be a member of A a little (μ_A close to 0), may more or less be a member of A (μ_A neither too close to nor too far from 1), could be strongly a member of (μ_A close to 1) or at the other extreme might fully be a member of (μ_A – 1) (Kaufmann, 1975). Therefore, the membership of an element within a fuzzy subset have a continuous characteristic function, and the set, therefore, cannot be assuredly delineated by a bounding line.

It would, however, be interesting to consider an agricultural region as a fuzzy subset. The preceding discussion has undoubtedly established the fact that an agricultural region cannot be clearly delineated as a set (an intersection set) whose elements must necessarily describe an exact territorial correspondence and thereby have a collinear termination. Some members are strongly present in a region, where others might have a rather weak membership. The predominance of a particular attribute within the region is variable. The demarcation of the boundary of a region is, therefore, highly conjectural and its validity can only be established by employing a suitable test of significance. But, mathematically, a region could well be conceptualised a fuzzy subset — a particular, distinct territory without a definite bound — that so much accords with our veteran empiric notion which frowns on the regional boundaries, but not on the region itself. This notion, in fact, does not invalidate the definition of a region given earlier, i.e.,

$$\text{or} \quad \bigcap_{i=1}^{n} A_i = A_i$$

but adds to it the criterion that the intersection set is a fuzzy subset. In the fuzzy intersection set, the members do not have the same grade or dominance and it is possible, therefore, to identify a core and a periphery. Thus a logical base for unification of the concept of uniform or homogeneous region and that of nodal region is provided. A uniform region is not fundamentally different from a nodal region, as on the basis of the notion of the continuous characteristic function, a uniform region can be rightly regarded as to have a core and a frame like the spatial structure of a nodal region. Planning regions, too, from this viewpoint, do not represent a basically different concept.

Conclusion

Due to the inadequacy of the interpretative tools, the formulation of the concept of region in general and that of agricultural region in particular has been a long-standing problem for the geographer. In 1951, when the conventional geography has already started to show the signs of senility, G.H.T. Kimble stated that to "spend our days 'regionalising' is to chase a phantom and to be kept continually out of breath for our pains". Later on, the geographers were divided on the edge of the two contesting philosophies on the question as to the real existence of a region. The application of the different statistical methods and that of the classical set theory have not succeeded to develop a logically sound and comprehensive concept of a region.

In the context of agricultural regions, however, it has been found that the theory of fuzzy subsets is quite well applicable to develop the concept from mathematical viewpoint. It not only provides a logical base for a notion that is empirically true, but it makes a ground for unification of such apparently different and diverging ideas as uniform region, nodal region and planning region. Of course, it is urgent for us to work on operational methods to deal the members of the fuzzy subsets and consider the cartographic problems that must arise in their representation. But it can be proclaimed that the application of the theory of fuzzy subsets ensures a positive solution of the regional problem, and it could enable the regional agricultural geographers to have a more realistic view of the territorial structure of agriculture. Moreover, it could formulate better planning strategies for the regional agricultural development by grafting to its framework the concept of the so-called nodal region that has proved its worth in the present-day regional planning.

REFERENCES

Alampiev, P. M., 1961 : The objective basis of economic regionalisation and its long range prospects, *Soviet Geography* II, pp. 64-74.

Alayev, E., 1986 : *Social and Economic Geography. An Essay in Conceptual-Terminological Systematisation*, Progress Publishers, Moscow, p. 58.

Alexander, J. W., 1977 : *Economic Geography*, Prentice-Hall of India Private Ltd., New Delhl, pp. 28-201.

Baker, O. E., 1926-32 : Agricultural Regions of North America, *Economic Geography,* 2, pp. 460-93, pp. 50-86, 309-39, 445-65, 4, pp. 44-73, 399-433; 5, pp. 36-69; 6, pp. 166-91, 278-309; 7, pp. 109-53, 326-64; 8, pp. 326-77.

Carter, H. and Thomas, J. G., 1969 : Referendum on Sunday Opening of Licensed Premises in Wales, *Regional Studies*, pp. 61-71.

Church, R. J. H., et al., 1964 : *Africa and the Islands,* Longmans, Green & Co. Ltd., London, pp. 168-70.

Cole, J. P. and King, C. A. M., 1968 : *Quantitative Geography*, John Wiley & Sons Ltd., London, pp. 30-43; 153-159; pp.574-587.

Davis, J. C., 1973 : *Statistics and Data Analysis in Geology*, John Wiley & Sons, Inc., pp. 442-56, 473-533.

Dickinson, R. E., 1964 : *City and Region : A Geographical Interpretation*, Routledge and Kegan Paul, London.

Dikshit, K. R., 1973 : Agricultural Regions of Maharashtra, *Geographical Review of India*, 35, No. 4, pp. 384-96.

Freeman, T. W., 1971 : *A Hundred Years of Geography*, Gerald Ducknorth & Co. Ltd., London, pp. 126-36.

Fruchter, B., 1954 : *Introduction to Factor Analysis*, D. Van Nostrand Co. Inc., Princeton, N. J., pp. 1-85.

Geogrescu-Roegen, N., 1972 : *The Entropy Law and Economic Process*, Cambridge, Mass., U. S. A.

Grigg, D., 1967 : *Regions, Models and Classes, in Chorley and Haggett, Models in Geography*, Methuen & Co. Ltd., London, pp. 461-509.

———, 1969 : The Agricultural Regions of the World — Review and Reflections, *Economic Geography*, 45, pp. 95-132.

Hartshorne, R., 1959 : *Perspective on the Nature of Geography*, Chicago, p. 31.

Indian Council of Agricultural Research, 1972 : *Proceedings of the Symposium on Cropping Patterns in India*, New Delhi, pp. 15-37.

Jones, C. F. and Darkenwald, G. G., 1965 : *Economic Geography*, 3rd edn., Macmillan, New York.

Jones, L. R. and Bryan, P. W., 1942 : *North America*, 6th edn., Methuen & Co. Ltd., London, pp. 195-96.

Kaufmann, A., 1975 : *Introduction to the Theory of Fuzzy Subsets*, Academic Press Inc., Harcourt Brace Jovanovich Publishers, Louvain, Belgium, pp. 1-25.

Kimble, G. H. T., 1951 : The Inadequacy of the Regional Concept, in Stamp and Wooldridge, London, *Essays in Geography*, Rodwell Jones Memorial Volume, Longmans, Green and Co. Ltd., London, pp. 151-74.

Knowles, R. and Wareing, J., 1976 : *Economic and Social Geography*, W. H. Allen, London, pp. 134-38.

Kolars, J. F. and Nystuen, J. D., 1974 : *Human Geography — Spatial Design in World Society*, McGraw-Hill Book Company, U. S. A., pp. 197-219.

Kostrowicki, J., 1968 : Agricultural Typology, Agricultural Regionalisation, Agricultural Development, *Geographia Polonica*, 14, Special Issue for the 21st IGC, New Delhi, PWN Polish Scientific Publishers, Warszawa (Warsaw), pp. 265-74.

Mamoria, C. B., 1975 : Geography of India (*Agricultural Geography*), Ist, edn., Shiva Lal Agarwal & Co., Agra-3, pp. 688-700.

Mitchell, C. W., 1973 : *Terrain Evaluation*, Longmans Group Ltd., London, pp. 23-26.

Morgan, W. B. and Munton, R. J. C., 1971 : *Agricultural Geography*, Methuen & Co. Ltd., London, pp. 126-28.

Mukhopadhyay, J., 1981 : Concept of Agricultural Region, in *Perspectives in Agricultural Geography*, V. 4, Regional Dimensions in Agriculture, ed. by Noor Mohammad, Concept Publishing Co., New Delhi, pp. 45-59.

Randhawa, M. S., 1958 : *Agriculture and Animal Husbandry in India*, Indian Council of Agricultural Research, New Delhi.

Spencer, J. E. and Horvath, R. J., 1963 : How Does an Agricultural Region Originate? *Ann. Ass. Am. Geogr.*, 53, pp. 74-92.

Stamp, Sir L. D. and Beaver, S. H., 1971 : *The British Isles*, 6th edn., Longmans Group Ltd., London, pp. 229-53.

Thoman, R. S., Conkling, E. O. and Yeates, M. H., 1968 : *The Geography of Economic Activity*, 2nd edn., McGraw-Hill Book Co., U. S. A., pp. 126-127.

Vince, S. W. E., 1951 : The Agricultural Regions of Belgium, in Stamp and Wooldridge, London *Essays in Geography*, Rodwell Jones Memorial Volume, Longmans, Green & Co., pp. 255-87.

Whittlesey, D., 1936 : Major Agricultural Regions of the Earth, *Ann. Ass. Am. Geogr.*, 26, pp. 199-240.

Wood, H. J., 1931 : *An Agricultural Atlas of Scotland*, George Gill & Sons, London. .

6

A HIERARCHY OF WORLD TYPES OF AGRICULTURE

JERZY KOSTROWICKI

THE FIRST, preliminary scheme of world types of agriculture, based on a broad discussion of the principles, criteria, methods and techniques of agricultural typology (Kostrowicki and Tyszkiewietz, 1970, 1979; Venzetti, 1972; Reed, 1975; Kostrowicki, 1977), was presented by the International Geographical Union (IGU) Commission on Agricultural Typology to the IGU Regional Conference, held in Hungary in 1971 (Kostrowicki, 1971), and then to the 23rd International Geographical Congress in Canada (Kostrowicki, 1955, 1972). A new, improved version of the scheme in which the discussion at the 5th Commission meeting, held in Hamilton, Canada (Reed, 1975), was taken into account, was elaborated and published in 1974 (Kostrowicki, 1975). The discussion at the 7th Commission meeting at Fontenay-aux-Roses, France, led to the elaboration of a further improved version of the scheme of world types of agriculture which was published in 1976 (Kostrowicki, 1976). The application of that version in several countries was presented and discussed at the 8th and last meeting of the Commission held in Odessa in 1976 (Kostrowicki and Tyszkiewiez, 1979).

Following all those discussions and many testing studies it has been decided to base agricultural typology on the four principal groups (social, operational, production and structural) of agricultural attributes. In the 1974 version each of these groups was represented by six-seven variables with the exception of the group of production attributes, which was described by four variables only. In order to balance the importance of each group the three variables representing production attributes were then doubled. That made the total number of variables characterizing each case rise from 24 to 27.

To make the variables fully comparable they had to be normalized. That was done by reducing the value of each of them to five classes, representing the world ranges of given phenomena. The 27 figures, representing those classes, were arranged in codes. Then, it was arbitrarily decided that codes which did not differ from each other by more than one-tenth of the total variance possible, i.e. by more than 10 variables (27×4 maximum distance between classes 1 and 5=108) could still be treated as representing the same type, i.e. that the types ought to differ from one another by at least 11 variables.

The codes based on this assumption for over 1,000 investigated cases--whether agricultural holdings or such aggregates as administrative or other units--were established and grouped into 61 model types for which codes were also elaborated.

As the attempt to group those types into types of a higher order was found not fully successful, the types of a higher order were not characterized by similar sets of variables, i.e. by the codes.

Such a characteristic became possible only after testing a number of methods of grouping the established model types into the types of a higher order.

At the same time a few minor changes were introduced to the set of the variables, as a basis of the codes. First of all, the three variables accepted representing the production attributes, were substituted by three new variables.

Finally it was decided that agricultural types were to be identified on the basis of the following attributes:

A. Social Attributes

1. The percentage rate of agricultural land which is owned, controlled, or held in common by a group of people under traditional customary rights of tenure.
2. The percentage rate of agricultural land operated under a labour or share tenancy or any other form of landed bondage.
3. The percentage rate of agricultural land, owned or held in ownerlike possession as an individual, joint or corporate private property.
4. The percentage rate of agricultural land operated by a consciously planned collective or State enterprises.
5. The size of holdings in terms of a number of actively employed

people per one agricultural holding.

6. The size of holdings in terms of the total amount of agricultural land (under temporary and perennial crops, cultivated and rough grassland, fallow land) in hectares per one holding.
7. The size of holdings in terms of gross agricultural output in conventional units[1] per one holding.

B. Operational Attributes

8. Inputs of labour in terms of the number of people actively employed in agriculture per 100 hectares of agricultural land.
9. Inputs of animal power in terms of the number of conventional/horse/draught units (Appendix-II).
10. Inputs of mechanical power in terms of the number of HP of tractors and other self-propelling machinery (combines, etc.) per 100 hectares of cultivated land (cropland, gardens, cultivated meadows and pastures).
11. Chemical fertilizing in terms of the amount of chemical fertilizers in kilograms used per one hectare of cultivated land.
12. Irrigation measured by the percentage rate of irrigated land (all systems of irrigation and entrapment) to the total of cultivated land.
13. Intensity of cropland used in terms of the percentage rate of harvested to the total land under temporary crops (including fallow).
14. Livestock population in the number of conventional (large) animal units (Appendix-III) per 100 hectares of agricultural land.

C. Production Attributes

15. Land productivity in terms of gross agricultural output in conventional units per one hectare of agricultural land.
16. (New) Productivity of cultivated land in terms of gross agricultural output from cultivated land per one hectare of actually cultivated land.
17. Labour productivity in terms of gross agricultural output in conventional units per one person actively employed in agriculture.
18. (New) Commercial labour productivity in terms of commercial

(sold or delivered off farm) production in conventional units per one person actively employed in agriculture.

19. The degree of commercialisation as the percentage rate of commercial to gross agricultural output.
20. Commercial production of land as the amount of commercial production in conventional units per one hectare of agricultural land.
21. (New) Degree of specialisation expressed as a coefficient of the degree to which the commercial part of agricultural production of an agricultural holding is concentrated on the least number of items (Szyrmer, 1975).

D. Structural Attributes

22. Land under perennial (trees, shrubs, vines) and semi-perennial crops (covering land without rotation for several years) as the percentage of the total agricultural land.
23. Permanent glasslands (including leys within field-grass systems and current fallows if used for grazing) as the percentage of the total agricultural land.
24. Land under primary food crops (edible grains, tuber, root and bulb crops, vegetables, fruits) as the percentage of the total agricultural land.
25. General gross production emphasis (orientation) as the percentage rate of animal products in the total agricultural output (in conventional units).
26. General commercial emphasis (orientation) as the percentage rate of commercialised animal products to the total commercial production of agriculture (in conventional units).
27. Production of industrial crops (to be used only or chiefly after industrial processing) as the percentage of gross agricultural output (in conventional units).

The classes resulting from the normalisation of indices representing individual attributes were not changed, but new classes were elaborated for three new attributes (Table 6.1).

When analysing the selected variables and their classes one has to take into account that most of them do not characterize only one attribute and that taken either individually or jointly they express many more attributes (Kostrowicki, 1976), which was overlooked by some authors (Kuzina and Yanvaryov, 1979).

TABLE 6.1: Classes of world ranges of individual attributes

No. of the class variable	Very Low	Low	Medium	High	Very High
1.	– 20	20-40	40-60	60-80	80-
2.	– 20	20-40	40-60	60-80	80-
3.	– 20	20-40	40-60	60-80	80-
4.	– 20	20-40	40-60	60-80	80-
5.	– 2	2-8	8-50	50-200	200-
6.	– 5	5-20	20-100	100-1000	1000-
7.	–100	100-1000	1000-10000	10000-100000	100000-
8.	– 3	3-15	15-40	40-150	150-
9.	– 2	2-8	8-15	15-30	30-
10.	– 6	6-15	15-35	35-90	90-
11.	– 10	10-30	30-80	80-120	200-
12.	– 10	10-25	25-50	50-80	80-
13.	– 10	10-30	30-70	70-130	130-
14.	– 10	10-30	30-80	80-160	160-
15.	– 5	5-20	20-45	45-100	100-
16.	– 5	5-20	20-45	45-100	100-
17.	– 40	40-100	100-250	250-800	800-
18.	– 20	20-60	60-180	180-600	600-
19.	– 20	20-40	40-60	60 80	80-
20.	– 3	3-12	12-30	30-80	80-
21.	– 0.1	0.1-0.2	0.2-0.4	0.4-0.8	0.8-
22.	– 10	10-20	20-40	40-60	60-
23.	– 20	20-40	40-60	60-80	80-
24.	– 20	20-40	40-60	60-80	80-
25.	– 20	20-40	40-60	60-80	80-
26.	– 20	20-40	40-60	60-80	80-
27.	– 20	20-40	40-60	60-80	80-

Such a deliberate selection made it possible to reduce the number of variables to the manageable quantity, below which it would be difficult to express all the important attributes of agriculture. On the other hand, with more variables, possible errors in their calculation lessen their importance.

To arrive at a proper method of grouping such multi-feature units a number of taxonomic methods, whether graphic or mathematical, have been tested by the IGU Commission on Agricultural Typology.

Among the graphic methods the following could be mentioned here:

- the Czekanowski diagram which, for quite a long time (Czekanowski, 1909) has widely been used in physical anthropology and, subsequently, in plant sociology, agricultural economics, etc.
- the Wroclaw dendrite (Perkal, 1953).
- the typogram method, i.e. a kind of a star diagram introduced to agricultural typology (Kostrowicki, 1970).
- the diagram proposed by J. Szyrmer (Szyrmer, 1973).

All of those methods have certain advantages and disadvantages.

The first two methods, based on the assumption that distances between objects under study are linear, have been discarded as considerably distorting the reality.*

The typogram technique which became quite popular in some countries (Bonnamour, Guermond and Gillette, 1971; Bonnamour, 1971, 1973; Chabaud and Champallert, 1973; Dongmo, 1973; Kampp, 1973) though very instructive and fairly effective when one has to deal with clear-cut differences between objects under study, requires too many arbitrary decisions when the differences are less evident.

The Szyrmer diagram is quite effective when units which are to be compared are not too numerous, otherwise it produces quite a complicated picture.

Out of numerous mathematical methods used in numerical taxonomy the following have been tested by a group of Polish mathematicians and geographers (Bielecka, Paprzycki and Paisecki, 1980).

- R and Q techniques of the factor analyses;
- the principal component analysis;
- the Berry grouping, based on either initial characteristics, or on the Hotteling components;
- the single-linkage method;
- the paired group method;
- the Hubert divisive methods A and B;
- the McQuitty elementary and restricted linkage analyses;
- the McQuitty maximum distance replacement;
- the Farell method; and
- the Ward method.

---

* See discussion above.

In addition, the following new taxonomic methods have been proposed by the same group (Bielecka, Paprzycki and Piasecki, 1980).

- the ORLINE method, elaborated by Z. Piasecki,
- the gravitation method, elaborated by M. Paprzycki,
- the modification of the FARELL method by M. Paprzycki,
- the IDWER method, elaborated by Z. Piasecki.

The consecutive stages of their studies and methods described in more detail in their papers were presented to the meetings of the IGU Commission on Agricultural Typology in Verona, Fontenay-aux-Roses, and Odessa (Kostrowicki and Tyszkiewicz, 1970, 1979; Vanzetti, 1972, 1975; Reed, 1975; and Kostrowicki, 1979).

One of the results of these investigations was the conclusion that while some methods could be quite correct from the formal, mathematical point of view and successfully used in various classifications, they could at the same time be little effective in the other. Therefore, the testing of formal effectiveness of the methods concerned should be supplemented by testing its cognitional (practical) effectiveness for a given purpose. While the first ought to be done by mathematicians, the second is a duty of the specialists in the subject in which the method is to be used.

TABLE 6.2: Symbols for agricultural typology

Sl.No.	Old Symbols	New Symbols
1	2	3
1.	Tsf	Eff
2.	Tsb	Efb
3.	Toc	Eth
4.	Ton	Enc
5.	Tnn	Enn
6.	Tem	Etm
7.	Tec	Etc
8.	Tds	Tir
9.	Tdh	Tmh
10.	Tia	Tio
11.	Tin	Tin
12.	Tii	Tii
13.	Tjn	—
14.	Tji	Tij
15.	Tjm	Tiu
16.	Tsv	Tss
17.	Tse	Tse

(Contd.)

TABLE 6.2 *(Contd.)*

1	2	3
18.	Tsc	Tsd
19.	Tmk	—
20.	TmG	Tme
21.	Tmm	Tmm
22.	Tml	Mmt
23.	Lcc	Tlc
24.	Lll	Tll
25.	Lpp	Tlp
26.	Mii	Mii
27.	Mss	—
28.	Mhg	Miv
29.	Mhf	Mif
30.	Mmc	Mmc
31.	Mmm	Mmm
32.	Mmn	Mmn
33.	Mml	Mma
34.	Mxm	Mml
35.	Mxg	—
36.	Mxc	Mem
37.	Mxl	Mmb
38.	Mcc	Mec
39.	Mvm	Mlc
40.	Mvh	Mlh
41.	Mvn	Mln
42.	Mpp	Mlp
43.	Mrr	Arr
44.	Muu	Add
45.	Mgg	Mig
46.	Sec	Sec
47.	Sem	Sem
48.	Smc	Smc
49.	Smm	Smm
50.	Scc	Scc
51.	Sll	Sme
52.	Srr	Aro
53.	Stc	—
54.	Sth	—
55.	Shv	Shv
56.	Shf	Shf
57.	Soo	Smi
58.	Sin	Sin
59.	Sii	Sii
60.	Sgg	Shg
61.	Suu	Ads

The formal (mathematical) effectiveness was tested by means of 14 criteria (Bielecka, Paprzycki and Piasecki, 1980).

In the light of the criteria adopted the sequence of effectiveness from the most to the least effective ones was established as follows : 1. gravitation, 2. FARELL-mod, 3. WARD, 4. ORLINE, and 5. IDWER.

To test the practical effectiveness of the abovementioned method for agricultural typology the results of grouping of the 61 types of world agriculture were compared with one anther by a graphic method. On the figures representing the distribution of those types obtained by means of the Szyrmer graphic method (Fig. 6.1) groupings obtained by each of those five methods were marked by lines (Fig. 6.2 to Fig. 6.6).

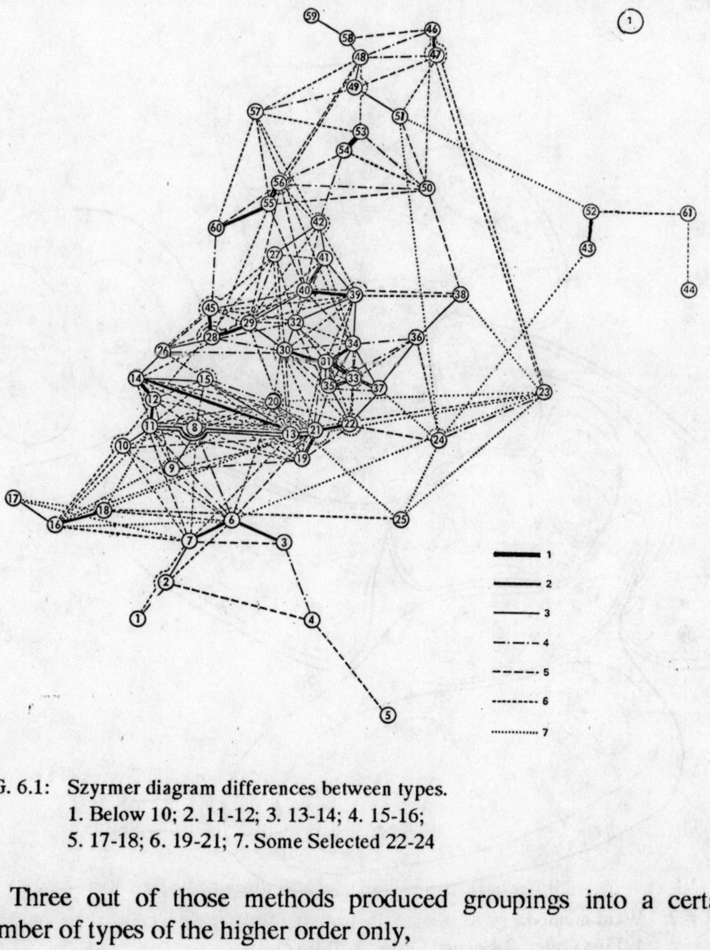

FIG. 6.1: Szyrmer diagram differences between types.
1. Below 10; 2. 11-12; 3. 13-14; 4. 15-16;
5. 17-18; 6. 19-21; 7. Some Selected 22-24

Three out of those methods produced groupings into a certain number of types of the higher order only,

ORLINE – into 17 types,
FARELL-mod – into 16 types,
IDWER – into 13 types

while the advantage of both the gravitation and Ward methods is that they make it possible to group the same types into several higher orders of types.

The gravitation method (Fig. 6.5), for example, grouped 61 units into 8 types of the highest order, which could be subdivided into the types of the second order and some of them into those of the third order.

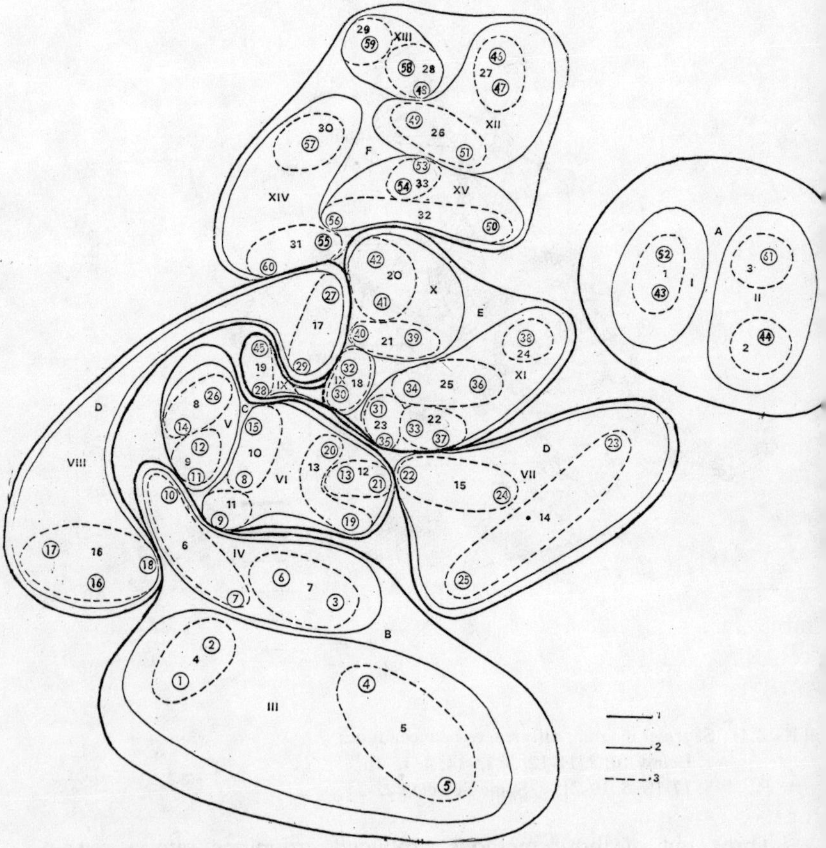

FIG. 6.2: Ward method.
1. First Order; 2. Second Order; 3. Third Order

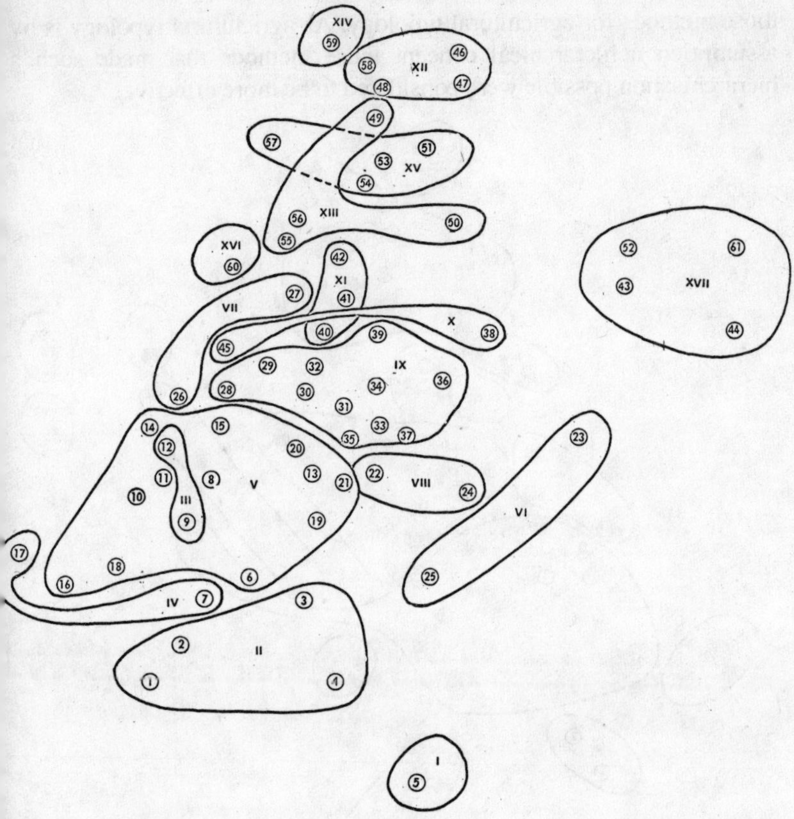

FIG. 6.3: **Orline method.**
1. First Order, 2. Second Order, 3. Third Order

The above diagram resulting from the Ward method could be used for the identification of any number of such orders depending on an arbitrary decision as to how many times and where exactly the diagram could be crossed, i.e. what degree of similarity is to be accepted for each order. In the present study it was arbitrarily decided to cross the diagram three times: first at the level of 200 steps, the second time at 1,000 steps, and the third time at the level of 5,000 steps.

A comparison of Fig. 6.2 to Fig. 6.6 reveals that the results to some of those groupings are more similar with one another, while the other produced more divergent results. The similarity of grouping and their consistence with empirical knowledge of the differentiation of world agriculture were used as the criteria of the practical effectiveness of those methods for agricultural typology. As agricultural typology is by assumption a hierarchical concept those methods that made such a hierarchisation possible were considered to be more effective.

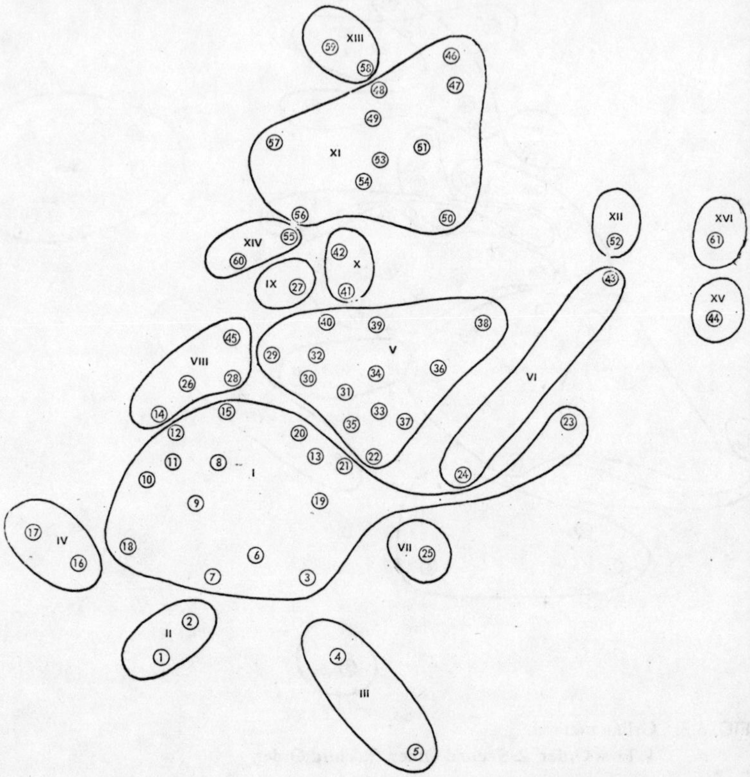

FIG. 6.4: Farell-mod method.

By use of such criteria the effectiveness for agricultural typology of the methods tested could be arranged in the following sequence: 1. Gravitation, 2. Ward, 3. Farell-mod, 4. Orline, 5. Idwer.

In the grouping of the 61 types considered as the types of the Third Order into types of the Second and First Order the results of all groupings were consulted while priority in the decision was given to those groupings that were corroborated by more methods.

Following the former assumption that the types of the Third Order ought to differ from one another by more than 10 per cent of the total variance, i.e. by at least 11 deviations, it was decided that the types of the Second Order should differ by more than 20 per cent of the total variance, i.e. by at least 22 deviations, and the types of the First Order by more than 30 per cent of the total variance, i.e. by at least 33 deviations.

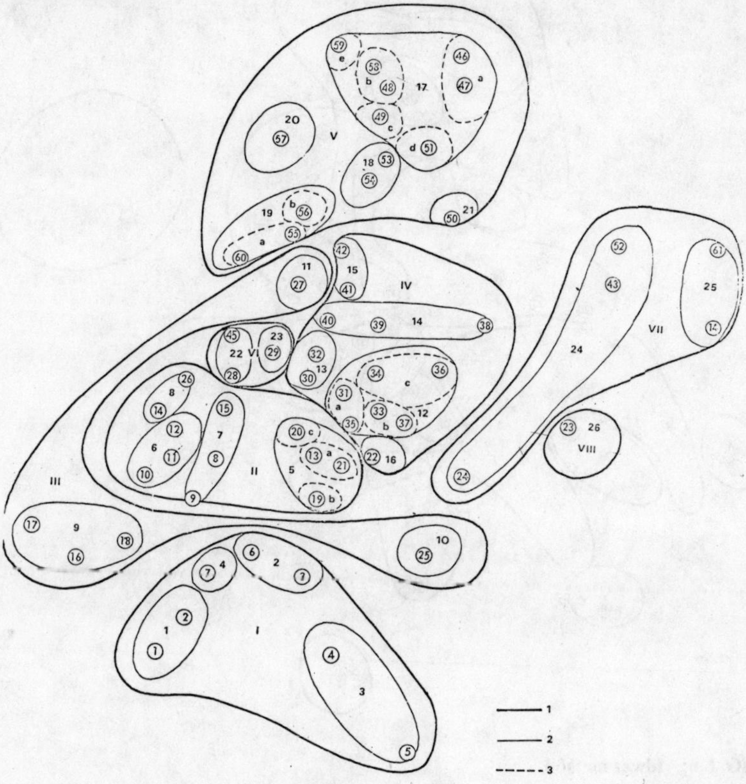

FIG. 6.5: **Gravitation method.**
1. First Order; 2. Second Order; 3. Third Order

The differences between each of the 61 types of the Third Order and every other were calculated by use of a computer according to the following programme, elaborated by M. Kopyt :

"Let's assume that we have *m* items, each of them described by *n* integer numbers, and that these numbers are in a ($m \times n$) matrix (called ITEMS).

We must calculate coefficients $V_{i,j}$ $1 \leq i, j \leq m$

$$V_{i,j} = \sum_{k=1}^{n} \left| \text{ITEMS}(i,k) - \text{ITEMS}(j,k) \right|$$

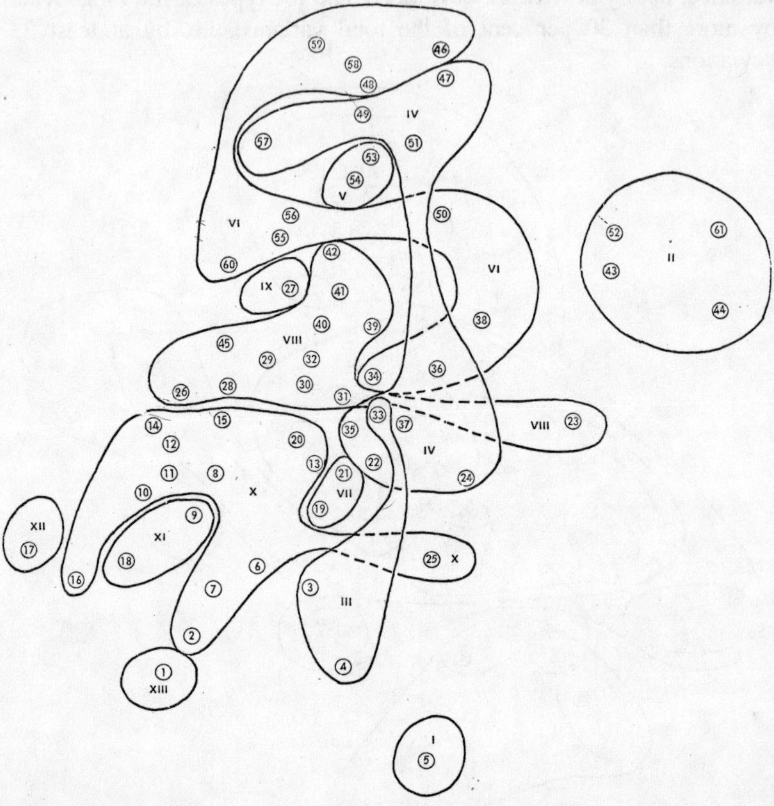

FIG. 6.6: Idwer method.

Our algorithm, in Fortran, has the form of a subroutine with parameters: ITEMS, IHELP, M, N.

```
SUBROUTINE TRANS (ITEMS, IHELP, M, N)
DIMENSION ITEMS (M, N), IHELP (M)
DΦ 100 II = 2, M
III = M — II + 2
IEND = III — 1
DΦ 200 IJ = 1, IEND
ISIGM = O
DΦ 300 IK = 1, N
ISIGM = ISIGM + IABS [ITEMS (III, IK) — ITEMS (IJ, IK)]
```

```
300 CONTINUE
IHELP (IJ) = ISIGM
200 CONTINUE
C    the auxiliary vector IHELP (I), 1 ≤ I ≤ IEND
C    ought to be printed here
100 CONTINUE
    RETURN
    END"
```

In cases when most of the methods applied produced similar groupings such a group was identified as the core of a given type of the Second Order. By use of the median method a code for each type was elaborated. In the same way based mainly on the gravitation and Ward methods as well as the Szyrmer diagram the types of the Second Order were grouped into the types of the First Order, for which codes were established in a similar way as for the types of the Second Order.

Then the codes for each type of the First, Second and Third Order were compared with one another once more by means of a computer according to the same programme as the types of the Third Order, first to find out whether the accepted thresholds of differences of the types of the Second and Third Order were maintained, and secondly, to decide what to do with the types where similarities to the types of a higher order varied considerably when different methods of grouping were applied.

As the 61 types of the Third Order certainly did not represent all possible types of that order, as the number of the types of the Second Order could also be increased with new investigations, and, individual types of any order did not represent the same number of cases, the codes obtained by means of the Median method were slightly modified with both the comparative weights of individual types of the lower orders and possible additions to each of the types of the higher order taken into account.

At the same time, the codes for the types of a lower order differing from those which were considered as the core types of various groups were compared with the codes established for the types of the higher order. Some of them turned out to be of transitional character between the types of a higher order, while the others represented individually, at least for the time being, separate types of a higher order. It was particularly true for the types representing poorly investigated territories.

TABLE 6.3: Hierarchy of world types agriculture

Order	Symbol	Code			
1	2	3	4	5	6
I	E	5321211	-1111121	-1221111	-133331
	T	1241222	-4412242	-4422221	-114221
	M	1151233	-2154343	-4455544	-122231
	S	1115555	-3243242	-3333433	-124221
	A	1133445	-3100003	-3045535	-151551
II	En	4211201	-1100121	-1221112	-141451
	Ef	5221221	-1111111	-1221111	-115111
	Et	2441211	-2211132	-2321211	-133221
	Ti	1241212	-5311452	-5511221	-114111
	Ts	2241222	-3212231	-2222325	-313113
	Tl	1351454	-3311132	-2333323	-123222
	Tm	1151222	-4423143	-3421321	-123231
	Ms	1151212	-3113341	-4434444	-411114
	Mi	1151212	-5145451	-5545555	-115111
	Mm	1151233	-3154143	-4444443	-132331
	Ml	1151345	-2154341	-4455545	-313113
	Me	1151254	-1152131	-2255524	-124221
	Se	1115443	-3232131	-2322321	-124221
	Sm	1115555	-2143143	-3434432	-132331
	Si	1115555	-5422343	-4522232	-114111
	Sh	1115545	-4155341	-4444555	-115111
	Ss	1115555	-2133442	-3344535	-411224
	Sc	1115555	-1142131	-1234515	-115111
	Ar	1133455	-1100001	-1034515	-151551
	Ad	1133455	-5100005	-5055555	-151551
III	Enn/Tnn/	5111000	-1100001	-1021122	-151551
	Enc/Ton/	4211211	-1111121	-1221112	-142451
	Eff/Tsf/	5111000	-1111111	-1321111	-214111
	Efb/Tsb/	4221221	-1111121	-2211111	-115111
	Etc/Tec/	2441211	-2111131	-2321211	-215111
	Eth/Toc/	3431221	-2211132	-1321211	-142331
	Etm/Tem/	2341211	-3311132	-2321211	-123221
	Tir/Tds/	1151211	-4411244	-2211211	-114211
	Tio/Tia/	2231212	-4211441	-3312222	-214111
	Tin/Tin	1241212	-4311142	-4411121	-214111
	Tii/Tii/	1251212	-5411452	-5511121	-115111
	Tij/Tji/	1152112	-5323452	-5522332	-114111
	Tiu/Tjm/	1151112	-4422454	-3322221	-214111
	Tse/Tse/	2241131	-1111131	-1123315	-312114
	Tsd/Tsc/	1351221	-3311332	-2312325	-323113
	Tss/Tsv/	2241222	-3112241	-2222325	-313113
	Tlp/Lpp/	1351444	-4411142	-3423435	-333213

(Contd.)

TABLE 6.3 *(Contd.)*

1	2	3	4	5		6
		Tlc/Lcc/	1351454	-2222132	-2333322	-124211
		Tll/Lll/	1251354	-2211132	-2333424	-141341
		Tmh/Tdh/	1251211	-4511343	-2411111	-122341
		Tmc/Tmc/	2241212	-4422243	-3423332	-224221
		Tmm/Tmm/	1151222	-3323143	-3423321	-123231
		Mii/Mii/	1151211	-4234551	-5534443	-114111
		Mif/Mhf/	1151222	-4134241	-4444545	-415111
		Miv/Mhg/	1151222	-4145352	-5534555	-115111
		Mig/Mgg/	1151313	-5155551	-5555555	-115111
		Mmt/Tml/	1151232	-2223143	-3433333	-131341
		Mmc/Mmc/	1151222	-4244242	-4434443	-213222
		Mmn/Mmn/	1151222	-3155342	-4445544	-112213
		Mmm/Mmm/	1151222	-3255144	-4444442	-123231
		Mma/Mml/	1151232	-3154145	-4444444	-131351
		Mml/Mxm/	1151344	-2154143	-4455443	-123231
		Mmb/Mxl/	1151243	-2143143	-3455434	-131451
		Mlc/Mvm/	1151244	-2154441	-4455542	-113112
		Mlh/Mvh/	1151345	-2154341	-4445545	-315111
		Mln/Mvn/	1151255	-2154242	-4455545	-311114
		Mlp/Mpp/	1151455	-3133341	-4434545	-411115
		Mem/Mxc/	1151254	-1152132	-3355534	-133331
		Mec/Mcc/	1151254	-1152131	-2255525	-115111
		Sec/Sec/	1115443	-3232231	-2322422	-124121
		Sem/Sem/	1115443	-3233132	-2333321	-133331
		Smm/Smm/	1115555	-2144143	-3434432	-132341
		Smc/Smc/	1115555	-3233242	-3433432	-223221
		Smi/Soo/	1115555	-2154443	-4455543	-122222
		Sme/Sll	1115555	-1142132	-2244423	-142441
		Sin/Sin/	1115555	-4322143	-4422222	-224221
		Sii/Sii/	1115555	-5422454	-5522232	-114111
		Shf/Shf/	1115555	-3133241	-4444545	-415111
		Shv/Shv/	1115545	-4154342	-4434555	-115111
		Shg/Shg/	1115545	-5155551	-5555555	-115111
		Scc/Scc/	1115555	-1143131	-2234525	-115111
		Arr/Mrr/	1151354	-1100001	-1044515	-151551
		Aro/Srr/	1115555	-1100001	-1034515	-151551
		Add/Muu/	1151323	-5151005	-5055555	-151551
		Ado/Suu	1115555	-5150005	-5055555	-151551

As a result, the following hierarchy of world types of agriculture was obtained. For their codes see Table 6.3, for their distribution on a diagram and deviations see Fig. 6.7 and for their final grouping see Fig. 6.8.

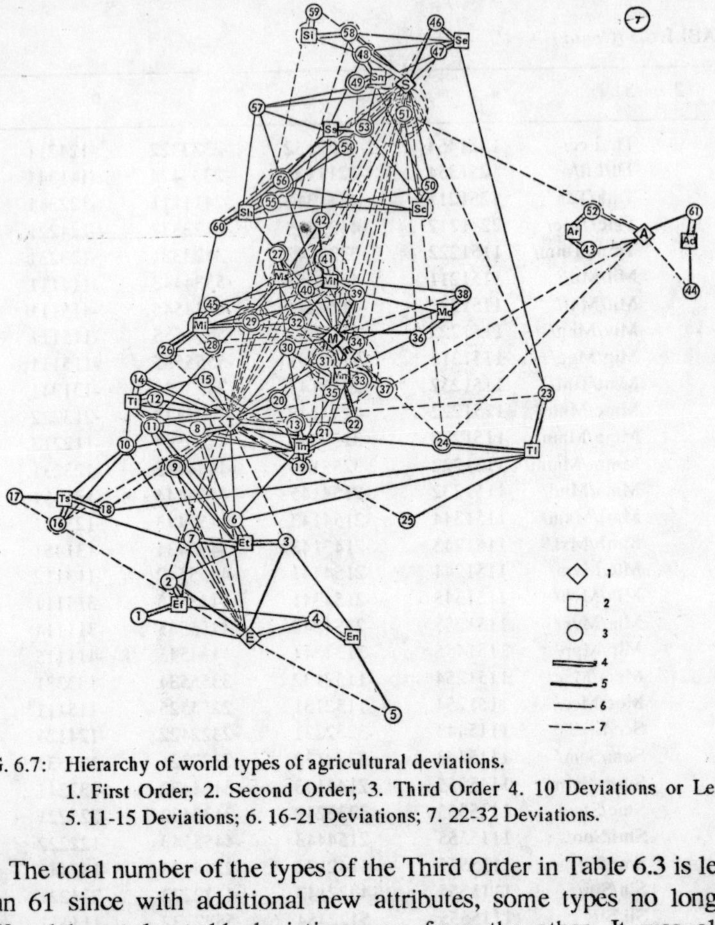

FIG. 6.7: Hierarchy of world types of agricultural deviations.
1. First Order; 2. Second Order; 3. Third Order 4. 10 Deviations or Less; 5. 11-15 Deviations; 6. 16-21 Deviations; 7. 22-32 Deviations.

The total number of the types of the Third Order in Table 6.3 is less than 61 since with additional new attributes, some types no longer differed by at least 11 deviations one from the other. It was also decided to eliminate a few types of the Third Order for which the codes were based on an insufficient number of cases, and to leave the further subdivision of the types of the Second Order open.

E. Traditional[2] Extensive Agriculture[3]

Land held in common or under servile or share tenancy. Small-scale agriculture with low inputs of labour, very low (if any), capital inputs, extensive use of land, very low to low land and labour productivity, very low commercialisation and specialisation, oriented chiefly toward food production whether vegetal or animal.

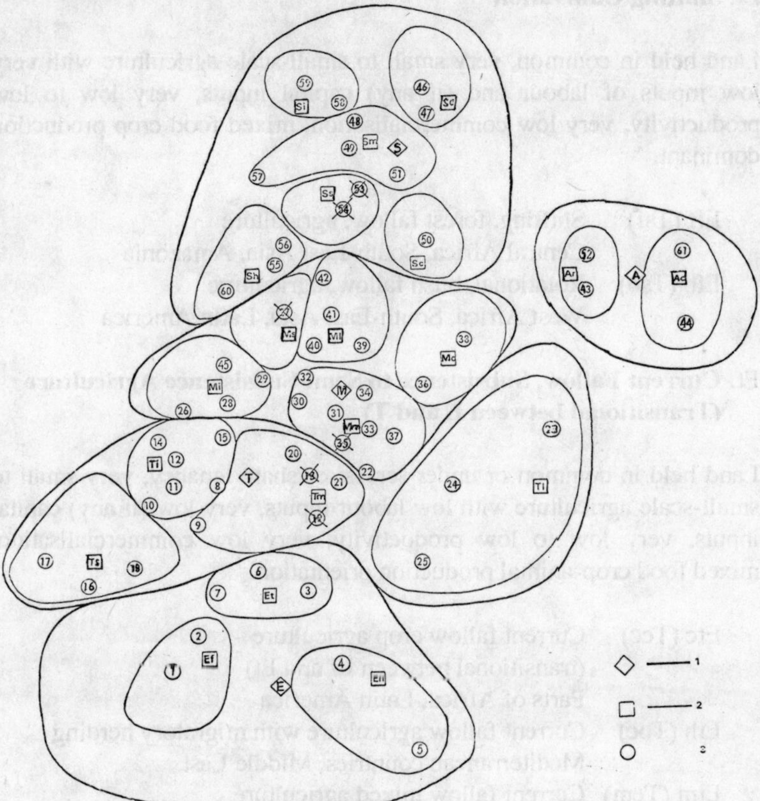

FIG. 6.8: **Hierarchy of world types of agriculture grouping**
1. First order; 2. Second order; 3. Third order.

En. Nomadic Herding

Land held in common. Small- to Medium-scale migratory herding, with or without supplementary crop growing. Very low labour and (if any) capital inputs, extensive use of land, very low to low productivity and commercialisation, livestock products dominant.

Enn (Tnn) Nomadic Herding
North Africa, Middle East

Enc (Ton) Nomadic herding with subsidiary crop growing
North Africa, Middle East

Ef. **Shifting Cultivation**

Land held in common, very small to small-scale agriculture with very low inputs of labour and (if any) capital inputs, very low to low productivity, very low commercialisation, mixed food crop production dominant.

Eff (Tsf)	Shifting, forest fallow, agriculture Central Africa, South-East Asia, Amazonia
Efb (Tsb)	Rotational, bush fallow, agriculture West Africa, South-East Asia, Latin America

Et. **Current Fallow, Subsistence to Semi-Subsistence Agriculture (Transitional between E and T)**

Land held in common or under servile or share tenancy, very small to small-scale agriculture with low labour inputs, very low (if any) capital inputs, very low to low productivity, very low commercialisation, mixed food crop-animal production orientation.

Etc (Tec)	Current fallow crop agriculture (transitional between Ef and Et) Parts of Africa, Latin America
Eth (Toc)	Current fallow agriculture with migratory herding Mediterranean countries, Middle East
Etm (Tem)	Current fallow mixed agriculture Some parts of Europe, Middle East

T. **Traditional Intensive Agriculture**

Land held by owner-operators, less often under customary rights of tenure, small-scale peasant or large-scale (latifundia) agriculture with high or medium inputs of labour, low labour capital inputs, medium to high land productivity, low labour productivity, low to medium commercialisation, oriented mainly toward food production whether crop or animal.

Ti. **Traditional, Small-scale, Labour-intensive Crop Agriculture**

Land operated by its owners, rarely under customary rights of tenure.

Very small to small-scale agriculture with high to very high inputs of labour and considerable inputs of animal power, very low mechanisation and chemical fertilisation, often with important irrigation, very high and productivity, very low labour productivity, low commercialisation, mixed crop production dominant.

Tir (Tds)	Semi-irrigated, low productive, semi-subsistence agriculture (transitional between Ti and Tm, closer to E). India.
Tio (Tia)	Irrigated, medium productive, semi-subsistence crop agriculture Arid lands of North Africa and the Middle East oases, inundated river valleys.
Tin (Tin)	Non-irrigated, highly productive, subsistence to semi-subsistence crop agriculture. South and East Asia.
Tii (Tii)	Highly intensive, irrigated highly productive, subistence to semi-subsistence crop agriculture. South and East Asia.
Tji (Tji)	Highly intensive, irrigated, highly productive, semi-commercial crop agriculture Parts of South and East Asia, Japan.
Tiu (Tjm)	Irrigated, medium productive, semi-subsistence to semi-commercial crop agriculture (transitional between Ti and Tm) Northern India.

Ts. Traditional Small-scale, Semi-commercial Specialised Crop Agriculture

Land operated by its owners, less often under various customary rights of tenure. Small-scale agriculture with low to medium inputs of labour, low to medium capital inputs, low land and labour, productivity, medium commercialisation, mixed food cash crop orientation, very high specialisation mostly in perennial industrial crops.

Tse (Tse)	Extensive, semi-commerical, specialised crop agriculture. Semi-arid countries of Latin America and Africa.

Tss (Tsv)	Semi-commercial, specialised crop agriculture West and East Africa, South-East Asia, Oceania, Latin America.
Tsd (Tsc)	Semi-commercial, semi-irrigated, specialised crop agriculture. Semi-arid and arid lands of north-west India, Pakistan, Sudan.

Tl. **Traditional, Large-scale Agriculture (Latifundia)**

Land privately owned, often operated under servile tenancy. Large-scale agriculture with medium inputs of labour and animal power, very low capital inputs; extensive use of land, with low to medium land productivity, medium labour productivity and commercialisation, medium specialisation in various crop or animal products.

Tlp (Lpp)	Traditional plantations Latin America, some other tropical countries.
Tlc (Lcc)	Traditional latifundia with crop-growing prevalent Southern Europe, Latin America.
Tll (Lll)	Traditional Latifundia with livestock breeding prevalent. Southern Europe, Latin America.

Tm. **Traditional, Small-scale Mixed Agriculture**

Land operated buy its owners. Small-scale agriculture with high inputs of labour and animal power, low-to-medium capital inputs, medium-to-high land productivity, low labour productivity, low-to-medium commercialisation, mixed crop and animal orientation.

Tih (Tdh)	Semi-irrigated, medium productive, subsistence agriculture with livestock grazing developed (transitional between Tm and Ti, closer to E) South Asian Mountains.
Tmm (Tmm)	Semi-subsistence to semi-commercial mixed agriculture Parts of Europe Latin America.
Tmc (Tmc)	Semi-subsistence to semi-commercial mixed

agriculture, with crop growing prevalent
Parts of Europe, Latin America.

M. Market-oriented Agriculture

Private ownership of land, low inputs of labour, high capital inputs, high land and labour productivity, high degree of commercialisation.

Ms. Small-scale, Specialised, Industrial Crop Agriculture

Very small to small-scale agriculture with medium inputs of labour, very low mechanisation, low-to-medium chemical fertilisation, often with irrigation, high productivity of land and medium labour productivity, high commercialisation and specialisation, mainly in perennial industrial crops.

For a time being no sub-division made because of insufficient information.
Japan, Taiwan, Malaysia, parts of Africa and Latin America.

Mi. Small-scale, Highly Intensive Crop Agriculture

Very small to small-scale agriculture with very high inputs of labour, high-to-very-high capital inputs, very high land productivity, high-to-very-high labour productivity, very high specialisation in the production of various food crops.

Mii (Mii)	Labour-intensive, irrigated, highly productive food crop agriculture (transitional between Mi and Ms) Japan, Taiwan.
Mif (Mhf)	Specialised fruit tree growing (transitional between Ma and Ms) Europe, Latin America.
Miv (Mhg)	Specialised vegetable growing (market gardening) Europe.
Mig (Mgg)	Highly industrialised horticulture (green houses important) Europe, North America.

Mm. **Mixed Agriculture**

Small- to medium-scale agriculture with medium labour and high-to-very-high capital inputs, high land and labour productivity, high commercialisation with mixed crop and animal orientation.

Mmt (Tml)	Medium-scale, semi-commercial mixed agriculture with livestock breeding prevalent (transitional between Tm and Mm) Mountains of North West Europe.
Mmc (Mmc)	Small-scale mixed agriculture with crop-growing prevalent Southern Europe, Latin America.
Mmn (Mmn)	Small-scale mixed agriculture with industrial crops prevalent (transitional between Mm, Ms and Ml) Parts of Europe.
Mmm (Mmm)	Small-scale mixed agriculture Central and Western Europe.
Mma (Mml)	Medium-scale, mixed, commercial agriculture with livestock breeding prevalent West Europe.
Mml (Mxm)	Large-scale mixed agriculture (commercial landed estates) West Europe.
Mmb (Mxl)	Large-scale, mixed agriculture with livestock breeding prevalent North America (dairy and corn belts) Australia, New Zealand.

Ml. **Large-scale Intensive Crop Agriculture**

Large-scale agriculture, low inputs of labour and high capital with inputs, high land and very high labour productivity, very high commercialisation and very high crop production specialisation.

Mlc (Mvm)	Large-scale, irrigated, mixed crop agriculture USA, South Africa.
Mlh (Mvh)	Large-scale horticulture (fruit and vegetable) USA, Latin America, South Africa, Australia.

Mln (Mvn)	Large-scale industrial crop agriculture USA, Australia, South Africa.
Mlp (Mpp)	Modern plantations (transitional between Ml and Ms) Latin America, parts of Africa, South Asia, Oceania.

Me. **Large-scale, Extensive, Specialised Grain-crop Agriculture**

Large-scale agriculture with very low inputs of labour, very high mechanisation, low land productivity, very high labour productivity, very high specialisation in grain-crop products, sometimes with subsidiary livestock breeding.

Mem (Mxc)	Extensive grain-crop growing and livestock breeding (transitional between Me and Mm) Australia.
Mec (Mec)	Extensive grain-crop growing USA, Canada, Australia, South Africa.

S. Socialised Agriculture

Socialised (collective or State) land ownership and operation. Large-scale agriculture of varying intensity and productivity with high degree and prevailing redistributive system of commercialisation.

Se. **Incipient Socialised Agriculture**

Large-scale agriculture with medium inputs of labour, low inputs of animal power, medium mechanisation, low chemical fertilisation, extensive use of land, low-to-medium land productivity, low labour productivity, low-to-medium commercialisation, mixed crop animal orientation.

Sec (Sec)	Incipient, socialised mixed agriculture with crop growing prevalent. South-east Europe, southern parts of the USSR.
Sem (Sem)	Incipient, socialised mixed agriculture East-Central Europe, USSR.

Sm. Socialised Mixed Agriculture

Very large-scale agriculture with low inputs of labour, high productivity, high commercialisation with mixed crop and animal orientation.

Mmt (Tml)	Medium-scale semi-commercial, mixed agriculture with livestock breeding prevalent (transitional between Tm and Mm)
Mmc (Mmc)	Small-scale mixed agriculture with crops growing prevalent Southern Europe, Latin America.
Mmn (Mmn)	Small-scale mixed agriculture with industrial crops prevalent (transitional between Mm, Ms and Ml) Parts of Europe.
Mmm (Mmm)	Small-scale mixed agriculture Central and Western Europe.
Mma (Mml)	Medium-scale, mixed, commercial agriculture with livestock breeding prevalent West Europe.
Mml (Mxm)	Large-scale mixed agriculture (commercial landed estates) West Europe.
Mmb (Mxl)	Large-scale, mixed agriculture with livestock breeding prevalent North America (dairy and corn belts) Australia, New Zealand.

Ml. Large-scale Intensive Crop Agriculture

Large-scale agriculture, low inputs of labour and high capital with inputs, high land and very high labour productivity, very high commercialisation and very high crop production specialisation.

Mlc (Mvm)	Large-scale, irrigated, mixed crop agriculture USA, South Africa.
Mlh (Mvh)	Large-scale horticulture (fruit and vegetable) USA, Latin America, South Africa, Australia.
Mln (Mvn)	Large-scale industrial crop agriculture

	USA, Australia, South Africa.
Mlp (Mpp)	Modern plantations (transitional between Ml and Ms) Latin America, parts of Africa, South Asia, Oceania.

Me. **Large-scale, Extensive, Specialised, Grain-crop Agriculture**

Large-scale agriculture with very low inputs of labour, very high mechanisation, low land productivity, very high labour productivity, very high specialisation in grain crop products, sometimes, with subsidiary livestock breeding.

Mem (Mxc)	Extensive grain-crop growing and livestock breeding (transitional between Me and Mm) Australia.
Mec (Mec)	Extensive grain crop growing USA, Canada, Australia, South Africa.

S. **Socialised Agriculture**

Socialised (collective or State) land ownership and operation. Large-scale agriculture of varying intensity and productivity with high degree and prevailing redistributive system of commercialisation.

Se. **Incipient Socialised Agriculture**

Large-scale agriculture with medium inputs of labour, low inputs of animal power, medium mechanisation, low chemical fertilisation, extensive use of land, low-to-medium land productivity, low labour productivity, low-to-medium commercialisation, mixed crop animal orientation.

Sec (Sec)	Incipient, socialised mixed agriculture with crop growing prevalent. South-east Europe, Southern parts of the USSR.
Sem (Sem)	Incipient, socialised mixed agriculture East-Central Europe, USSR.

Sm. Socialised Mixed Agriculture

Very large-scale agriculture with low inputs of labour, high productivity, high commercialisation with mixed crop and animal orientation.

Mmt (Tml)	Medium-scale, Semi-commercial, mixed agriculture with livestock breeding prevalent (transitional between Tm and Mm) Mountains of North-west Europe.
Mmc (Mmc)	Small-scale mixed agriculture with crop-growing prevalent Southern Europe, Latin America.
Mmn (Mmn)	Small-scale mixed agriculture with industrial crop prevalent (transitional between Mm, Ms and Ml) Parts of Europe.
Mmm (Mmm)	Small-scale mixed, commercial agriculture Central and Western Europe
Mma (Mml)	Medium-scale, mixed, commercial agriculture with livestock breeding prevalent West Europe
Mml (Mxm)	Large-scale mixed agriculture (commercial landed estates) West Europe.
Mmb (Mxl)	Large-scale, mixed agriculture with livestock breeding prevalent North America (dairy and corn belts) Australia, New Zealand.

Ml. Large-scale Intensive Crop Agriculture

Large-scale agriculture, low inputs of labour and high capital with inputs, high land and very high labour productivity, very high commercialisation and very high crop production specialisation.

Mlc (Mvm)	Large-scale, irrigated, mixed-crop agriculture USA, South Africa.
Mlh (Mvh)	Large-scale horticulture (fruit and vegetable) USA, Latin America, South Africa, Australia.
Mln (Mvn)	Large-scale industrial crop agriculture

	USA, Australia, South Africa.
Shg (Sgg)	Highly industrialised horticulture (green houses important)
	Various socialist countries.

Ss. Socialised, Specialised Industrial-Crop Agriculture

Very large-scale agriculture with low-to-medium inputs of labour, medium-to-high capital inputs, medium land productivity, high productivity of labour, high-to-very-high commercialisation and very high specialisation in perennial, industrial crop growing.

No subdivision made, for the time being, because of insufficient information.

USSR (cotton, tea), Cuba (sugarcane, etc.)

Sc. Socialised, Extensive, Specialised Grain-Crop Agriculture

Very large-scale agriculture with very low inputs of labour, high mechanisation, low chemical fertilisation, extensive use of land, very-low-to-low land productivity, medium-to-high labour productivity, high commercialisation and very high specialisation in grain crops.

Scc (Scc)	Extensive specialised grain crop growing
	Southern part of the USSR.

A. Highly Specialised Livestock Breeding

Types of agriculture in which common technological characteristics combined with a large scale of operation and a high degree of specialisation in livestock breeding prevail over other agricultural attributes.

Ar. Extensive Commercial Herding

Market-oriented or socialised large-to-very-large-scale agriculture, with very low inputs of labour and capital, very low land productivity, medium-to-high labour productivity, high commercialisation and very high specialisation in animal products.

Arr.	Market-oriented livestock grazing (ranching)
	Western USA, Australia, South Africa.

Aro. Socialised livestock grazing (otgon)
Southern parts of the USSR, Mongolia.

Ad. **Highly Industrialised (Feed-Lot), Livestock Breeding**

Market-oriented or socialised large-scale agriculture with very high inputs of labour and capital, very high land and labour productivity, very high commercialisation and specialisation in various animal products.

Add. Market-oriented, highly industrialised livestock breeding.
Various West European countries, North America.

Ado. Socialised highly industrialised livestock breeding.
Various socialist countries.

The present hierarchy deals with the three highest orders of the types only. If needed, it is possible to subdivide each type of the Third Order into types of still lower orders. To do so 10 classes instead of five as proposed in the 1974 version of the agricultural typology (Kostrowicki, 1974) could be used and some additional variables relevant for a given territory could be introduced, such as, for example, irrigation systems, the role of part-time farmers, the number and average size of crop fields per one holding, proportion of agricultural production delivered off farm but not sold (share cropping), etc., under the condition that the balance between four groups of attributes is maintained, or else by using five per cent differences, i.e. at least deviations from the model types.

The first attempt to present such a subdivision without introducing any additional variables or more classes was made by W. Tyszkiewicz (Tyszkiewicz, 1979) in her study of agriculture of the Thracian Basin, Bulgaria, which appeared to be very monotonous. Two methods were tested in that study: the Szyrmer diagram and the deviation method with the conclusion that the second one was more effective for that purpose.

Transitional cases are another problem. Here, the technique of successive products—somewhat reciprocal to successive quotients technique (Kostrowicki, 1970, 1976; and Szyrmer, 1976) was applied successfully to measure the distances between various transitional cases and two or more model types of agriculture (Stola; and Tyszkiewicz), particularly when larger aggregates were used as basic units of investigation.

In cases where differences between two or more cases, and the model code are around 10, but one is slightly below 11 and another 11 or slightly more, a 50 per cent clause was proposed to eliminate possible identification of the first case as a pure type, and the repudiation of the other. Such clause states that in such a situation, besides the case which falls within the accepted limit, those that are up to 50 per cent higher should also be taken into account. Practically it applies only to the cases with, 8, 9 and 10 deviations from the model codes, which plus 50 per cent are respectively 12, $13^1/_2$ and 15 deviations.

The scheme of the world agricultural types is an open system. While it is doubtful that more types of the First Order, differing sufficiently from the five listed above, could be identified at least nowadays, it is quite certain that many more types of the Second Order and particularly of the Third Order could still be discovered, particularly for the countries or periods for which data or literature are scarce. Described in the same way that could supplement the present scheme.

In fact numerous new types were already described in a number of studies carried on, or encouraged by the Commission. Most of those descriptions, however, were based on the former versions of the typology, differing in some variables, with codes elaborated only for the types of the Third Order, Rearranged in a proper way they could be included into the present scheme.

At the same time a great number of other studies have recently appeared that bring quantitative data characterizing agriculture of various areas or holdings. If expressed by codes, they could also be used for the extension of the present scheme to cover agricultures for which no sufficient data had earlier been available. Another problem is a temporal extension of the scheme to cover agricultures of the past. Many data for their characterisation could be found in numerous countries, both in statistics and in some historical studies.

Their use and presentation in a comparable way could greatly contribute to the knowledge of the processes and direction of transformation of agriculture in time, endowing the agricultural typology with a more dynamic character. Some studies of that kind have already been completed, none of them, however, covers a period longer than one-three decades (Stola, 1975; Szczesny, 1979). Besides a cognitional interest such comparative dynamic studies could be of practical importance for forecasting or programming agricultural development (Kostrowicki, 1972, 1976, 1977).

APPENDIX-I

CONVENTIONAL UNITS TO MEASURE AGRICULTURAL PRODUCTION

The proposed list of conventional units was based on both the wheat equivalent units elaborated by FAO and published every year in the *FAO Agricultural Yearbook*, and the grain units, elaborated originally by German agricultural economists and then adapted by various authors to differing local conditions and extended to the crops not characteristic of Germany. The first ones are based on the relation of the average value of individual agricultural products to the average value of wheat, the second on the relation in the starch and protein content in individual products to the starch and protein content of cereal grains. Both those kinds of units have certain advantages and disadvantages. Wheat equivalent units are based on values of particular crops but those values are estimated not from their current prices, which solely would make it possible their use in studying such important problems as net production, agricultural income, etc. On the other hand, as current prices undergo constant changes, their use in agricultural typology would be pointless as they could not serve as a basis for any comparisons, both in time and space, which is possible when either wheat equivalent or grain units are used.

Therefore, since many products are not aimed for either starch or protein, both these categories of conventional units should be considered artificial.

At present, when prices undergo rapid changes, grain units as a measure of agricultural production would guarantee better comparability in time and space. As, however, grain units have never been worked out for many products, particularly for tropical crops, and wheat equivalent units for fodder crops, it was necessary to combine both those units in order to establish a full list of units for measuring agricultural production. Since both wheat equivalent units and grain units are based on the value of the unit of grain, they do not differ greatly from each other — as one can see from below.

Products that are not listed below should be assessed from the relation of their value to the value of wheat.

Proposed Simplified Weight Units: Grain Units: 100 kg of a given Crop is Equivalent to 100 kg of cereal grain

A. Crops

1. Cereals	
Wheat	1.00
Rye	0.90
Barely	0.90
Oats	0.80
Maize	0.80
Millets	0.80
Sorghum	0.70
Buckwheat	0.80
Rice, paddy, rough	0.80
clean	1.10
Other cereals	0.70
Mixed grain	0.70
2. Pluses	1.20
3. Starchy roots	
Potatoes	0.40
Sweet potatoes, yams, etc.	0.30
Cassava (manioc)	0.20
Oca, ulluco, mashua	0.40
4. Vegetables	0.50
Onion, garlic, asparagus	0.40
Green beans, green pea, cauliflower	0.30
Roots, lettuce	0.25
Cabbages, celery, kohlrabi	0.20
Carrots, rape, spinach	0.15
Melons, pumpkins	0.10
Lettuce, chicory	0.25
Tomatoes, cucumbers, sweet peppers	0.20
Seeds	3.00-8.00

(Contd.)

5. Fruit, fresh	
Citrus fruit	0.70
Bananas	0.40
Grapes	0.50
Other fresh fruit	0.70
Dates, figs	1.50
Nuts in shell	3.00
Small fruits, berries	0.50
Raisins	2.50
6. Sugar crops	
Sugar, centrifugal, raw	1.10
Sugar, non-centrifugal	0.55
Sugar beets	0.25
Sugarcane	0.20
Seeds	3.00
7. Beverage crops	
Cocoa, beans, fermented and dried	4.50
Coffee, green beans	5.50
Tea, made	9.00
Wine	1.00
8. Spices, condiments	
Hops	5.00
Cummin	2.00
9. Tobacco	
Farm sales weight	10.00/2.00
10. Oil crops	
Olive oil	5.00
Olive fruit	2.00
Palm kernels	1.00
Palm oil	1.50
Soybeans	1.20
Groundnuts, in shell	1.50
shelled	2.00

(Contd.)

	Rapeseed	1.80
	Mustard seed	1.80
	Sesame seed	1.80
	Sunflower seed	1.60
	Coconuts, shelled	1.40
	unselled	1.10
	Castor beans	1.20
	Poppy seeds	2.50
	Cottonseed	0.90
	Hempseed	2.00
	Linseed	1.70
11.	Fibres	
	Cotton lint	8.00
	raw	1.20/1.00
	ginned	6.50
	unginned	3.00
	Flax fibre raw	6.00
	Flax fibre straw	1.00/0.70
	Hemp fibre raw	6.00
	straw	0.70/0.50
	Jute, kenaf, mesta, etc.	2.00
	Fibre agavas	2.00
	Abaca	2.20
	Other hard fibres	1.50
12.	Rubber	5.00
13.	Feeds	
	Fodder roots	0.12
	Maize, green	0.13
	Sugar bet leaves,	0.11
	thrash, bagasse	0.70
	Clover, lucerne green	0.15
	hay	0.50
	Other hay crops green	0.13
	hay	0.40
	Meadow green	0.10

(Contd.)

hay	0.30
Pasture	0.10
Mixed cereal	
crops green	0.13
hay	0.50
Pulses straw	0.25
Cereal straw	0.10
Seeds	4.00

B. **Livestock Products**

1. Meat (live weight — 100 kg)	
Cattle, calves, buffaloes	6.00
Pigs	6.00
Sheep, lambs	6.00
Horses	5.00
Poultry	7.00
Other animals	6.00
Rabbits	8.00
Fish	6.00
2. Milk	1.00
3. Eggs	6.00
4. Honey	6.00
5. Fibres : Wool (greasy) raw	18.00
Wool (greasy)	40.00
pure	100.00
Cocoons (fresh)	14.00
6. Manure	0.03

APPENDIX-II

CONVENTIONAL DRAUGHT (HORSEPOWER UNITS)

Camels	2
Horses	1
Buffaloes	0.9
Oxen	0.5
Cows	0.2
Mules	0.8
Asses	0.3

5 hP in tractor power equals 1 horsepower unit.
1 horsepower unit equal 5 men.

APPENDIX-III

CONVENTIONAL (LARGE) ANIMAL UNITS
(Simplified for Typology of World Agriculture)

Camels	2.0
Cows, buffaloes, horses	1.0
Other cattle	0.8
Cattle without specification	0.6
Llamas, reindeer, asses	0.3
Mules	0.6
Pigs	0.25
Sheep and goats	0.1
Geese, turkeys	0.02
Laying hens, ducks	0.001
Chicken	0.003
Poultry non-specified	0.01

The units are based essentially on animal weight and 500 kg of live weight was taken as one unit. If weight of animals in a specific region differs heavily from that average, the units applied could be corrected on the basis of their real weight. At the same time, in case the number of slaughtered animals per year is higher than the number recorded by an agricultural census, equivalent multiplier should be applied.

APPENDIX-IV

DEGREE OF SPECIALISATION COULD BE ASSESSED BY MEANS OF THE FOLLOWING FORMULA*

$$W_i = a_{i1}^2 + a_{i2}^2 + a_{i3}^2 + \cdots + a_{in}^2, \quad w_i = \sum_{j=1}^{n} a_{ij}^2$$

where W — index of degree of specialisation
a_{ij} — part of given product in total commercial production
n — number of products

In case the data are not available, the degree of specialisation can be estimated in the following way:

Class 1. Full dispersal, no single product surpasses 10 per cent of commercial production.

Class 2. First five products no more than 20 per cent of commercial production each, or first four products about 25 per cent each, the remaining less, or first three products about 30 per cent each, the remaining less, or first product about 40 per cent, the second less than 20 per cent, the remaining less than 10 per cent each.

Class 3. Two first products about 45 to 50 per cent commercial production each, the remaining less than 10 per cent or first product about 60 to 75 per cent, the second 25 to 35 per cent, the remaining less than 10 per cent.

Class 4. First product 75 to 90 per cent of commercial production, the second 20 to 25 per cent or the remaining dispersed less than 10 per cent each.

Class 5. First product over 90 per cent of commercial production, the remaining less than 10 per cent each.

NOTE: **It has been decided that in case the degree of commercialisation (variable 19) is very low (Class 1) the degree of specialisation should be considered as insignificant and accidental and, therefore, marked 1.**

(— Editors)

---

* **In the calculation decimal or common fractions (1/2, 1/15, 0.5, etc.) should be used rather than percentages.**

NOTES

1. For the list of conventional units, see (Kostrowicki, 1974), Appendix-I (100 kg of wheat is accepted as one conventional unit).
2. In the present version of typology, the symbols have been changed in such a way that they symbolise certain characteristics, for example, a — animal, c — crop, E, e – extensive, i – intensive, l — large-scale, M — market-oriented, m – mixed, S — socialised, s — specialised, T — traditional, etc. The codes from the former version are presented in parentheses.
3. As it is impossible for the names of types to represent all the 27 characteristics, only some of those characteristics that differentiate best a given type from the other have been used to form a name.

REFERENCES

Bielecka, K., Paprzycki, M. and Piasecki, Z., 1980 : "Applicability of Numeric Taxonomy Methods in Agricultural Typology. Problems, Criteria and Methods of Evaluation", in Mohammad, N. (Ed.) *Perspectives in Agricultural Geography*, Vol. I, New Delhi.

Bonnamour, J., Guermond, Y. and Gillette, Ch., 1971 : "Typologie des Systems d'exploitation agricole utilises en France", *Annales de Geographie* 438, Mars, Avril, pp. 144-166.

———, 1971 : "Les systems regionaux d'exploitation agricole en France — Methods d'analyse typologique". *Etudes Rurales*, 43-44, pp. 78-169.

———, 1973 : *Geographie Rurale — Methodes et Perspectives*, Paris, pp. 44-48.

Chabaud, M. and Champalbert, A., 1973 : "Methode d'analyse de l'espace agricole : typogramme et matrice ordonnable. Examples du Minervois et de al vallee du Rhone". *Bulletin de la Societe Languedocienne de Geographie*. Tome 7, fasc. 2. Montpellier.

Czekanowski, J., 1909 : "Zur Differentialdiognose der Neandertalgruppe", in *Korrespondenz-Blatt der Deutschen Gesell schaft fur Antropologie, Ethnographie und Uhrgeschichte*, V.XL, pp. 44-47.

Dongmo, J.L., 1973 : "Typologie de l'agriculture camerounaise. Essai d'application a l'Afrique Noire de la methode statistico-geographique du professeur Kostrowicki. Universite de Yaounde", in *Annales de la Faculte des Lettres et Sciences Humaines*, 5, pp. 19-40.

Kampp, Aa. H. 1973 : Et forslag til landbrugstypologi', in *Geografisk Orientering* 3, pp. 134-137.

Kostrowicki, J. and Tyszkiewicz, W. (eds), 1970 : "Essays on Agricultural Typology and Land Utilization", *Geographia Polonica*, 19, p. 290.

Kostrowicki, J., 1970 : "Some Methods of Determining Land Use and Agricultural 'Orientations' as Used in the Polish Land Utilization and Typological Studies", *Geographia Polonica*, 18, pp. 93-110.

———, 1970 : "Types of Agriculture in Poland: A Preliminary Attempt at a Typological Classification", *Geographia Polonica,* 19, pp. 99-110.

———, 1971 : "The Methodological Bases for the Typology of World Agriculture", in Krajko G. et al. (eds), *Symposium on Agricultural Typology and Agricultural Settlements*, Szeged-Pecs, pp. 141-173.

———, 1972 : "A Preliminary Attempt at a Typology of World Agriculture", in

Adams, W.P. and Helleiner, F.M. (eds), *International Geography*, V. 2, Montreal, pp. 1097-1100.

_______, 1974 : *The Typology of World Agriculture: Principles, Methods and Model Types*, Warsaw, p. 74 (Mimeographed).

_______, 1975 : "The Typology of World Agriculture: A Preliminary Scheme", in Reeds L.G., *Agricultural Typology and Land Use*, Hamilton, Ontario.

_______, 1976 : *World Types of Agriculture*, Warsaw, p. 49, (Mimeographed).

_______, 1976 : "On the Methods of Determining Land Use, Crop and Livestock Combinations as Used in the Polish Studies", in *Essays in Applied Geography*. In Memory of the late Professor S.M. Ali, Sagar, M.P., India, pp. 111-120.

_______, 1977 : "Agricultural Typology, Concept and Method", *Agricultural Systems*, 2, pp. 33-45.

_______, 1976 : "Agricultural Typology as a Tool in Planning Spatial Organization of Agriculture", *Geoforum*, 7, pp. 234-259.

_______, 1979 : "Agricultural Typology", *Geographia Polonica*, 40, p. 260.

Kuzina, I.M. and Yanvaryova, L.F., 1979 : "Types of World Agriculture Map for Higher Schools", *Geographia Polonica*, 40, pp. 17-27.

Perkal, J., 1953 : Taksonomia Wroclawska (Wroclaw Taxonomy): *Przeglad Antropologiczny* 19, pp. 82-96.

Reeds, L.G. (ed), 1975 : *Agricultural Typology and Land Use*, Hamilton, Ontario, p. 350.

Stola, W. : "Essai d'application des methodes typologiques a l'etude comparee sur development des agricultures belge et polonaise". *Geographia Polonica*.

_______, 1975 : "Changements dans les types de l'agriculture belge dans les annees 1950-1970", in Vanzetti C. (ed.) *Agricultural Typology and Land Utilization*, Verona, pp. 336-349.

Szyrmer, J., 1973 : "Propozycja zastosowania nowej metody taksonomicznej do typologii rolnictwa", (Sum: The Proposal of the New Taxonomic Method to be Applied in Agricultural Typology), *Przeglad Gerograficzny*, 45, pp. 739-756.

_______, 1975 : Stopien specializacji rolnictwa (Sum: The Degree of Agricultural Specialization), *Przeglad Geograficzny*, 47, 1, pp. 117-135.

Szyrmer Przemiany, J., 1976 : "w strukturze przestrzennej produktywnosci ispecjalizacji w rolnictwie indywidualnym w Polsce w latach 1960-1970" (Sum: The spatial Changes of the Productivity and Specialisation of Individual Farming in Poland in 1960-1970). *Dokumentacja Geograficzny*, 4-5, pp. 50-61.

_______, 1960 : "On the Analysis of a Set Characteristics", *Zastosowania Matematyki*, 5, pp. 35-45.

Szczesny, R., 1979 : "Changing Types of Austrian Agriculture, 1960-1970", *Geographia Polonica*, pp. 40.

Tyszkiewicz, W., 1979 : "Agricultural Typology of the Thracian Basin, Bulgaria, as a Case of the Typology of World Agriculture", *Geographia Polonica*, pp. 40.

_______, "Types of Agriculture in Macedonia as a Sample of the Typology of World Agriculture", *Geographia Polonica*.

Vanzetti, C. (ed.), 1972 : *Agricultural Typology and Land Utilization*, Verona, p. 448.

_______, 1975 : *Agricultural Typology and Land Utilization*, Verona, p. 498.

7

TYPOLOGY OF AGRICULTURE
Procedure and Techniques

JACEK H. SZYRMER

THE BASIC principles and methods applied for the Types of Agriculture Map of Europe, described by Kostrowicki (1991), could equally be used for the map of world types of agriculture.

The elaboration and publication of the Map of Europe did not put an end to the works on the typology of agriculture. The extension of the works on all the world was subject of the scientific conference organised by UNESCO and IGU which was held in Rabat, Morocco, (Troughton, 1986). After the discussion it was assumed, that the map of world types of agriculture could be elaborated, provided that adequate funds are available.

Meanwhile, a more modest project has been undertaken in 1986 at the Institute of Geography and Spatial Organisation, Polish Academy of Sciences, directed by J. Kostrowicki, to elaborate a general map of types of world agriculture on the scale of 1:15 million and to prepare an appropriate monograph on its basis. The scholars of some countries, like France, Soviet Union, India, China, Thailand and Cameroon are taking part in the execution of this project. We hope other scholars will help us.

Idea of Agricultural Types

Agriculture is understood as a branch of material production, which consists on the cultivation of plants and breeding of animals by man, which in a conscious way interferes in the biological processes, and transforms the natural environment, producing raw plant and animal goods. Based on this definition, agriculture includes neither collecting, hunting, forestry, fishing and fish-breeding nor processing of agricultural goods. Thus to agricultural production belong raw products: milk but not butter, or cheese; grapes but not wine; olives but not oil.

Each agricultural holding or the agriculture of the given area may be characterised by variables expressing its components and the relations between them. Theoretically each combination of given values of these variables is possible, but in practice certain combinations are very common, others seldom, or some do not exist at all. Those groups of similar holdings characterised by combinations of similar variable values are types of agriculture.

The above conclusions may be presented in a geometrical way: each agricultural holding constitutes a point in multidimensional space, where each dimension, each axis, is one of the adopted variables. Knowing the values of these variables for the given holding, and ascribing them respectively to the axes, we may determine the position of the point-holding in the multidimensional space. If in our space we would situate points representing all the holdings existing presently in the whole world, then in certain places of our space we could see big concentrations, clusters, while other places would be empty or almost empty. Those concentrations of points representing agricultural holdings are types of agriculture, and the coordinates of concentration centres are models of types.

The typology of agriculture ought to be basically understood as the typology of agricultural holdings. However, very frequently, specially in studies covering larger areas, aggregates of holdings, are investigated instead of agricultural holdings. Consequently, dominant types over a given territory are identified. Quite often, however, there is no single type dominating an area, but two or more types coexist simultaneously, differing more or less from each other. An example may be constituted by the coexistence of small and large holdings, traditional and modern, private and state owned. In such a situation, if possible, groupings of differing holdings should be investigated separately.

The determination of types are not an easy task. The uneven distribution of points-holdings in space does not allow us to divide this space a priori into regular polyhedrons. Thus we must determine the types and their central points on the basis of empirical data. The Method of Agricultural Typology, considered in such a way, must at least embrace four elements:

1. Set of diagnostic variables,
2. Rules of expressing diagnostic variables in a uniform way (normalisation),
3. Method of measuring of differences between units (holdings or other territorial units), or between units and models of types, and
4. Rules of ascribing units to the individual types.

Diagnostic Variables

The set, currently applied in the Method of World Agricultural Typology, comprises 28 diagnostic variables. They are secondary quantities, calculated on the basis of certain primary quantities. All diagnostic variables with one exception are ratios. Their values are obtained by dividing one out of the quantities listed below by another one (see Appendix). These primary quantities are basic variables:

A. Number of agricultural holdings.
B. Total agricultural land (in hectares).
C. Agricultural land owned, controlled or held in common under traditional customary rights of tenure (in hectares).
D. Agricultural land operated under servile, labour or share tenancy (in hectares).
E. Agricultural land being a private property (in hectares).
F. Agricultural land operated by collective and state enterprises (in hectares).
G. Cultivated land (in hectares).
H. Total area of temporary crops including fallow (in hectares).
I. Harvested land under temporary crops (in hectares).
J. Land under perennial and semi-perennial crops (in hectares).
K. Permanent grassland (in hectares).
L. Land under primary food crops (in hectares).
M. Irrigated land (in hectares).
N. Number of farm animals (in conventional animal units).
O. Number of herbivorous farm animals (in conventional animal units).
P. Number of draught and pack animals (in conventional draught units).
R. Number of people engaged in agriculture (in conventional units i.e. converted into full time employees).

S. Mechanical draught power (in Metric horsepower, hp).
T. Use of chemical fertilisers (in kilograms of pure content of NPK).
U. Gross agricultural output (in conventional units).
V. Gross animal output (in conventional units).
W. Gross output of industrial crops (in conventional units).
Y. Total commercial agricultural production (in conventional units).
Z. Commercial animal production (in conventional units).

Although the method of estimating values of the diagnostic variables does not usually raise any doubts, the description of basic variables is much more difficult. This is due to the variety of agriculture in different countries of the one hand, and to a different understanding of terms by researchers coming from various disciplines and schools on the other. Therefore, it is necessary to define precisely how to calculate the values represented by those terms. The rules for calculating these variables will be described below.

Normalisation

The identification of types is impossible without defining in a uniform way the differences between territorial units, i.e. taxonomic distance. This is not possible without expressing diagnostic variables in a uniform way, viz. by the means of the same measures. This is achieved in the Method by applying the "normalisation" of values of those variables, the empirical value being transformed into the "normalised" one. All possible values of each diagnostic variable are divided into five classes. The first class contains the very low existing values of the variable; second — low; third — medium; fourth — high; while the fifth class — very high. The lower limit of the ranges of all variables is 0. The upper limit of a range of occurrence, consists of half the variables : in thirteen variables (no. 1–4, 12, 19, 22–28) expressed in percentages, it is equal to 100 per cent, in variable 21 only — 1. To accommodate the classification to changes in world agriculture, the remaining variables do not have the upper limit of their range, and thus in their case the fifth class is left open (Kostrowicki, 1991) A "normalised" value of a given variable ascribed to the unit under study, constitutes the number of the class of its empirical value.

In consequence, each unit under study (agricultural holding, group of holdings, or agriculture of a territorial unit), may be presented in the form of a *code* which consists of 28 digits, each from 1 to 5, for example the code of self-governed agriculture of the Algiers wilaya, Algeria, in 1977-1978:

1115443 — 4154342 — 3322432 — 2131115

The digit 4 at the beginning of the second group indicates that the value of diagnostic variable no. 8 *input of labour per 100 ha of agricultural land,* in this agriculture is contained in class 4, i.e. between 40 and 150 persons per 100 ha of agricultural land. Further, the digit 1 in the middle of the fourth group indicates that the value of diagnostic variable no. 25 *percentage of animal produces in gross agricultural output* is contained in this case in class 1, i.e. amounts to less than 20 per cent.

The code of a unit is a set of coordinates enabling its location in our 28-dimensional space where each variable is one dimension of this space. With the help of codes formulated in such a way, we may also describe central points of each type. Codes of central points may be considered as model types.

Determination of Types

A typology of world agriculture consists of three levels of types. This system comprises until now six types of the first order, 20 of the second order, and over a 100 of the third order. Types classified in three levels are arranged in a hierarchical way, i.e. each type of the third order belongs to one of the types of the second order, and each type of the second order — to one of the types of the first order.

Each type is marked by a symbol containing one (types of the first order), two (types of the second order), or three letters (types of the third order). In case of a symbol of the third order types, the first letter represents the symbol of the type of the first order, whereas the first and the second taken together represent a symbol of the second order type, to which it belongs. For example, the symbol *Mel* informs that the type of the third order represented by it, belongs to types: *M* (first order), *Me* (second order) and *Mel* (third order). Each of those types is represented, by its model in the form of a code comprising 28 digits, representing 28 variables (Kostrowicki, 1991). The codes are coordinates of central points of types. The unit under study is ascribed to the closest type of the first, second and third order.

The change of the value of diagnostic variables from empirical to comparable "normalised" ones, enables an evaluation of differences between each given pair of real units or between each real unit and model of each type, i.e. the taxonomic distance of this unit from central points of all types.

From various taxonomic distance which are in use, the one expressed by the following formula was chosen:

$$D_{kl} = \sum_{j=1}^{n} \left| a_{kj} - a_{lj} \right|$$

where: D_{kl} – taxonomic distance between unit k and type l,
a_{kj} – normalised value of j variable for k unit,
a_{lj} – normalised value of j variable for model of l type,
n – number of variables (28).

The *taxonomic distance* is thus evaluated as the sum of absolute values of differences within all individual variables. Each variable has the same weight. The distance understood in such a way may have values from 0 to 112 (4 × 28 variables).

It has been arbitrarily decided that the maximum deviation (taxonomic distance) from the model type (centre) to limits of the type of the third order should equal 11, to limits of the type of the second order — 22, and to limits of the first order type — 33. It corresponds with 10 per cent, 20 per cent and 30 per cent of the maximum distance.

If the unit under study is situated outside the limits of all hitherto described agriculture types of the given level, this indicates that it represents a new world type, not described as yet.

Calculation of Basic Variables

Land Ownership and Land Tenure

Land ownership is being defined as the broadest right accepted in the given society for the disposal of land and obtaining profits from it. However, in many traditional societies the term of land ownership does not exist, i.e. nobody who can transfer his rights to the land, and thus is not its rightful owner. In some cases there is a belief that land is owned by ancestors, gods, etc. A certain group of people (tribe, lineage, family, etc.), or a single person, is entitled only to use this land, even to grant permission for its temporary use by another person. However, he or she cannot dispose of it permanently.

If the term of land ownership is to be universal, i.e. so that it would always be possible to determine the owner of each piece of land in agricultural use, the term should be applied more widely. In that case a land owner ought to be considered as a natural person or a group of persons, which is entitled to decisions concerning the selection of a land user, or to authorise somebody else to select such a user.

Among various forms of land ownership the following could be singled out:

1. *Traditional common ownership*: In this case the land owner, often in the broader sense, is a group of people wider than a family. This group is usually bound by ties such as kinship, tradition of descending from a common ancestor, common dwelling. Contrary to the case of private group property, the right of belonging to a community is usually a consequence of tradition, birth, etc. A member of community who wants to leave that community, cannot "withdraw his share", or turn it over to a selected purchaser. Such "ownership" is commonly described as a "use right" i.e. the right to use the land is owned, not the land itself. In this case the ownership is indivisible.

2. *Mortmain ownership*: Land belongs to religious institutions and foundations. To this group belongs above all the property of churches, religious associations, and religious communes. In Christian countries, a classic example consists of lands owned by the Catholic church, both belonging to orders, or a dowry of bishops or parsons; and in the Muslim countries — lands of habous (*waqf*). Similarly to traditional communities, the right to transfer land rights does not exist here or is substantially limited. Many people consider God himself as a formal owner of land. Moreover, disposing of profits from the given land is also usually, at least formally, limited. It may be destined for requirements of the cult, keeping of monks, charitable activity . . . . Property of foundations, being more common in recent times, is of a similar character as the ownership of religious institutions, and that is why it may be considered as mortmain ownership.

3. *Private individual ownership*: Land belongs to a natural person. Characteristic for this form of ownership is a considerably high degree of freedom left by the law to the owner. He or she may, according to his or her wish, use the land by him- or herself, pass it over to

somebody else in return for rent, transfer all rights in regard to the land to others (e.g. sell and land), and dispose it off in case of, his or her death.

4. *Private group ownership*: This is a more rare case of private ownership. This form comprises the ownership of various kinds of corporations, joint stock companies etc, as well as group holdings.

5. *Collective ownership*: This is another form of a group ownership, either established by the state or organised by individuals. It includes agricultural productive cooperatives. In this form of group ownership, contrary to the private group ownership, entering and leaving the cooperative, purchasing and ceiling of land are being limited and regulated by the authorities. This is the main difference of this form of ownership from the traditional common ownership, in which also members of the group have no full freedom of disposing of land belonging to them, but the source of those limitations is the tradition, customs, and not decisions of administrative bodies.

6. *Public ownership* embraces the ownership of the central and local governmental bodies.

The basic variable "C" concerns traditional common ownership and the basic variable "E" (private land ownership) embraces both the individual and the group ownership.

Land tenure usually concerns the right to use land. The various existing forms of land tenure may be classed into the same six basic groups like land ownership. Obviously, not always the owner of land is utilising it, and, therefore, the form of tenure may differ from the form of ownership. This should be kept in mind while calculating the value of basic variables C, D, E, and F, and as a consequence of diagnostic variables 1, 2, 3, and 4. Basic variable F embraces all cases of collective (5) and public (6) land tenure.

Forms of land tenancy: As land tenancy in a broad sense may be understood all cases of transferring the right of landuse for a determinate or indeterminate period, by agreement by the owner or by a person authorised by him, to a third party. This agreement may be written or oral; it may be concluded in various forms, according to the local customs or laws. In return for the right of landuse, the tenant is obliged to provide the owner with rent. This rent may be provided in different forms:

1. Rent in work (labour tenancy),
2. Rent in nature,
3. Rent in money.

Tenancy may also have various forms depending on way of determination of the amount of a rent:

1. Land rent depending on the size of harvest. It consists of definite part or share of the harvest. This form is described by the term *share tenancy* or *share cropping*. In French it is called *merayage* (from *meitie*, the previous form of *moitle* — half) which indicates that the owner and the user share the crops half by half. However, in the Arab countries it is named *khamessal* from five — *khamsa*, as the tenant is being left with only one-fifth of the crops, while the other shares go to the land owner or to the party supplying the seeds, draught animals and water. There exists a considerable variety of share tenancy systems.
2. Land rent fixed in advance, without any consideration to the amount of harvest.

Rent in nature is usually connected with the share tenancy, and rent in money with fixed payments. There could be, however, other kinds of combinations, like a fixed rent payed in nature, particularly when prices of agricultural goods are not stable.

By estimating the basic variable D, *agricultural land operated under servile, labour or share tenancy*, one should take into consideration all cases of share tenancy, whether the rent is to be paid in work, nature or in money as well as all kinds of labour tenancy.

Agricultural Holding

An agricultural holding is a unit distinguished in an organisation and technical sense, and not in the legal sense. Who owns the land, or the production means, is here of no importance. The holding is separate when current decisions concerning production are undertaken by one person, or jointly by a group of persons. The same persons may use lands belonging to two or more owners. On the other hand, lands belonging to one owner, but used separately by a few persons (e.g. leased to many tenants) are to be treated as separate holdings.

When some lands are used by a group of people larger than a family, they are to be considered as one holding, provided that the whole labour input in agriculture and profits from it, or at least their bigger part, come from lands used in common.

Landuse

Agricultural land (B) is considered in a broad sense. These lands embrace all areas directly or indirectly used for agricultural production, and namely:

– Land under temporary crops,
– Land under perennial and semi-perennial crops,
– Permanent grassland,
– Current fallow,
– Long fallow,
– Lands under farm buildings, and farm yards.

On the other hand, fish breeding is not included in agriculture, and consequently fish ponds are not a part of agricultural land. It also does not include other inland waters, forests (apart from those used as pastures), built-up areas (apart from buildings used for agricultural purposes), and wasteland.

Land under temporary crops comprises all cultivated annual crops such as cereals, root crops, industrial annual crops and fodder crops, including some fodder crops left on field for few years such as clover, alfa-alfa etc.

Land under perennial and semi-perennial crops (J) includes all area under cultivations of trees and shrubs. It also includes such plants, as cotton or lavender, which occur both as shrubby and as herbaceous plants, as well as those herbaceous plants, which may cover the land for several consecutive years, as sugarcane, hops, pineapple, bananas, agaves and strawberries. On the other hand, they do not include herbaceous fodder crops.

Permanent grassland (K) includes the land covered with grass occurring naturally or cultivated for pasture or hay for at least three consecutive years, including land with scarce vegetation used as pastures or livestock passage tracks in semi-arid or polar areas. They also include fallows and forest areas which are regularly grazed.

Current fallow is a cultivated land which is being allowed to rest uncropped for a short period established in advance (1–3 years) in order to re-establish soil fertility. Also some sown land, the crops from which are destined for ploughing-in (green manure), could be included here.

Long fallow is uncultivated and unsowed land, left for a longer, usually undefined, period of time in order to re-establish soil fertility. Regularly grazed long fallows are not to be included in this category, but ought to be considered as pastures. On the other hand, lands on which cultivation has been given up completely, and which are not being grazed, should not be considered as long fallows, and therefore, are not to be included in agricultural land. They should be considered as wasteland.

Cultivated land (G) This is a more narrow category than agricultural land. Here only land under temporary, perennial and semi-perennial crops, as well as cultivated meadows and pastures should be included. Uncultivated meadows and pastures, land under long and current fallow, and land under farm buildings should not be included in this category.

Total area of temporary crops including fallow (H) embraces land under temporary crops as well as current and long fallows even if grazed. Additionally here also cropped lands, from which there were no harvests, e.g. as a consequence of drought or frost should be included. This group embraces also vegetable gardens as well as glass-houses, plastic-sheet tunnels etc. Here the real physical area is to be taken into consideration, without any regard to the number of harvests obtained from that area during the year.

Harvested land under temporary crops (I). Each plot of land is to be counted as many times as there is a number of harvests obtained from it during the year. The areas of current fallow are not being taken into consideration here (even those with plants destined for ploughing in), nor sown areas from which no crops came.

Food crops (L) are plants yielding products for direct consumption without much transformation. It includes food grains, edible leguminous crops and edible bulbs and roots, as well as vegetables and fruits.

Farm Animals

Number of farm animals, in *conventional animal units* (N) is obtained by multiplying the number of heads of the given animal species (sometimes species, category and age) by the appropriate coefficient. Coefficients are calculated on the basis of an average live weight for the given-species. It is assumed that one conventional animal unit (CAU) equals 500 kg of live weight. For example the coefficient for cows and horses equals 1.0, for pigs-0.25, for sheep and goats-0.1

However, in some countries and regions the weight of animals differs substantially from the average values adopted here. In such cases the adopted coefficients ought to be modified in such a way that one conventional animal unit really constituted 500 kg of live weight. For example, when in a given region cows on average weigh not 500 kg, but 300 kg, the coefficient applied for them should be 0.6 and not 1.0. A similar approach may be assumed in studies concerning agricultural holdings or little groups of holdings, provided that data regarding the weight of animals are available.

Herbivorous (or herd) farm animals (O) include cattle, buffaloes, yaks, goats, sheep, horses, mules, donkeys, camels, llamas and reindeer; while non-herbivorous include pigs, poultry and fur-bearing animals, as well as bees and silk-worms.

Draught and pack animals (P) are those which are engaged in some work in the process of agricultural production, such as cultivation, irrigation or draining of agricultural lands, transport of agricultural production tools and agricultural products. This variable includes horses, mules, donkeys, camels, oxen, buffaloes and yaks.

If possible, those animals which do not work directly in agriculture should be eliminated, e.g. sports horses or animals used for non-agricultural transport, especially in those cases where their share in the given species is important.

The number of draught and pack animals of various species is calculated in conventional units of animal power by multiplying number of heads by co-efficients representing conventional draught units. The basic unit is the power of one draught horse. For example the coefficient for camels equals-2.0, buffaloes-0.9, donkeys-0.3.

Agricultural Production

In the World Agricultural Typology, agricultural production is characterised by two categories: gross agricultural output and commercial production. *Gross agricultural output* (U) is a sum of crop and animal production, while *Crop production* embraces all raw (unprocessed) products of vegetal origin.

Animal production (V) embraces slaughter animals and other raw (unprocessed) materials of animal origin. The production of slaughter animals includes basically animals sold or slaughtered in a holding, and the increase (decrease) in livestock in the period with which the study is concerned. In case of problems in the determination of this last element, it may be omitted, provided that in the period under study no substantial changes in animal stock took place.

The amount of production of slaughter animals is described by live weight. However, sometimes the available data contain only the number of animals sold or slaughtered. In that cases their weight should be calculated on the basis of the average live weight during slaughter which was usual in the given region and given character of agriculture.

Raw stuffs of animal origin include milk, eggs, honey, wax, wool, cocoons of silk-worms, feathers, hair etc. In microscale study manure should also be included, with animal products. It could be omitted, however, in more general studies.

In accordance with the adopted concept of agriculture, agricultural production does not include fish, neither sea nor fresh-water, nor hunted game animals.

The Method of World Agricultural Typology considers *Commercial production* (Y) as gross commercial production. It consists of all goods produced in all branches of an agricultural holding that are sold or delivered off the holding in any other way. Commercial production does not include goods turned over to employees as a part of their wages, or to members of a cooperative or collective farms as a part of their share in profits of the cooperative as well as products used for consumption of employees, also in canteens. Land rent in nature going to the land owner is not to be included in commercial production either.

Commercial production of the given year embraces goods produced during that year, even if they are sold after the year has ended. On the other hand, goods sold in the given year, but produced in the previous years, are not included in the production of given year.

Commercial production could additionally embrace derivative products from the processing of basic agricultural goods within the holding, such as meat, leather, milk products, oil, wine, fibres, fruit or vegetable products. According to the method, they are not agricultural products *sensu stricto*. However, in case of lack of data they could take the place of basic goods. Therefore, they should be converted into unprocessed basic goods and included in commercial production.

Industrial crops (W), are those, the produce of which could not be used without previous considerable processing. Industrial crops include sugar-yielding plants (sugarcane, sugar beet), fibre crops (cotton, flax, hemp, jute, agaves and other), oil crops (soybeans, coconuts, oil palms, groundnuts, olives, rape, sunflower, poppy, mustard, sesame, castor oil plant and others), stimulants (coffee, cocoa, tea, tobacco, and others), hop, dye plants (henna, indigo and others), aromatic oil plants (rose, lavender and others), condiments, medical plants, tanning crops as well as rubber and gum yielding plants.

Conventional units of production: Both the gross output and the commercial production are expressed in conventional units, based on both grain units and the units elaborated by FAO. *Grain units* are determined on the basis of the starch and protein content in various products. On the other hand, FAO *units* are based on the price relations of various products to wheat.

By way of example here are some equivalents in conventional units for one quintal: wheat — 1.0, rice — 0.8, all pulses — 1.20 potatoes — 0.4, most fresh fruits — 0.7, pigs — 7.0, cattle — 6.0, milk — 1.0.

Employment

The number of people engaged in agriculture (R) may be evaluated in two ways, depending on the situation in the agriculture under study and on the available data:

1. Estimating the number of people engaged in agriculture in conventional units, i.e. full-time employees. One full-time employee equals 300 workdays per year. If the number of workdays is available, estimation of this number is simple. If only the number of people engaged is available (e.g. number of people engaged exclusively, mainly or part-time in agriculture), then for each category a coefficient is to be adopted, namely the approximate average number of workdays during the year for these categories of employees.

   The method of calculating the number of employees should be applied especially in those cases where seasonal or part-time agricultural workers are of importance, or if there is an important number of part-time farmers working both inside and outside their holdings.
2. Taking into account only those people who are employed exclusively or mainly in agriculture: i.e. since work in agriculture is often combined with work in a household, all those, who work exclusively or mainly in the household and participate in agricultural work only occasionally, should be

excluded from the total number engaged in agriculture.

Mechanical Draught Power (S)

It mainly consists of tractors and other self-propelled machinery (combines). Only those should be taken into consideration, the technical condition of which enables their utilisation.

The resources of mechanical power should be evaluated in *Metric Horsepower* (hp) — by summing the power of all tractors and self-propelling machines. In the case of a lack of accurate data, the resources should be estimated by the help of an average power of various groups of tractors and machines in the given form of agriculture and in the given region.

Chemical Fertilisation (T)

Here the nitrogen, potassium and phosphorus fertilisers are to be taken into account, both single — and multi-component, converted into pure content (kg and NPK).

Generally the mass of fertilisers used during the years should be taken into consideration, and not those purchased. However, if there is a lack of such data, information concerning purchases in the given year may be applied, providing that these are purchases approximate to average purchases in a number of years, or an average purchase from a few successive years must be assumed.

Irrigation (M)

The area of irrigated land (M) ought to embrace only those areas which are truly irrigated. Areas on which irrigating facilities are out of order or are not functioning for other reasons, should not be included here.

The percentage of irrigated land in relation to the total land under cultivation is to be calculated here. In case of cultivation of land irrigated for some weeks or months only — the whole cultivated area has to be taken as 100 per cent irrespective of the period when land is left uncultivated. On the other hand in case when the irrigation is seasonal, and in another season crops are grown without irrigation (e.g. *rabi* and *kharif* crops in India), the percentage ratio of irrigated land to the total harvested land in all seasons, should be estimated.

Degree of Specialisation in Agriculture

The degree of specialisation is calculated by the help of the following formula:

$$| W_i = \sum a_{ij}^2 |$$

where: a_{ij} – share of the *j* item in the commercial production, estimated in conventional units, of the *i* unit (holding or aggregate of holdings).

It is the sum of squares of shares of all commercialised products in the commercial production of holding or aggregate of holdings under study. Each item of the commercial production should be treated separately.

As already has been mentioned, diagnostic variables representing attributes are ratios obtained from the division of one of the 24 basic variables, computed in accordance with the presented principles, by another of them. They are calculated as in Appendix.

APPENDIX

CALCULATION OF DIAGNOSTIC VARIABLES

I. *Social Attributes*

1. Percentage of agricultural land owned, controlled or held in common under traditional customary rights of tenure:

$$X_1 = (C : B) * 100\%$$

where: *B* – total agricultural land (in hectares),
C – agricultural land owned, controlled or held in common under traditional customary rights of tenure (in hectares).

2. Percentage of agricultural land operated under servile, labour or share tenancy:

$$X_2 = (D : B) * 100\%$$

where: *B* – total agricultural land (in hectares),
D – agricultural land operated under servile, labour or share tenancy (in hectares).

3. Percentage of agricultural land owned as a private property:

$$X_3 = (E : B) * 100\%$$

where: *B* – total agricultural land (in hectares),
E – agricultural land owned as a private property (in hectares).

4. Percentage of agricultural land operated by collective and state enterprises:

$$X_4 = (F : B) * 100\%$$

where: *B* – total agricultural land (in hectares),
F – agricultural land operated by collective and state enterprises (in hectares).

5. Size of operation in terms of the number of people engaged in agriculture per one agricultural holding:

$$X_5 = R : A$$

where: *A* – number of agricultural holdings,
R – number of people engaged in agriculture (in conventional units).

6. Size of operation in terms of the agricultural land per one agricultural holding:

$$X_6 = B : A$$

where: A – number of agricultural holdings,
B – total agricultural land (in hectares).

7. Size of operation in terms of gross agricultural output per agricultural holding:

$$X_7 = U : A$$

where: A – number of agricultural holdings,
U – gross agricultural output (in conventional units).

II. *Operational attributes*

8. Inputs of labour:

$$X_8 = (R : B) * 100$$

where: B – total agricultural land (in hectares),
R – number of people engaged in agriculture (in conventional units).

9. Inputs of animal power:

$$X_9 = (P : G) * 100$$

where: G – cultivated land (in hectares),
P – number of draught and pack animals (in conventional draught units).

10. Inputs of mechanical power:

$$X_{10} = (S : G) * 100$$

where: G – cultivated land (in hectares),
S – mechanical draught power (in Metric Horse-power).

11. Chemical fertilising:

$$X_{11} = T : G$$

where: G – cultivated land (in hectares),
T – use of chemical fertilisers (in kg of pure content of NPK).

12. Irrigation:

$$X_{12} = (M : G) * 100\%$$

where: G – cultivated land (in hectares),
M – irrigated land (in hectares).

13. Intensity of cropland use:

$$X_{13} = (I : H) * 100\%$$

where: H – total area of temporary crops including fallow (in hectares),
I – harvested land under temporary crops (in hectares).

14. Intensity of livestock breeding:

$$X_{14} = (N : B) * 100$$

where: B – total agricultural land (in hectares),
N – number of farm animals (in conventional animal units).

II. *Production attributes*

15. Land productivity:

$$X_{15} = U : B$$

where: B – total agricultural land (in hectares),
U – gross agricultural output (in conventional units).

16. Productivity of cultivated land:

$$X_{16} = U : G$$

where: G – cultivated land (in hectares),
U – gross agricultural output (in conventional units).

17. Labour productivity:

$$X_{17} = U : R$$

where: R – number of people engaged in agriculture (in conventional units),
U – gross agricultural output (in conventional units).

18. Commercial labour productivity:

$$X_{18} = Y : R$$

where: R – number of people engaged in agriculture (in conventional units),
Y – total commercial agriculture production (in conventional units).

19. Degree of commercialisation:

$$X_{19} = (Y : U) * 100\%$$

where: U – gross agricultural output (in conventional units),
Y – total commercial agricultural production (in conventional units).

20. Commercial land productivity:

$$X_{20} = Y : B$$

where: B – total agricultural land (in hectares),
Y – total commercial agricultural production (in conventional units).

21. Degree of specialisation: see above.

IV. *Structural attributes*

22. Percentage of land under perennial and semi-perennial crops:

$$X_{22} = (J : B) * 100\%$$

where: B – total agricultural land (in hectares),
J – land under perennial and semi-perennial crops (in hectares).

23. Percentage of permanent grassland:

$$X_{23} = (K : B) * 100\%$$

where: B – total agricultural land (in hectares),
K – permanent grassland (in hectares).

24. Percentage of land under primary food crops:

$$X_{24} = (L : B) * 100\%$$

where: B – total agricultural land (in hectares),
L – land under primary food crops (in hectares).

25. Percentage of animal production in gross agricultural output:

$$X_{25} = (V : U) * 100\%$$

where: U – total gross agricultural output (in conventional units),
V – gross animal production (in conventional units).

26. General commercial emphasis:

$$X_{26} = (Z : Y) * 100\%$$

where: Y – total commercial agricultural production (in conventional units),
Z – animal commercial production (in conventional units).

27. Percentage of industrial crops in gross agricultural output:

$$X_{27} = (W : U) * 100\%$$

where: U – total gross agricultural output (in conventional units),
W – gross output of industrial crops (in conventional units).

28. Percentage of herbivorous animals in the total number of farm animals:

$$X_{28} = (O : N) * 100\%$$

where: N – number of farm animals (in conventional animal units),

O – number of herbivorous farm animals (in conventional animal units).

REFERENCES

Kostrowicki, J., 1982 : *The Types of Agricultural Map of Europe.*

Mohammad, N., 1981 : *Perspectives in Agricultural Geography*, Vol. I., Concept Publishing Company, Delhi.

Scott, 1983 : The Typology of Australian Agriculture, *Geogrphia Palonica*, 46.

Troughton, M.J., 1986 : Report and Proceedings of IGU/UNESCO Seminar on Methods of Comparative Studies in Agricultural Development; Rabat, (March 18-21st).

8

TYPES OF AGRICULTURAL MAP OF EUROPE

Concept, Method, Techniques

JERZY KOSTROWICKI

THE CONCEPT of the Types of Agriculture Map of Europe is based on the general concept of agricultural typology elaborated by the IGU Commission on Agricultural Typology during its 12 years' activity — 1964-1976 (Kostrowicki 1979).

The tasks of the Commission were originally determined as follows : "1. to establish common principles, criteria, methods and techniques of agricultural typology; 2. to initiate, promote and co-ordinate regional studies on agricultural types; 3. to elaborate the typological and regional classification of world agriculture (Kostrowicki 1977; Kostrowicki and Tyszkiewicz 1970)".

In fact all those tasks served one another : the principles, criteria, methods and techniques, elaborated on the basis of two questionnaires distributed to numerous geographers of the world, were discussed in the Commission meetings, subsequently tested in regional and local studies, then discussed and tested again and again. In consequence already in 1971 the first draft of the typology of World Agriculture was presented for discussion (Kostrowicki 1971, 1972, 1974), then improved and discussed again. Finally, on its meeting in Odessa (1976) the concept was accepted by the Commission. Nevertheless some methods and techniques of agricultural typology were still further developed and modified (1980).

Already in 1972 the elaboration of the types of agriculture map of the world was included in the Commission programme. However, lack of sufficient means delayed considerably the whole work and made impossible its completion, at least during the period of the

Commission's activity. However, the idea has not been abandoned. With the aid of a number of European geographers (Kostrowicki 1984) the types of agriculture map of Europe on 1 : 2.5 million scale was completed and published in September 1984.

The cartographic aspects of the map as well as the results obtained, in terms of the typology of European agriculture and its evolution, were presented elsewhere (Kostrowicki, 1982, 1984). A number of typological studies of several European countries have also been published marking the evolution of the concept, method and techniques of agricultural typology (Bonnamour and Gillette, 1980, Galczynska, 1984, Gorbunova et al., 1979, Rikkinen and Vapaoksa, 1983, Stola, 1975, Szczesny, 1981 a, b, Tschudi, 1983, Tyszkiewicz, 1976, 1980, 1982).

As the types of Agriculture Map of Europe is based on the general concept of agricultural typology, some most essential characteristics of that concept have to be recalled.

The Concept of Agricultural Typology

"As agriculture as a whole should not be considered as a simple sum of its components but as a set of highly interconnected and interrelated phenomena and processes, it can be treated as a complex or a system in terms of a system approach. Individual agricultures understood as such systems could be compared with each other and then grouped into types according to their similarities.

Following these assumptions, the type of agriculture is understood :

1. As a more or less established form of crop growing and/or livestock breeding for production purposes, characterised by a set or association of its attributes (characteristics, features, properties).
2. As a supreme and overall concept in agricultural classification comprising all other concepts used in classifying agriculture, such as land tenure systems, landuse systems, cropping systems, systems of livestock breeding, farming systems, types of farming, etc.
3. As a hierarchical concept encompassing types of varying orders, from types of farms based on a study of individual holdings, through several intermediate orders to the highest order — types of world agriculture.

4. As a dynamic concept, with types changing in an evolutionary or revolutionary way along with a change of their basic attributes.

Typology is often confused with regionalisation. Although both concepts are meant to synthetise a complicated reality in order to make it more comprehensible, they belong to two distinct categories. A type is a systematic or taxonomic concept, and its definition is based essentially on similarities between various individuals. As individuals, characterised by similar sets of attributes, may occur repeatedly both in time and space, the same types can be identified in various periods or territories. As agricultures with similar sets of their attributes are often distributed in space in a mosaic-like pattern, the distribution of resulting types does not necessarily form a contiguous area, but agricultures of the same types are usually dispersed and intermingled with some others.

By contrast, the region is a spatial or territorial concept. It is delimited on the basis of differences between places, rather than similarities between individuals. Consequently, the region should be understood as a fraction of the earth's surface, extending within definite limits and characterised by a peculiar association of features that render its character unique and differentiate it from all other territorial units.

Both the type and the region are hierarchical concepts. On the basis of their similarities, types of a lower order may be grouped into types of a higher order, irrespective of their distribution in space and time, while regions of a lower order always form territorial parts of regions of a higher order. By its very significance, regionalisation is then a static concept, while typology is a dynamic one, involving all possible changes.

If typology has already been established, agricultural regions can easily be delimited by generalisation of a more complicated typological pattern to a simpler regional picture, based on dominance or co-dominance of individual types over a given territory (Kostrowicki, 1977).

Criteria of Agricultural Typology

In accordance with the logic of any classification, the identification of agricultural types ought to be based on internal (inherent or endogenous) attributes of agriculture : external (or exogenous) attributes, or, rather, conditions in which agriculture develops, should not be used as a basis for agricultural typology, important though they might be for explaining why, in a particular place and time, individual types of agriculture have developed.

The simultaneous use of such exogenous, natural and other conditions alongside endogenous attributes of agriculture is futile, since it presupposes rather than proves their impact on the formation of agricultural types. This can be proved much better by the study of agricultural characteristics and their associations, independently of the conditions of their development, and then by a subsequent analysis of their inter-relationships by means of correlation calculus.

On the other hand, the external or exogenous conditions, such as natural, locational, transportation and market conditions, the effect of supply and demand on agricultural products, prices, the agricultural policy etc. certainly play an important role in the formation of agricultural types and their separate attributes, which change with a change of those conditions. Their study is, therefore, necessary for both understanding and interpretating the development and spatial distribution of agricultural types.

An agricultural holding is the best basic unit in agricultural typology, as it is the only real unit of operation. At the same time, however, irrespective of all their deficiencies, other units (administrative areas or whatever is convenient) can be used, at least in macro-scale typologies and particularly when dealing with a great number of small holdings for which no separate data are available.

This is particularly true of the so-called village agricultures, with plots belonging to different holdings, scattered throughout the village territory. But even in the countries where farms are larger, farm data are often confidential and, therefore, available only in an aggregate form, which may contain a considerable variety of inter-farm differences.

Therefore, when studying such aggregates, it should always be kept in mind that the data do not refer to real units, but are averages for certain territories, with a more or less unknown internal differentiation.

This is why detailed sample studies are highly recommended, whenever possible, not only to verify the range of these differences but also to assess the accuracy of the statistical data.

In fact, nowhere in the world are agricultural statistics fully accurate, or provide all necessary material for agricultural typology.

Therefore, even in those countries where agricultural statistics are relatively good, there are always certain gaps that can be filled only by using estimates. In some countries, however, the scarcity of data, or their unreliability, makes it necessary to base typological studies on estimates, rather than on the statistical data. When the problem and area under study are sufficiently known to the scholar, his estimates might even yield better results than the direct use of unreliable statistics (Kostrowicki, 1977).

Internal attributes that may serve as a basis for the identification of types can be grouped as follows : 1. social attributes, 2. operational attributes, 3. production attributes and 4. structural attributes.

The social attributes provide answers to questions who is the land owner, holding operator or decision maker and what is the scale of operation.

The operational attributes explain what the labour and capital inputs are, what intensity of farming is and how a holding, or holdings, are operated.

The production attributes disclose how much is produced and for what purpose.

The structural attributes answer the question about the proportion of various branches of agricultural production i.e., about the enterprise combinations in terms of landuse, livestock breeding, gross agricultural output and commercial production.

To attain the comparability of results in space and time, the identification of agricultural types should always be based on the same criteria, irrespective of the order, scale of research as well as time and area concerned.

Methods and Techniques of Agricultural Typology

There are two important methodological problems with which any scholar is faced when starting to work an agricultural typology of any order or area, namely : 1. the choice of variables to represent various agricultural attributes and their adequate expression and 2. the choice of a technique for comparing individual basic units of study, characterised by sets or associations of such variables, and grouping them according to their similarities.

To attain comparability of results in time and space the selected variables should be expressed quantitatively. Although it is true that a good expert, with a deep and intimate knowledge of the problems and area concerned, can produce excellent typology without using any quantitative methods (like D. Whittlesey) it is also true that nobody else, not even the same scholar, is likely to obtain comparable results after some lapse of time, because the line of thinking and the way of interpreting facts cannot be repeated. It would be even more difficult or even impossible to obtain in this way comparable results for another area or another period of time. Only the use of quantitative data and techniques can guarantee that the same method, when applied to the same data, will always yield the same results, irrespective of when and by whom the data are processed. Similarly only quantitative methods could give comparable results when applied to another area or another period of time. The agricultural attributes could be expressed by an innumerable quantity of different variables.

In fact, some views were expressed at the Commission meetings that the greater the number of variables covering all aspects of agriculture, the better the typology.

However, following the discussion, the method of the selection of the variables, based on the purposeful choice of a limited number of the variables of synthetic or composite character, as many as possible universal, significant and representative, was chosen out of several possible ways and recommended by the Commission.

The synthetic character of these variables implies that each variable comprises a number of elementary attributes of agriculture. Their universal character indicates that the selected variables are relevant in describing most, if not all, possible types of world agriculture. The significant character will ensure that the selected variables represent only the important attributes of agriculture. The representative character means that the selected set of variables represent the most important aspects of agriculture in a balanced way.

Preference was given to such an approach over that based on an unlimited number of the variables not only because those might be too numerous. With computer techniques the problem can easily be solved. However, when the number of the variables is unlimited, the degree of coverage of all important aspects of agriculture by those variables as well as their influence on the type formation can hardly be assessed. Most of them would be of local, regional but not universal character, their importance as far as the type formation is concerned could vary greatly and finally the variables of elementary character would certainly dominate over those of synthetic character much less numerous. Consequently, the use of a great number of variables, representing only one aspect of agriculture, with a much smaller number of the variables that could represent another aspect, may result in exaggerating the type-forming influence of one aspect, reducing the influence of another. But even within a set of the selected variables of synthetic character certain differences in their type-forming importance could persist. The only way to solve that problem is the weighing of individual variables. However, the greater the number of variables their weighing is more and more difficult, if not impossible at all, while with a smaller number of the variables it is easier to balance their respective significance. As a leading French agricultural economist, Jean Klatzmann pointed out, if the variables are not weighed at all they are in fact weighed because the same value is ascribed to them. In fact, in the earlier versions of the agricultural typology structural or operational attributes, easy to be pointed out, but of less synthetic character, clearly dominated production attributes much less numerous but of much more synthetic character. However, when put to test, it appeared that the elimination of all, but very important production attributes, had not changed much the resulting typology. That was the reason why certain rearrangement of the variables was made, first by the elimination of some variables from the operational group and doubling, at the beginning, the value of production attributes, further substituted by some new ones. At the same time to keep the number of the variables as low and balanced as possible a few quite important variables had to be eliminated.

Finally the set of 27 variables (Table 8.1) was recommended, out of which 7 represent social attributes, 7—operational ones, 7—production attributes and 6—structural attributes. Only recently, after the Types of Agriculture Map was sent to the publishers, has the 7th variable been added to the last group representing orientation of animal breeding, an important attribute, formerly overlooked, though several attributes representing the orientation of crop growing were applied.

TABLE 8.1: **Variables used in agricultural typology**

A. *Social Attributes*

1. **Percentage rate of land held in common in the total agricultural land (arable land including fallow + perennial crop + permanent grassland).**
2. **Percentage rate of land in labour and share tenancy (share cropping) in the total agricultural land.**
3. **Percentage rate of land owned by private persons (irrespective of the land tenure system) in the total agricultural land.**
4. **Percentage rate of land operated by the consciously planned collective or state enterprises in the total agricultural land.**

(Contd.)

TABLE 8.1 *(Contd.)*

5. Number of people actively employed in agriculture per 1 agricultural holding.
6. Amount of agricultural land in hectares per 1 agricultural holding.
7. Amount of agricultural gross production in conventional units per one agricultural holding.

B. *Operational Attributes*

8. Number of people actively employed in agriculture per 100 hectares of agricultural land.
9. Number of draugh animals (horses, mules, asses, oxen, buffaloes — if used in agricultural work) in conventional draught units (see appendix) per 100 hectares of cultivated land (arable land without fallow + perennial crops + cultivated grassland without uncultivated meadows and pastures).
10. Number of tractors and other self-propelling machinery in HP per 100 hectares of cultivated land.
11. Amount of chemical fertilisers in pure content (NPK) per 1 hectare of cultivated land.
12. Percentage rate of irrigated land in the total cultivated land.
13. Percentage rate of harvested land in the total arable land (including follow).
14. Number of farm animals in conventional (large) animal units per 100 hectares of agricultural land.

C. *Production Attributes*

15. Gross agricultural production in conventional units per 1 hectare of agricultural land.
16. Gross agricultural production in conventional units per 1 hectare of cultivated land.
17. Gross agricultural production in conventional units per 1 person actively employed in agriculture.
18. Commercial (delivered off farm) agricultural production in conventional units per 1 person actively employed in agriculture.
19. Percentage rate of commercial (delivered off farm) agricultural production in gross agricultural production.
20. Commercial agricultural production in conventional units per 1 hectare of agricultural land.
21. Degree of specialisation in commercial agricultural production.

D. *Structural Attributes*

22. Percentage rate of perennial (trees, shrubs, vines) and semi-perennial (covering land without rotation for some years such as hops, cotton, sugarcane etc.) crops in the total agricultural land.
23. Percentage rate of permanent grassland (including leys within field-grass system and current fallow if used for grazing) in the total agricultural land.
24. Percentage of land under food crops (food grains, tuber, root and bulb crops, vegetables and fruits) in the total agricultural land.
25. Percentage rate of animal products in gross agricultural production.
26. Percentage rate of animal products in commercial agricultural production.
27. Percentage rate of industrial crops (such as fiber, oil, sugar crops, tobacco etc.) in gross agricultural production.

Irrespective of the fact that the type — forming importance of such synthetic variables does not differ very much, an attempt was made to weigh them though not directly, but indirectly by their balanced selection, normalisation or adaptation to certain subdivisions, either natural or known from literature.

As those variables are often expressed in different ways or by different measures, to be compared, they have to be normalised. World ranges of each phenomenon, expressed by a variable, were adopted as a basis of such a normalisation. Those ranges each subdivided into 5 classes: 1. very low, 2. low, 3. medium, 4. high, 5. very high represent a relative value of each variable. In most variables expressed in percentage proportion such classes represent each 20 per cent of their world ranges. For the others as well as those expressed in different measures an attempt was made to distinguish by such classes the most significant subdivisions or thresholds.

With the exception of those expressed in percentages the lowest and the highest classes were left open. Every one of such classes are marked by figures from 1 to 5 (Table 8.2) with 0 applied only when a given attribute in a given case is completely irrelevant or virtually non-existent.

The use of the variables representing world ranges rather than world averages has several advantages : 1. World averages are hardly calculable and their use would require constant adjustment of the classes to changing averages, 2. World ranges can easily be calculated from the world agricultural statistics and the adopted classes, if properly selected, would not require adjustment at least for several decades.

The adoption of broad classes as variables instead of indices facilitates greatly the typological classification of agriculture for the countries where adequate or reliable statistics are not available. In fact, in such cases the use of estimates instead of statistical data may less distort reality provided that the person carrying on such research knows intimately the area and the problem concerned. The same can be said about possible omission of some variables because of the lack of data. With the method adopted even the broadest estimate would less distort the effect than the lack of such a variable.

On the other hand, many adopted variables when combined with the others express more than one of such variables.

For example, the variables representing land ownership and tenure systems (1-4), when combined with one another and the variables representing a scale of operation (5-7), can provide an additional information with what systems of landownership and land tenure we have to deal. Variable 3 combined with 2, 5 and 6 as well as with low 10 and 11 provides the information about traditional large-size landed estates operated by serfs, servile tenants or share-croppers (155155) while variable 3 combined with 5 and 6 only and high 10 and 11 informs about a large-scale modern holdings. The same variable 3 but combined with low 5 and 6 (115155) informs about small or medium-sized farm operated by owner-operators or fixed-rent tenants. On the other hand, high variable 5 combined with high 1 says about the large tribe, clan or extended family holding, while low variable 6 when combined with high 3 indicates a small-scale holding or owner-operators or fixed-rent tenants, etc.

TABLE 8.2: Classes of world ranges of individual attributes

No of the variable	CLASSES				
	Very low 1	low 2	medium 3	high 4	very high 5
1	–20	20-40	40-60	60-80	80-
2	–20	20-40	40-60	60-80	80-
3	–20	20-40	40-60	60-80	80-
4	–20	20-40	40-60	60-80	80-
5	–2	2-8	8-50	50-200	200-
6	–5	5-20	20-100	100-1000	1000-
7	–100	100-1000	1000-1000	10000-100000	100000-
8	–3	3-15	15-40	40-150	150-
9	–2	2-8	8-15	15-30	30-
10	–6	6-15	15-35	35-90	90-
11	–10	10-30	30-80	80-200	200-
12	–10	10-25	25-50	50-80	80-
13	–10	10-30	30-70	70-130	130-
14	–10	10-30	30-80	80-160	160-
15	–5	5-20	20-45	45-100	100-
16	–5	5-20	20-45	45-100	100-
17	–40	40-100	100-250	250-800	800-
18	–20	20-60	60-180	180-600	600-
19	–20	20-40	40-60	60-80	80-
20	–3	3-12	12-30	30-80	80-
21	–0.1	0.1-0.2	0.2-0.4	0.4-0.8	0.8-
22	–10	10-20	20-40	40-60	60-
23	–20	20-40	40-60	60-80	80-
24	–20	20-40	40-60	60-80	80-
25	–20	20-40	40-60	60-80	80-
26	–20	20-40	40-60	60-80	80-
27	–20	20-40	40-60	60-80	80-

Among operational attributes the first two variables (8 and 9) represent separately or together traditional agriculture while the following two (10 and 11) modern industrialised agriculture. On the other hand, the classes representing the intensity of cropland use (variable 13) are so arranged that they reflect 5 following forms of rotated cropland use (without perennial crops).

1. up to 10 per cent of land under crops — long /forest/ fallow
2. 10-30 per cent under crops — short /bush/fallow
3. 30-70 per cent under crops — current fallow /al tercio, two year or three field systems
4. 70-130 per cent — full use of land with after-crops
5. Over 130 per cent — multicropping.

Among production attributes the differences between the variables 15 and 16 are such that in the 15th total gross production is calculated per hectare of total agricultural land, whether actually cultivated or not, while in the 16 gross production from cultivated only — per hectare of cultivated land (without fallow and permanent grassland) what could be controlled with variables 13 and 23.

The variables 18, 19 and 20, contrary to 15, 16 and 17, are referred to production delivered off farm only (not used on place either for self consumption or for feeding animals).

A very important but difficult to calculate is the variable 21 — degree of specialisation, that could be estimated by use of either a mathematical formula (Szyrmer, 1973) or a simpler evaluation (Kostrowicki, 1974). However, the results obtained by those two techniques sometimes are not identical. This variable introduced to the first set of characteristics, subsequently for some time eliminated and then reintroduced a new has also another shortcoming. It is perfect, when used to measure the specialisation of a single holding. When applied to larger areas, in which agriculture is highly specialised but in different products, it may show a low degree of specialisation.

To reduce the number of variables representing structural attributes the proportion of two out of three, principal systems of landuse only (22 — perennial crops and 23 — permanent grassland), was adopted, while the higher proportion of the third one (arable land) is shown by low degree of variables of the first two. Similarly within crop production the proportion of food and industrial crops is represented in a different way by variables 24 and 27 while the proportion of fodder crops is treated as the remainder.

The importance of livestock breeding is shown by the intensity or rather density of animals (variable 14) as well as by the proportion of animal products within gross and commercial production (25 and 26), while nothing is said about the orientation of livestock breeding. This gap has been filled recently by the introduction of the 26th variable representing the proportion of herd (or herbivorous) animals to the total number of animals. As non-herbivorous animals (pigs, poultry) usually do not survive from one to another annual census their proportion has been somewhat exaggerated.

The second methodological problem in agricultural typology is the selection of the best possible method for comparing and grouping into types individual multi-feature of multi-variable units according to their similarity.

In the early stages of agricultural typology some taxonomic graphic methods were tested, developed or adapted and introduced in some studies (Kostrowicki, 1980, 1981) such as the typogram method (Kostrowicki, 1970, 1972 b) and a diagram, proposed by Szyrmer (1973). The first has gained some popularity particularly in France and some other countries (Bonnamour, 1971, 1973; Bonnamour, Guermond, Gillette, 1971; Chabaud and Champalbert, 1973; Dongmo, 1973; Kampp, 1973; Tschudi, 1983). Though very instructive particularly in teaching, and fairly effective when one deals with clear-cut spatial differences, it requires too many arbitrary decisions where differences are less evident. On the other hand, the Szyrmer diagram is quite effective when units to be compared are not too numerous; otherwise, it produces a too complicated picture. In the past, it was effectively used to compare the types of agriculture already described in order to group them into types of a higher order (Kostrowicki, 1980).

Later on, a number of taxonomic quantitative methods applied to date were evaluated and some new proposed (Kostrowicki, 1980) by a group of geographers and mathematicians (Bielecka, Paprzycki, Piasecki, 1975, 1979, 1980; Bielecka, Paprzycki, 1979).

As it has been proved all of them are more or less effective for a single set of units, while any addition or withdrawal of a unit may change the whole grouping. Therefore, none of them can fulfil the principal requirement of the agricultural typology and a number of other classifications, i.e. to make their results comparable, both in time and space, unless that single sets of units would cover the whole world.

But even in that case the comparability in time could not be secured. In such a situation the deviation method, which is one of the methods implying a frame of reference, was applied. By this method the sets of variables arranged as codes characterising given units of study are compared with the model codes of individual types established earlier. Originally they were based on the study of over one thousand cases covering the whole world.

In the earlier stages of their introduction the codes were elaborated for the types of the 3rd order only. This procedure brought about difficulties when the codes for agricultures differing from those applied as the basis of model codes were compared (Scott, 1983, Troughton, 1979). The establishing of the model codes for the types of the 1st and 2nd orders eliminated that problem since any code representing any agriculture could now be compared and classified effectively to one or another type of the 1st or 2nd order.

In any case the way of establishing the model codes is until now one of the weaker points in the typological method since it involves some arbitrary decisions. On the other hand, even the most sophisticated quantitative methods are not entirely free from arbitrary decisions.

With such model codes the codes for any unit under study have to be compared and taxonomic distances from the model codes evaluated by summing up the deviation of every variable from the same variable of the model code. When such a taxonomic distance expressed as the sum of such deviations between the code of the unit under study and one or several model codes does not exceed the arbitrarily adopted maximum, the agriculture of such a unit could be considered as being of the type represented by such a model code. As the whole system is open, the cases that deviate from the model types by more than the adopted maximum are used as the basis for the description of new types. In fact, a further analysis and codification of thousands of new cases have enriched the original world typology of agriculture (Kostrowicki, 1980) with a number of additional model-types.

Since the whole system is hierarchical, it has been arbitrarily decided, on the basis of the material collected, that the maximum deviation from the model codes should be 10 per cent of the total variance, i.e. 11 deviations for the types of the 3rd order; 20 per cent, i.e. 22, for the types of the 2nd order, and 30 per cent, i.e. 33, for the types of the 1st order.

When the units under study are not numerous, such a deviation calculation could easily be made by a visual confrontation of distances between the unit under study and the model codes. When, however, a greater number of units is involved such a comparison can successfully be made only by using a computer with a special programme. The programme used in Poland was published (Kostrowicki, 1980) but it changes with the change of the type of the computer.

By this procedure 6 types of world agriculture of the 1st order were identified and described by model codes, over 20 types of the 2nd order and over 100 types of the 3rd order. While there is a little possibility that more types of the 1st order could still be identified, it is certain that the number of the types of the 2nd order and particularly of the 3rd order could still be singled out, particularly in the countries that have not as yet been more thoroughly studied.

The Types of Agriculture Map of Europe — Typological Procedure

As most of the Thematic Maps of Europe, published to day or being under elaboration, including the land-use maps published under the editorships of Vanzetti (1969) and Csati (1980) are on 1 : 2,5 million scale, to ensure the comparability with those maps, the same scale has been adopted. The consequence of such a scale is the adoption of the units of the highest order, whether administrative or agricultural (if available). However, the size of the administrative units of the highest order varies greatly among the European states. In the USSR (oblast) they are much larger than elsewhere, however, since the internal differentiation of the Great Russian Plain is not great, the reality is not greatly distorted. The exception is the Caucasus and particularly the Transcaucasian republics, too large and internally too differentiated to be treated as single units, even within the adopted scale. The administrative units for most European countries as well as agricultural regions for Belgium, Austria and Finland used for that purpose are close to the optimum size. On the other hand, the highest administrative units of the Federal Republic of Germany (Lands) are greatly differentiated and some of them are too large. Therefore they were substituted with the units of 2nd order (Regierungsbezirken). A special case is Yugoslavia where there are no intermediate administrative units between the federal republics and communes (obstinas), the last being too small for the adopted scale. Recently, some kind to intermediate statistical units (zajednica) have been introduced to some but not all Yugoslav republics adopted for the map, while in the other republics the communes had to be grouped into larger units.

For all those units (altogether over 870) the material was collected, the adopted indices elaborated and transformed into codes. Those codes (Table 8.3) together with the model codes, were sent to the computer, where codes for those units were compared with the earlier established model codes

TABLE 8.3: Model Codes

I.	E	5321311	—	1111121	—	1221111	—	133331	4
	T	1241222	—	4412242	—	4422221	—	114221	3
	M	1151233	—	2154343	—	4455544	—	122331	3
	S	1115555	—	3243242	—	3333433	—	124221	4
	A	1133445	—	3000003	—	3045535	—	151551	3
	L	1551354	—	2211132	—	2334422	—	132342	4
II.	En	4211201	—	1100121	—	1221112	—	141451	5
	Ef	5321221	—	1111111	—	1221111	—	115111	4
	Et	2441211	—	2211132	—	2311211	—	132221	4
	Ll/Tl	1551354	—	2211132	—	2333323	—	124331	4
	Lp/Tp	1351444	—	4411142	—	3323434	—	333213	3
	Ti	1241212	—	5311452	—	5511221	—	114111	3
	Ts	2241222	—	3212231	—	2222325	—	313113	2
	Tm	1151222	—	4423143	—	3321321	—	124231	4
	Ms	1151333	—	3113341	—	4433444	—	511114	2
	Mi	1151212	—	5145451	—	5545555	—	215111	2
	Mm	1151222	—	3154143	—	4444443	—	122331	3
	Ma	1151143	—	2153144	—	3455535	—	151551	4
	Ml	1151355	—	2154341	—	4455544	—	213223	4
	Me	1151254	—	1152132	—	1255525	—	134221	4
	Se	1115343	—	3232131	—	2222321	—	124221	4
	Sm	1115555	—	2143143	—	3333432	—	133331	3
	Sa	1115455	—	3255144	—	4444445	—	151551	4
	Si	1115555	—	5422343	—	4522232	—	114111	2
	Sh	1115545	—	4155441	—	4434554	—	215111	2
	Ss	1115555	—	2133442	—	3344535	—	411214	4
	Sc	1115555	—	1142131	—	1234515	—	115111	4
	Sg	1115554	—	4434342	—	2312423	—	141233	5
	Ar	1133455	—	1000001	—	1033515	—	151551	5
	Ad	1133415	—	5000005	—	5055555	—	101551	3
III.	Etm	1341221	—	3311132	—	2321221	—	133221	4
	Llm/Tll/	1251354	—	2211132	—	2333424	—	143341	4
	Tir/Tmr/	1151211	—	4411244	—	2211211	—	114111	5
	Tiu/Tmu/	1151112	—	4432454	—	3322321	—	214111	4
	Tmh	1251211	—	4511343	—	2311111	—	122341	5
	Tme	1151111	—	3423134	—	2211111	—	123331	5
	Tmb	1151222	—	3412133	—	3332232	—	122341	4
	Tmm	1151222	—	3323143	—	3323321	—	113231	3
	Tmo	1151111	—	3223142	—	4422133	—	113331	3
	Tme	1241212	—	4422243	—	3423332	—	224221	4

(Contd.)

Tmy/Tmj	1151111	—	4434244	—	3311223	—	214222	5
Tmg	1151121	—	3243132	—	2322432	—	332232	4
Tmf	1151112	—	4254141	—	3322332	—	513112	4
Tmk	1151112	—	4423144	—	4421233	—	132451	4
Mso	1151122	—	2143132	—	3334534	—	411115	2
Mif	1151222	—	3154341	—	4445544	—	515111	3
Min	1151112	—	4455353	—	5533553	—	515221	3
Miv	1151223	—	3145351	—	5534554	—	114111	2
Mig	1151314	—	5155551	—	5555555	—	115111	1
Mmr	1151233	—	2154142	—	3343433	—	123221	4
Mme	1151222	—	4254242	—	4434443	—	213222	4
Mmf	1151112	—	3243343	—	3433433	—	324221	4
Mmv	1151123	—	2255243	—	5555552	—	114121	3
Mmp	1151111	—	4254454	—	4422342	—	313341	3
Mmz	1151122	—	2243342	—	2333322	—	141232	4
Mmt	1251122	—	3153243	—	4443342	—	313231	3
Mmm	1151222	—	3255144	—	4444442	—	122331	2
Mmh	1151112	—	4454254	—	5533442	—	142451	4
Mmg/Tma/	1151232	—	2223143	—	3433333	—	131351	5
Mmw	1151122	—	3155145	—	3343333	—	151251	5
Mmi	1151222	—	3155355	—	5545553	—	212451	3
Mam/Mma/	1151233	—	2154144	—	4544444	—	141451	4
Mab/Mmb/	1151143	—	1143142	—	2355424	—	141451	5
Mai	1151223	—	3255145	—	5555555	—	131551	3
Mae	1151142	—	1155542	—	1544514	—	151551	5
Mlm	1151355	—	2154143	—	4455542	—	123331	3
Mlc	1151254	—	2154441	—	4455543	—	113112	4
Mel	1151132	—	1143132	—	2244522	—	222222	4
Mem	1151254	—	1152132	—	2355533	—	133331	4
Sem	1115443	—	3233133	—	2333321	—	133331	4
Sec	1115443	—	3232231	—	2222422	—	124121	4
Sep	1115343	—	2142132	—	2333422	—	222222	3
Smw	1115555	—	2142142	—	2223423	—	124331	3
Smm	1115555	—	2144143	—	3434432	—	132341	3
Smy/Smj/	1115353	—	1153241	—	1433413	—	153111	4
Smg	1115354	—	2154142	—	2434422	—	151341	4
Smc	1115555	—	3233242	—	3333432	—	223221	4
Smu	1115555	—	3144143	—	3344534	—	114222	4
Smi	1115555	—	2154443	—	4455543	—	122222	2
Sad/Smd	1115454	—	3155144	—	4444443	—	113351	3
Sae	1115555	—	2254142	—	2323425	—	151551	5
Sai	1115555	—	2155145	—	3444445	—	131551	3
Sgt	1115555	—	4444342	—	2312423	—	141233	5
Shy/Shj/	1115444	—	3144341	—	3333534	—	323111	4
Shv	1115555	—	4154342	—	4433545	—	115111	2
Shg/Shh	1115545	—	5155551	—	5555555	—	115111	1
Ssf/Shf	1115555	—	3133241	—	3344535	—	515111	4
Ssc	1115555	—	2154442	—	3334535	—	332114	5
Scm	1115555	—	1141131	—	1333413	—	154441	4
Scc	1115555	—	1142131	—	2233525	—	115111	4
Arr	1151354	—	1000001	—	1022515	—	151555	5
Aro	1115555	—	1000001	—	1034515	—	151551	5
Add	1151323	—	5050005	—	5055555	—	101551	3
Ado	1115555	—	5050005	—	5055555	—	101551	3

The computer presents the deviations between the model codes of any order and every unit under study in the form of prints of figures. The next step was to select and write on a special table all the deviations that did not exceed 14 differences (11 + possible 25 % tolerance clause). There were however some cases when the distances between the given unit and two or more model codes were lower than the adopted maximum. Such units have to be treated as transitional between two or more types. The degree of such transition could be various and was measured by means of the successive products technique (Kostrowicki 1979, Tyszkiewicz 1980) with no more than 4 products adopted (e.g. such a transition with four products could be 3 : 1, 2 : 2, 1 : 1 : 1 : 1 or 1 : 3).

For example, let's assume that distances between a given unit and four model codes ABCD are 2, 7, 10 and 12. The procedure is then as follows (Table 8.4).

TABLE 8.4: Type of Model Codes

Types	A	B	C	D
Distances	4	7	10	13
multiplied by 2	8	14	20	26
multiplied by 3	12	21	30	39
multiplied by 4	16	28	40	52

In such a situation the first four successive products are 4, 7, 8, 10 i.e. AABC or $A_2B_1C_1$; D being too far.

However, when the deviations from more than one model code are slightly less or slightly more than the adopted maximum the 25 per cent tolerance clause was applied in order not to eliminate very similar units. (Table 8.5).

TABLE 8.5: Model Code

Types	B	C	D
1st product	10	12	16
2nd product	20	24	32

i.e. B_2C_2, D — being too far.

In fact such a clause applies only to the cases when the distance between the unit under study and one model code is more than 9 and than less 14 from another (9 + 25 % = 11.25, 10 + 25 % = 12.5, 11 + 25 % = 13.7).

There are also some situations with the not very accurate date when the codes for some greater, internally very diversified, units exceed as a whole the adopted maximum and could not be grouped in any sensible way into types. Then some greater tolerance should be adopted, if closer investigation proves that several types of very different character occur therę side by side. Otherwise, the study ought to be made of the occurrence of similar deviations that could lead to the identification of new types. However, to do so the deviations of such units from the model codes of the 2nd and then, if necessary, of the 1st order have to be analysed to learn about the possible common direction of such deviations.

Some examples of such a procedure taken from Portugal are given below.

To interpret such a situation another table (Table 8.6) is presented into which only the distances below 14 (including tolerance clause) are included. The distances 11 and below are marked = those between 12 and 14. The second column represents the interpretation : 1. those units with distances below 11 to one or more model codes are singled out, 2. unit no. 10 represents the application of the 25 per cent tolerance clause, 3. nos. 4, 14, 15 — with distances slightly above the adopted maximum, 4. the problem of nos. 4, 11 with the same distances to more than the adopted number of products has to be solved finally on the basis of the types distinguished in the neighbouring units.

As one can see from the symbols, a variety of types have been identified in Portugal, starting from small-scale traditional (T), peasant agriculture (Tmg), transitional between L (Traditional Large-Scale or Latifundia) and M (Market Oriented), large-scale agriculture (Mel), through various types of small-scale Market-Oriented mixed agriculture (Mmc, Mmp, Mmt, Mmf) differing by the relative importance of crop growing, including fruit crops and animal breeding, to the incipient socialised agriculture (Sep) that represents the collectives spontaneously organised in some southern provinces of Portugal during the "Carnation Revolution". In some units one type of agriculture clearly dominates, while in the other two, three or four types co-exist in various proportions (2 : 2, 2 : 1 : 1, 1 : 1 : 1 : 1).

In order to produce a full typology of European agriculture such comparisons had to be made several times : those units that differed from the already established model codes by more than 14 deviations should be eliminated and grouped into new types, then compared with a supplemented list of model codes again, eliminating again those that still deviated too much etc. In fact the establishment of new types is another weak point of the present typological method, particularly when a new type is described on the basis of a few units only. With a further addition of new units it happens that such model code ought to be changed. This is the reason why certain codes of the third order were withdrawn from the original list (Kostrowicki, 1980) until a greater number of cases representing such a type could make a better description of such a type possible.

By means of the methods and techniques, described above, 6 types of the 1st order, 22 types of the 2nd order and over 60 types of the 3rd order are identified in Europe. Some of them were found in their residual or transitional forms only, some identified in Europe could not be marked on the map due to their scattered occurrence. Finally, the 2 new types of the 2nd order and a few types of the 3rd order added when the map had already been sent to the publishers augmented the total number of the types of the 2nd order to 24 and those of the 3rd order to 66 (Table 8.3). Their characterisation together with the conclusions concerning the evolution of the European agriculture in the light of the Types of Agriculture Map of Europe are be presented elsewhere (Kostrowicki, 1984 b).

In spite of certain shortcomings, that could possibly be eliminated, the method has proved to be effective and could be used in any scale, from the lowest applied to individual agricultural holdings, to the highest — the typology of world agriculture.

It is thus believed that the Types of Agriculture Map of Europe could serve as a model for the elaboration of comparable maps of other continents, perhaps in a different scale (1 : 5 million) and limited to two first orders of the typology, and then of the Types of Agriculture Map of the World.

TABLE 8.6: The typological procedure on the example of Portugal

Nos. of unit	No of Type/distances	Nos. of types selected	Symbols of types and the proportions between them (no of products)
1	54/8 ====	54	Mmp_4
2	68/11 72/10 ===== =====	68-72	Mel_2Sep_2
3	54/2 ====	54	Mmp_4
4	42/13 55/12 68/12 ________________	42/55-68	$Tmg_1/Mmz_{1-2}Mel_{1-2}/$
5	42/11 55/14 68/14 ===== ===== =====	42	Tmg_4
6	43/14 54/6 _____ ====	54	Mmp_4
7	68/13 72/6 _____ ====	72	Sep_4
8	42/11 =====	42	Tmg_4
9	42/10 14/13 ==== _____	42	Tmg_4
10	51/14 54/11 56/13 _____ ==== _____	54-56	Mmp_2Mmt_2
11	38/14 51/11 52/11 _____ ===== =====	$/51_{1-2}$-	$Mmc_{1-2}Mmf_{1-2}Mmt_1$
	54/12 56/10 _____ =====	$—52_{1-2}$-56/	
12	68/11 72/10 ===== =====	68-72	Mel_2Sep_2
13	52/14 54/5 _____ ====	54	Mmp_4
14	52/13 _____	52	Mmf_4
15	52/14 _____	52	Mmf_4
16	54/5 ====	54	Mmp_4
17	40/14 41/4 52/8 _____ ==== ==== 54/13 _____	52	Mmf_4
18	45/17 52/14 ==== _____	54	Mmp_4

REFERENCES

Aitchison, J. W., 1983 : Model types of world agriculture. Problems of definition and case identification. *Geographia Polonica* 46, pp. 175-186.

Bielecka, K. M. Paprzycki, Z. Piasecki, 1975 : An evaluation of the applicability of selected mathematical methods for the typology of world agriculture. In : C. Vanzetti (ed) *Agricultural Typology and Land Utilisation*, Verona pp. 55-70.

Bielecka, K., M. Paprzycki, Z. Piasecki, 1979 : Proposal of new taxonomic methods for agricultural typology. *Geographia Polonica* 40, pp. 191-200.

Bielecka, K., M. Paprzycki, Z. Piasecki, 1980 : Applicability of numeric taxonomy methods in agricultural typology. Problems, criteria and methods of evaluation. (in) Noor Mohammad (ed) *Perspectives in agricultural Geography*, N. Delhi, 1981, v.i, pp. 381-416.

Bielecka, K., M. Paprzycki, 1979 : Evaluation of taxonomic methods from the point of view of comparability of results in space and time — in optimisation aspects. *Geographia Polonica* 40, pp. 187-190.

Bonnamour, J., 1973 : Geographie rurale. *Mēthodes et perspectives*. Paris, pp. 44-48.

Bonnamour, J., Ch. Gillette, 1980 : Les types d'agriculture en France, 1970. *Essai methodologique*. Paris 184 p.

Bonnamour, J., Y. Guermond, Ch. Gillette, 1971 : Typologie des systēmes d'exploitation agricole utilisēs en France. *Annales de Geographie, 438, Mars-Avril*, pp. 144-166.

Bonnamour, J., Y. Guermond, Ch. Gillette, 1971 : Les systēmes rēgionaux d'exploitation agricole en France. *Mēthodes d'analyse typologique Etudes rurales* 43-44, pp. 78-169.

Chabaud, M., A. Champalbert, 1973 : Methode d'analyse de l'espace agricole : typogramme et matrice ordonnable. Exemples du Minervois et de la vallēe du Rhōne. *Bulletin de la Sociētē Languedocienne de Gēographie* 7.2.

Christians, Ch., 1975 : La typologie de l'agriculture en Belgique : mēthodes, problems, rēsultats. In : *Agricultural Typology and Land Utilisation*. Verona, pp. 93-110.

Csati, E., 1980 : Land Utilisation map of Europe. *Budapest Cartographia*.

Dongmo, J. L., 1973 : Typologie de l'agriculture camerounaise. Essai d'application a l'Afrique Noire de la mēthode statistico-geo-graphique du professeur Kostrowicki. Universities de Yaunde. *Annales de la Faculte des Letters et Sciences Humines*. 5, pp. 19-40.

————, 1975 : Typologie de l'agriculture camerounaise. In : C. Vanzetti (ed) *Agricultural Typology and Land Utilisation*. Verona pp. 115-145.

Galczynska, B., 1984 : *Agricultural Typology of Bulgaria*. Geographia Polonica 50 (in print).

Gorbunova, L. I., M. V. Komleva, L. V. Shishkina, 1979 : Agricultural Typology of the USSR. *Geographia Polonica* 40, pp. 83-92.

Kampp, A. H., 1973 : Et forslag til landbrukstypologi. *Geografisk Orientiring* 3, pp. 134-137.

Kostrowicki, J., 1970 : Types of agriculture in Poland. A preliminary attempt at a typological classification. *Geographia Polonica* 19, pp. 99-110.

————, 1971 : the methodological bases for the typology of world agriculture. In : G. Krajko et. al. (eds) *Symposium on Agricultural Typology and Agricultural Settlements*. Szeged-Pecs pp. 141-173.

————, 1972a : A preliminary attempt at a typology of world agriculture. In: W. P.

Adams and F. M. Helleiner (eds) *International Geography* v. 2, Montreal pp. 1097-1100.

———, 1972b : The typology of world agriculture. A preliminary scheme. In : L. Reeds (ed) *Agricultural Typology and Land* Use. Hamilton Ontario.

———, 1974 : The typology of world agriculture. Principles, methods and model types. Warsaw, 74 p. Reprinted in : C. Vanzetti (ed) *Agricultural typology and Land Utilisation*. Verona, 1975, pp. 429-479.

———, 1976 : "On the methods determining landuse, crop and livestock combinations as used in the Polish studies. In : *Essays in Applied Geography*, pp. 111-120.

———, 1977 : Agricultural typology. Concept and method. *Agricultural Systems* 2, pp. 33-45.

———, 1979 : Twelve years' activity of the IGU Commission on Agricultural Typology. *Geographia Polonica* 40, pp. 235-253.

———, 1980 : A hierarchy of world types of agriculture. In Noor Mohammad (ed) 1981. *Perspectives in Agricultural Geography* v. I, New Delhi, pp. 165-208.

———, 1982 : The types of agriculture map of Europe. *Geographia Polonica* 48, pp. 79-91.

———, 1984a : Types of agriculture in Europe. *Geographia Polonica* 50, (in print).

———, 1984b : *Transformation de l'agriculture europeenne a la lumiere de la carte des types agricoles de l'Europe*. 33 p. (duplicated).

Kostrowicki, J., and W. Tyszkiewicz, 1970 : Agricultural Typology Selected Methodological Materials. *Dokumentacja Geograficzna* 1, 60 p.

Rikkinen, K., T. Vapaoksa, 1983 : The application of world agricultural typology to Finland. *Geographia Polonica* 46, pp. 93-106.

Scott, P., 1983 : The typology of Australian agriculture. *Geographia Polonica* 46, pp. 7-20.

Stola, W., 1975 : Changements dans les types de l'agriculture belge dans les annees 1950-1970. In : C. Vanzetti (ed) *Agricultural Typology and Land Utilisation*, pp. 339-356.

———, 1983 : Essai d'application des methodes typologiques a l'etude comparee sur le development des agricultures belge et polonaise. *Geographia Polonica* 46, pp. 159-173.

Szczesny, R., 1981a : Changing types of Austrian agriculture 1960-1970. *Geographia Polonica* 40, pp. 161-169.

———, 1981b : Transformations of agricultural types in Poland 1970-1973. In : Noor Mohammad (ed) *Perspectives in Agricultural* Geography v. I, New Delhi pp. 209-228.

———, 1982 : Type rolnictwa Szwajcarii (Sum : Agricultural types of Switzerland). *Przeglad Geograficzny* 54, 4 pp. 511-532.

———, 1984 : *Agricultural typology of the Alpine* areas. (duplicated).

J. Szyrmer, 1973,: Propozycja zastosowania nowej metody taksonomicznej do typologii rolnictwa (The proposal of the new taxonomic method to be applied in agricultural typology). *Przeglad Geograficzny* 45, 4, pp. 739-756.

———, 1975 : Stopien specjalizaeji rolnictwa (Sum : The degree of Agricultural specialisation). *Przeglad Geograficzny* 47, 1, pp. 117-135.

Troughton, M., 1979 : Application of the revised scheme for the typology of world agriculture to Canada. *Geographia Polonica* 40, pp. 96-112.

Tschudi, Aa. B., H. Johnansen, 1983 : Types of agriculture in Norway by the typogram method : Notes on the problem of establishing commercialisation. *Geographia*

Polonica 46, pp. 83-92.

Tyszkiewicz, W., 1975 : Types of agriculture in Poland as a sample of the typology of world agriculture. *Geographia Polonica* 40, pp. 171-186.

———, 1980 : Types of agriculture in Macedonia as a sample of the typology of world agriculture. *Geographia Polonica*, pp. 163-185.

———, 1982 : Zastosowanie metod typologiezuych do badan rolnictwa Szwecji (Sum : An application of typological methods to the study of Swedish agriculture). *Prezeglad Geograficzny* 53, 4, pp. 551-570.

Vanzetti, C. (ed), 1969 : *World Atlas of Agriculture*. Novara.

9

TYPES OF AGRICULTURE OF SMALL TROPICAL ISLANDS

MACIEJ JEDRUSIK

THE PAPER is an effort of setting up the typology of agriculture of 16 small tropical Islands. Typology was obtained with the known method of definition of agriculture types elaborated by the Commission of Agricultural Typology of the International Geographic Union, under the leadership of Professor Jerzy Kostrowicki. The essence of the method resides in the fact that each agricultural system considered can be described with the 27-digit code. Particular digits of this code represent normalised values of social, ownership, organisational technical, productive and structural features of agriculture and take values from 1 to 5, these values corresponding to classes adopted for each feature. Class code 1 corresponds to very low values of the feature, class code 5 — to very high values. Once the overall code established, it is compared to the previously established codes of agricultural model types of the world on three hierarchical levels, and, on the basis of similarity of the code of the unit considered to the codes of model types this unit is assigned to one of the previously described types or a new type is established.

The method outlined has been initially applied to the study of agriculture in moderate zone, but recently it has also been started to be applied in typology of agriculture of tropical areas, in such countries as Malaysia, India, Cameroun, Cuba and Mexico (Kostrowicki, 1988). An attempt of determining the agriculture types on small tropical islands is contained in this direction of study and is, simultaneously, the first attempt of this kind.

The study encompassed 16 tropical islands. They were: five Pacific islands — Abemama, Christmas, Nanumea, Samui and Tutuila, then six Atlantic islands — Barbados, Basse-Terre, Desirade, Grande-Terre, Martinique and Montserrat, and finally five Indian Ocean islands — Mahe, Majotta, Mauritius, Reunion and Silhouette (Table 9.1). This is very few, as compared to the multitude of tropical islands in the world, but a very significant limitation was set by the requirement of reliability and comparability of cartographic, and, even more so, statistical material, being the basis of the study undertaken.

TABLE 9.1: The island surfaces and geographical coordinates of the island centres

Island	Ocean	Surface in sq. kms	Latitude	Longitude
Reunion	Ind	2512	21°05′ S	55°30′ E
Mauritius	Ind	1865	20°05′ S	57°50′ E
Martinique	Atl	1100	14°30′ N	61°W
Basse-Terre	Atl	815	16°10′ N	61°40′ W
Grande-Terre	Atl	585	16°20′ N	61°25′ W
Barbados	Atl	430	13°10′ N	59°20′ W
Majotta	Ind	374	12°50′ S	42°50′ E
Christmas	Pac	338	2°N	157°30′ W
Samui	Pac	247	9°30′ N	100°E
Mahe	Ind	145	4°40′ S	55°25′ E
Tutuila	Pac	140	14°20′ S	170°45′ W
Montserrat	Atl	102	16°35′ N	62°05′ W
Desirade	Atl	30	16°2+′ N	61°W
Abemama	Pac	27	0°20′ N	173°55′ E
Silhouette	Ind	25	4°25′ S	55°10′ E
Nanumea	Pac	3, 6	5°40′ S	176°10′ E

The data used for determination of agriculture system code values for the islands considered come from the period after 1977.

Characterisation of Agriculture of the Islands Studied

The study proved that agriculture of the islands selected is differentiated both in terms of features of agriculture itself and in terms of conditions in which this agriculture developed. The values of codes for all the features proposed in the typological system adopted, excepting feature no. 21, are presented in Table 9.2 values of the code for feature no. 21 were not presented because of the different manner of determination of values for this feature. Likewise, data for the island of Samui were omitted because of the different method of source data collection and determination of the code.

TABLE 9.2: Values of feature codes for agriculture of the islands studied

Feature No.	Abe	Bar	B-T	Chr	Des	G-T	Mah	Maj	Mar	Mau	Mon	Nan	Reu	Sil	Tut
	Island name abbreviations, in alphabetical sequence														
1	0	0	0	0	6	0	0	17	0	0	0	0	0	0	0
2	0	0	19	0	0	19	13	0	6	4	66	0	21	13	3
3	100	100	100	100	94	100	76	23	92	100	94	100	100	100	85
4	0	0	0	0	0	0	0	0	0	0	0	0	0	0	0
5	3	5	2	4	3	2	1	4	1	5	1	4	1	1	1
6	3	8	2	77	4	2	2	4	2	6	1	2	3	15	1
7	48	414	144	44	92	141	59	13	99	988	43	32	352	154	38
8	103	57	78	5	88	88	78	86	30	83	86	165	26	8	73
9	0	0	6	0	0	17	0	0	9	4	122	0	1	0	8
10	0	78	65	0	0	66	19	0	112	11	127	0	80	0	16
11	0	176	73	0	0	73	0	0	335	249	0	0	206	0	0
12	0	2	9	0	0	3	0	0	19	20	0	0	10	0	0
13	0	65	100	0	100	100	47	31	94	100	74	13	93	100	86
14	16	62	129	1	97	216	134	28	139	55	800	67	71	4	116
15	15	50	61	1	24	71	31	3	47	172	39	14	115	10	28
16	15	65	83	1	36	110	31	4	64	182	109	14	122	10	34
17	15	88	77	12	28	81	40	4	155	208	45	9	449	129	38
18	9	82	71	5	4	72	38	1	145	203	22	1	423	63	3
19	58	93	92	42	14	89	94	28	93	98	49	8	94	49	8
20	9	46	56	0	3	64	30	1	44	168	19	1	108	5	2
21															
22	100	47	59	100	0	54	87	17	36	95	12	100	73	100	51
23	0	47	32	0	33	40	56	9	47	8	73	0	5	0	6
24	32	4	8	43	67	5	13	14	40	2	11	87	17	40	79
25	14	24	9	27	52	11	60	19	19	6	83	39	12	8	41
26	0	19	6	0	0	8	62	0	16	5	84	0	8	0	1
27	58	74	52	42	0	86	22	28	28	93	0	8	86	49	0

Social Features

Land Ownership and Administration

On the majority of islands studied virtually all agricultural land is privately (individually) owned. There are islands where private ownership is accompanied by state property or property of rural communes (Tutuila, Martinique, Montserrat, Mahe and Majotta). It is only on the last two of the islands mentioned that the share of land not owned by private farming exceeds 20 per cent, with the highest share characterising Majotta and reaching there 75 per cent of the total surface of agricultural land. Most of the agricultural land on this island (56.5%) belongs to the so called territorial commonality, i.e. in fact — the state. Then, almost 18 per cent of agricultural land belongs to village communities ("terre de villages" in French). In reality, however, the boundaries between the two ownership forms are not too precisely defined. This results, for instance, from the lack of regular cadastral measurements, and also from difficulties in application of European law in the areas where Muslin law dominated for centuries, originated from customary law, according to which land ownership appeared with taking of the land, ceased when a farm was left, and inherited land was given over to the heirs without divisions (Fig. 9.1).

Majotta is also quite specific in terms of land administered by village communities, for it is only in this island that their share is quite high, although even there the share of land administered by the traditional tribal, kin or village communities does not exceed the threshold of 20 per cent, taken as the threshold of significance in the approach here applied.

The land ownership share of agricultural cooperatives or state farms appears on only a few islands and is never any how significant. State land is usually given for tenancy to private farmers or is not cultivated by anybody. Some state farms were created on Mahe, but there is lack of information as to existence of state farms on any of the other islands.

Tenancy. There is relatively little land used according to the traditional principles of tenancy, that is, tenancy for labour or half-half splitting or other, similar forms. On three of the islands only, i.e. on Reunion, Samui and Montserrat the shares of the three forms of tenancy in agricultural land exceed 20 per cent. Data concerning Montserrat, however, are not too precise, for the results of the

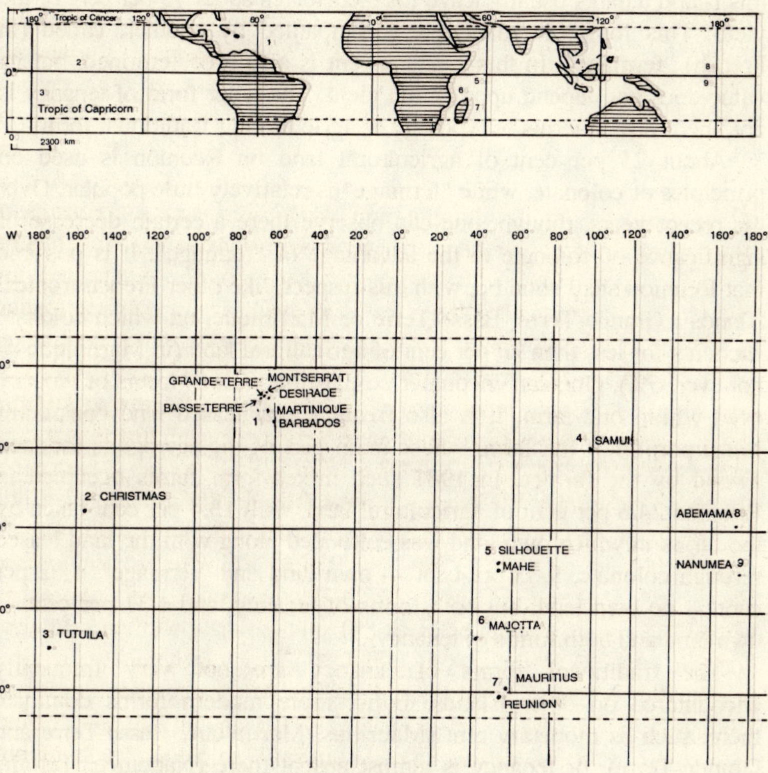

FIG. 9.1: Islands studied in the cartesian system of orthogonal coordinates

agricultural census of 1984, sent from this island, encompass only the totals of agricultural land under tenancy (66% of the total) without distinction into various forms of tenancy. On Reunion the traditional and dominating form of tenancy is colonate (French: "colonage"). It consists in leasing of the land of an owner of a bigger estate, most often a plantator, for a rent which is a predetermined part of crop yield. Thus, the absolute value of the rent depends upon the actual yield. On this island tenants usually leave for themselves about 75 per cent of the yield. This form of tenancy is accompanied by another, called (in French) "fermage". In this case the rent is also predetermined, but its value does not depend upon actual yield. The latter form of tenancy is not, therefore, in terms of typology of agriculture, a traditional form.

About 25 per cent of agricultural land on Reunion is used on principles of colonate, while "fermage" is relatively little popular. Over the recent years, though, one can observe there a certain decrease of significance of colonate to the advantage of "fermage". It is possible that Reunion shall soon be, with this respect, like other French tropical islands : Grande-Terre, Basse-Terre or Martinique, on which colonate accounts for less than 20 per cent of agricultural land (in Martinique — not even 6%). One can encounter co-existing various forms of tenancy even within one farm. It is also frequent that leased land constitutes just a portion of the farm, whose main part of the acreage is the land owned by the farmer. In 1981 such mixed-form farms occupied in Reunion 24.8 per cent of agricultural land, with 13.4 per cent taken by the farms in which own land was cultivated along with the land leased through colonate, 9.03 per cent — own land and "fermage", 1.02 per cent — no own land, but both forms of tenancy, and 1.31 per cent — own land and both forms of tenancy.

The traditional forms of tenancy are not very frequently encountered on other islands. Either more modern forms dominate there, such as monetary rent (Mauritius, Martinique, Basse-Terre and Grande-Terre), or tenancy is almost absent there (Nanumea, Tutuila, Barbados, Abemama, Christmas and Desirade). An interesting, though non-traditional, form of land tenancy exists of Mahe. Some 200 farmers of this island live and work on the land rented from the government of Seychelles for a very low rent. An additional, quite specific manner of paying back the rent is the obligation of cultivating the land in tenancy. This form of tenancy accounts for more than 10 per cent of agricultural land of the island.

Farm Acreage

All the 16 islands studied are characterised by domination of small and very small farms, their magnitude understood both in terms of pure acreage and the number of persons employed, or, finally, output volume.

In reality, however, the structure of farms on the islands in question is much more complex. Thus, usually, an important share of agricultural land on an island is taken by a handful of farms of large surface, accompanied by a large number of small farms. This, for instance, is the case of Mauritius, where there are 55 large farms of more than 200 hectares each, taking together 60 per cent of agricultural land, and, simultaneously, there are also more than 28000 farms of less than 5 hectares, including more than 40 per cent of those having less than 0.5 hectare. Small farms account for less than 25 per cent of agricultural land of the island.

A similar, though perhaps not so contrasted situation is observed on Reunion. In 1981 some 90 per cent of farms existing there had surface of less than 5 hectares, taking up together 41 per cent of agricultural land, while barely 0.32 per cent of farms, having surfaces of more than 100 hectares accounted for some 30 per cent of agricultural land of this island.

Available source materials for Mahe do divide farmers into two groups — those having bigger and smaller farms ("large farmers" and "small farmers"). The average acreage of a large farm in Mahe is at 130 hectares, and there are 16 such units there, accounting for approximately 60 per cent of agricultural land. On the other hand there are more than 500 of small farms, accounting for the most part of the rest of agricultural land. To this, we may add nearly 1200 farms almost without land, where livestock raising, mainly of pigs and poultry, is conducted in farm-adjacent gardens.

On Silhouette there exist only just a few large farms which take up almost entirely the area of agricultural land of this island. Several dozens of very small farms have been created on very limited portions of land bordering the large farms. These very small farms are conducted by tenants employed on plantations.

Out of the islands of Indian Ocean only Majotta features relatively smaller differences in farm acreages. The surface of most of farms does not exceed 1-2 hectares. It is, anyway, hard to get precise data, because of inaccuracy of cadastral measurements on this island and because of very popular there burn-out technology.

Concentration of agricultural land is also strongly advanced on the islands of the Caribbean Sea, and especially so on Barbados. Thus, barely one per cent of all farms, with surface of more than 200 hectares, occupy there as much as 76 per cent of agricultural land, while 97 per cent of all farms, the small ones, with up to 4 hectares — occupy only 12 per cent. On Grande-Terre and on Basse-Terre the classes of farms of less than 10 hectares and of more than 50 hectares take similar areas of agricultural land. Still, while the former constitute more than 97 per cent of all the farms, the latter — only 0.5 per cent. The data available, though, concern not only true farms, but also appropriated agricultural land, which may not be cultivated at all. Such an "estate" may in fact be just a residence with a vegetable garden, or it may be used for commercial or industrial purposes.

There are, on both these islands, very few farms with average acreage. The same occurs in Martinique. Average farms (5 to 20 hectares) are almost non-existent there, while small farms (up to 5 hectares) and large farms (more than 50 hectares) take, respectively, 36.1 and 33.8 per cent of agricultural land surface, with the small ones accounting for 93.9 per cent of the total number of farms and the large ones — only for 0.6 per cent. There is a smaller differentiation of magnitudes of farms of Desirade, presumably because of the fact that the total area of agricultural land is quite limited on this island.

Generally, differentiation of farms acreages is lower for Pacific islands, although small farms dominate on each of them. An exception is constituted by Christmas, where there exist some large coconut plantations and the overall number of farms is relatively low, so that the average magnitude of a farm is the highest among all the islands considered, and it exceeds 70 hectares. On Abemama farms are small and in view of lack of plantations they are also little diversified as to their acreage. There is a similar phenomenon on Nanumea.

Differentiation of farm acreages on Tutuila is much less dramatic than on most islands of other oceans. Larger farms (of more than 20 hectares) do constitute there less than one per cent of the total number of farms and occupy 7 per cent of agricultural land. Small farms (below 4 hectares) do account for 94 and 69 per cent respectively.

Own observations of the present author concerning agriculture of Samui make it possible to assume that farms of rather small acreages dominate, as to their quantity, on this island. Such small farms are composed of fields of up to one hectare. These farms do also encompass the fallowed fields between the wild growing, but harvested, coconut palms. Such fields, lying fallow, are sometimes used as pastures. Their boundaries, however, are not well defined. In the South-Western and Northern parts of the island a dozen of large farms developed oriented mainly at rice growing. Thus, the magnitude structure of farm population seems similar to the one encountered on Mahe.

Organisational and Technical Features

One can observe, within this group of features, a much greater differentiation than in the case of social features. This concerns a majority of particular features, but for some of them from the organisational-technical group, agriculture of the islands studied displays important similarities.

Labour Intensity

Labour intensity of agriculture is, on most of the islands considered, rather high. On Nanumea the number of agriculturally employed per 100 hectares of agricultural land is, though, even higher, which may result from errors in the evaluation of the overall employment in agriculture. On the other hand, lower employment in agriculture is observed in Reunion and Martinique, and the lowest — on Christmas and Silhouette.

Live Pulling Power Intensity

Likewise, low intensity of use of live traction power in agriculture is typical for a majority of islands studied. Somewhat higher values of this feature are attained on Mauritius, Basse-Terre, Tutuila, Martinique and Grande-Terre.

Mechanical Traction Power Intensity

Within the set of islands studied one can distinguish two clearly distinct subsets: islands with very low mechanisation in field work, and islands on which agriculture relies heavily upon machines. The first of these subsets contains: Majotta, Abemama, Christmas, Nanumea, Desirade, Samui and Silhouette, while the second: Montserrat,

Martinique, Reunion, Barbados, Basse-Terre and Grande-Terre. The remaining three islands, i.e. Mahe, Tutuili and Mauritius cannot, however, be classified into any of these two groups of islands.

Fertiliser use

On as many as 10 islands out of 16 studied (Majotta, Mahe, Montserrat, Abemama, Christmas, Nanumea, Tutuila, Desirade, Samui and Silhouette) mineral fertiliser use does not exceed 10 kgs of NPK per one hectare of cultivated land. Somewhat higher (average) fertiliser use can be observed on Basse-Terre and Grande-Terre, while the highest values — on Barbados, Reunion, Mauritius and Martinique.

Irrigation

The shares of agricultural land subject to irrigation in total agricultural land are, on the other hand, quite similar on the islands considered. In 13 cases the share of irrigated land does not exceed 10 per cent of agricultural land, and irrigation has a bit bigger significance only on Mauritius, Martinique and Samui.

Intensity of use of Arable Land

Most islands are characterised by high or average intensity of use of arable land. One should, though, remember, that this indicator, when used with respect to agriculture of tropical islands is not the most adequate in view of big difficulties in determining its values. By virtue of assumption, this feature should account, for instance, for the number of harvests per year. With almost inobservable climatic changes in the annual cycle on small tropical islands the number of harvests is difficult to define. This number is even not given in agricultural censuses and other reports being the statistical basis of the present paper. Thus, the value of the feature considered was obtained for purposes of the present study from one crop yield during a year. Differences among islands result also from magnitudes of surfaces taken as non-cultivated and fallowed.

Accepting the above reservations, one can still state that while the intensity of use of arable land is on the majority of islands either high or average, it is very low on Abemama, Christmas and Samui, and only slightly higher on Nanumea.

Numbers of Animals Raised

There are much greater differences among the islands as to the numbers of animals raised per 100 hectares of agricultural land. The lowest numbers of animals are observed on Christmas (several hundred pigs and some fifteen hundred poultry), on Silhouette (approximately 100 pigs and several hundred poultry), and on Samui where, besides pigs and poultry, cattle is also raised, but numbers of all these animals are low. Animal husbandry has somewhat bigger significance on Majotta (cattle, goats, sheep, poultry and even donkeys) and on Abemama (pigs and poultry). Quite some pigs and poultry, along with about 20 goats and sheep are raised on small Nanumea, due to which this tiny island has a similar animal raising indicator as Reunion with its almost 20,000 of cattle, tens of thousands of pigs, goats and sheep, and hundreds of thousands of poultry, as Mauritius, where, along with cattle, pigs, goats and sheep there are almost one million and a half of hens, and as Barbados with similarly differentiated structure of animal husbandry. Even more animals per 100 hectares of agricultural land are on Desirade, although only 300 cattle and not even 2000 goats and sheep are being raised there. This agricultural activity is similarly intensive on Mahe (with almost 2000 cattle, 6600 pigs and over 100000 poultry), on Tutuila (88 cows, 7320 pigs, more than 20,000 poultry and 105 horses), on Basse-Terre (22,000 cattle, almost the same number of pigs, 15,000 goats, 1000 sheep and several hundred thousand hens), and on Martinique (more than 40,000 cows, over 22,000 pigs, almost 60,000 goats and sheep, almost 400,000 poultry, more than 300 horses and 500 mules, as well as several thousand rabbits).

The highest livestock raising indicator values are obtained, though, on Grande-Terre, where livestock raised is dominated by cattle (some 45,000), and where there are also some 17,500 pigs, 15,000 goats, 1000 sheep and more than 300,000 poultry, and on Montserrat, although there are only 3500 cattle, some 11,000 goats and sheep and limited numbers of poultry on this island, but, due to small area of agricultural land, the resulting indicator value is the highest there.

Productive Features

Land Productivity

Agriculture of most of the islands is characterised by high productivity of agricultural and cultivated land. Both these magnitudes are in most cases very similar on each of the islands studied. On a majority of islands perennial crops, and especially coconut palm, are, simultaneously, fallowed land, but, in accordance with assumptions adopted, they should be treated as cultivated land. It is only on Grande-Terre and Montserrat that there are important divergences between land productivity and cultivated land productivity. This fact results from existence, on these islands, of large surfaces of not cultivated pastures and meadows, which importantly diminishes the area of cultivated land. The lowest land and cultivated land productivity is on Majotta and Christmas, somewhat higher — on Silhouette, Nanumea and Abemama. Islands of Samui, Desirade, Tutuili, Mahe and Montserrat feature average land productivity level, with Montserrat having clearly higher cultivated land productivity. On remaining four islands productivity of cultivated land, although somewhat higher than land productivity, does not exceed the average level. Even higher land productivity is observed on Martinique, Barbados, Basse-Terre and Grande-Terre. On all these islands productivity of cultivated land is, of course, higher than that of land in general, but only on Grande-Terre this difference is highly significant. The highest productivities are observed on Mauritius and Reunion.

Labour Productivity

There are much smaller differences among the islands studied when labour productivity is considered. On most of these islands labour productivity is low or even very low. This group encompasses such islands as: Majotta, Nanumea, Christmas, Abemama, Desirade, Tutuila, Mahe, Montserrat, Samui, Basse-Terre, Grande-Terre and Barbados. Labour productivity is average on Silhouette, Martinique and Mauritius, and it is high on Reunion.

Similar differences are observed for the value of commercial product per one person employed in agriculture. Here, however, there are a bit more islands on which the value of commercial product per one person employed in agriculture is average or high. Such is the situation on Silhouette, Samui, Basse-Terre, Grande-Terre, Barbados and Martinique. Then, the highest values of this indicator are observed on Mauritius and Reunion.

Degree of Commercialisation

Agriculture of almost half of the islands studied is characterised by the very high degree of commercialisation. Commercialisation of agriculture is the highest on Mauritius, Reunion, Mahe, Basse-Terre, Martinique, Grande-Terre and Barbados. At the other extreme along the axis of this indicator we find Nanumea, Tutuila and Desirade. Agriculture of Majotta is also little commercialised. On the remaining five islands : Montserrat, Abemama, Christmas, Samui and Silhouette the degree of commercialisation is average.

Commercial Production and Specialisation in Agriculture

The value of commercial production calculated per one unit in agricultural use is most often referred to as level of commercialisation. Differences with this respect among the islands studied belong to the greatest ones. Islands with low level of commercialisation are : Christmas, Majotta, Nanumea, Tutuila, Desirade, Silhouette, Abemama and Samui. Average and high values of this feature are observed on Maontserrat, Mahe, Martinique, Barbados, Basse-Terre and Grande-Terre.

It is, on the other hand, characteristic for agriculture of a majority of islands studied, that the degree of agricultural specialisation is high or very high. Degree of specialisation of agriculture is understood here in such a way that it corresponds to degree of concentration of agricultural commercial production on a definite number of products. Since the notion of specialisation, as used here, concerns only commercial production, the highest values of the indicator considered do appear not only on Reunion and Grande-Terre, where plantations of sugarcane dominate, but also on Christmas, Silhouette, Nanumea and Abemama, where agriculture is poorly developed, and where virtually the only commercial product is copre, whose yield is, anyway, very modest. There is, as well, high specialisation of commercial agricultural production on Mauritius (sugarcane and tea), on Majotta (first of all ylang-ylang), on Desirade (corn), on Barbados (sugarcane), and on Basse-Terre (sugarcane and bananas). Islands with average degree of commercial specialisation of agriculture are: Samui (coconut palm and rice), Montserrat (mainly beef and pork meat production), Martinique (bananas, sugarcane, pineapples), and Mahe (coconut palm, cinnamon tree, tea and some products from animal husbandry). The lowest degree of specialisation is observed on Tutuila, where the structure of commercial production is quite differentiated.

Structural Features

Perennial and Semi-perennial Crops

When turning towards the production structure of agriculture, one may notice that differences among the islands studied are with that respect much lower than for the previously considered group of features. In a vast majority of cases perennial and semi-perennial crops dominate in the surface structure of crops. They take a lower share only on Desirade, Montserrat and Majotta, and, slightly lower — also on Martinique.

Permanent Grasslands

Important similarities among islands appear also with respect to the share of land taken by permanent grasslands in the total surface of agricultural land. On the majority of islands studied this share is relatively low. Somewhat greater shares of permanent grassland in total agricultural land are encountered on Basse-Terre and Desirade. On Grande-Terre, Barbados, Martinique and Mahe, on the other hand, permanent grasslands take approximately as much as half of agricultural land. They are also quite significant on Montserrat, where 12 per cent of the overall surface of permanent grasslands of almost 418 hectares is occupied by cultivated meadows and pastures.

Food Crops

Similarly as permanent grasslands, also food crops take on the majority of islands considered a relatively small share of agricultural land. From among the food crops most important are tuberiferous plants and coconut palm, as well as vegetables and fruits. The lowest shares of these cultures occur on such islands as Mauritius, Barbados, Grande-Terre, Basse-Terre, Montserrat, Mahe, Majotta and Reunion. There are a bit larger surfaces under food crops on Abemama, Martinique, Silhouette, Samui, Christmas, Desirade and Tutuila. The highest share of food crops occurs on Nanumea.

Production Direction and Commercial Attitude

Agriculture of a majority of islands is directed primarily on crop production. Animal production has greater significance only on Tutuila and Desirade, and — the highest — on Mahe and Montserrat.

General commercial attitude on as many as 14 islands (excepting Mahe and Montserrat) is also towards crops. Commercial attitude is most oriented towards livestock-based production on Montserrat.

Industrial Crops

The shares of industrial crops in gross production of insular agriculture are quite differentiated. Industrial crops attain importance greater than average on such islands as Christmas, Silhouette, Samui, Basse-Terre and Abemama, and highest importance on Barbados, Reunion, Grande-Terre and Mauritius.

* * *

According to the system of typology proposed for agriculture world-wide by Kostrowicki (1980), agriculture of the units studied can be ascribed the following codes, corresponding to values 27 typological attributes, after appropriate standardisation is made:

1. Abemama	1151211	4111112	2211325	512113
2. Barbados	1151222	4144133	4423544	431214
3. Basse-Terre	1151112	4243144	4423544	421113
4. Christmas	1151231	2111111	1111315	513213
5. Desirade	1151211	4111144	3311124	124311
6. Grande-Terre	1151112	4443145	4523545	431115
7. Mahe	1141111	4131134	3322533	531342
8. Majotta	1121211	4111132	1111214	211112
9. Martinique	1151111	3355244	4433543	332112
10. Mauritius	1151222	4225243	5534554	511115
11. Montserrat	1251111	4551145	3522333	241551
12. Nanumea	1151211	5111123	2211115	515211
13. Reunion	1251112	3145143	5544555	511115
14. Samui	1251222	4111211	3323323	413113
15. Silhouette	1151122	2111141	2233325	513113
16. Tutuila	1151111	4331144	3311112	414311

Conclusions

On the basis of typological procedure it could be concluded that agriculture of the islands studied can be assumed to belong to three types of agriculture on the hierarchical level of IInd order: T_s (little intensive, small-scale, semi-commercial, mixed agriculture), T_m (traditional, small-scale, mixed agriculture), and M_s (small-scale agriculture, specialised in industrial crop production). Classification of islands among these three types of agriculture is as follows: type T_s: Abemama, Christmas, Majotta, Nanumea, Samui and Silhouette; type T_m: Desiderade, Mahe, Montserrat and Tutuila; type M_s: Barbados, Basse-Terre, Grande-Terre, Martinique, Mauritius and Reunion.

Among the IIIrd order types determined for this set of data it is only the *Tmy* type corresponding to Desiderade and Tutuila that already exists in the existing hierarchical system of agriculture types of the world. Three other types determined are relatively close to existing types *Mso*, *Tss* and *Tmk*, while one, corresponding to agriculture of Mahe is quite distant from the IIIrd order types existing previously. The study reported here made it possible to distinguish two new types of the IIIrd order, namely: *type Ms1* (small-scale agriculture with high intensity of labour and production means, high productivity, high commercialisation, and clear specialisation in industrial crops), described with the code:

1151112 4244143 4533544 421115,

to which Barbados, Basse-Terre, Grande-Terre, Martinique, Mauritius and Reunion belong, and *type Tsj* (traditional, small-scale, with high labour intensity, low intensity of production means, low productivity, specialised in perennial crops with high importance of food crops), described with the code :

1151221 4111131 2212214 413112,

to which Abemama, Christmas, Majotta, Nanumea, Samui and Silhouette belong.

It is, though, difficult, on the basis of individual cases, and with reservations raised here, to accept the other two types appearing.

The agriculture types determined for particular hierarchical levels of typology may not exhaust all agriculture types encountered on small tropical islands. Since, however, the method of setting up agricultural typology, accepted in the present study, provides an open, rather that closed, formula, it is possible to include in it a new type, if it is determined.

On the basis of the reasoning presented it is possible to state that the method outlined and applied, admitting certain reservations concerning definition of values of features characterising agriculture — especially with regard to the definition of intensity of use of arable lands, proved to be useful in geographical studies of agriculture on small tropical islands.

REFERENCES

Kostrowicki, J., 1980 : A hierarchy of world types of agriculture, *Przeglad Geograficzny* LII, Vol. 2, pp. 271-302.

Kostrowicki, J., 1986 : Transformations de l'agriculture europêenne a la lumiere de la carte des types agricoles de l'Europe, *Geographia Polonica*, Vol. 52, pp. 191-208.

Kostrowicki, J., 1988 : Comparative research in world agriculture. Methodological review, *Przeglad Geograficzny*, Vol. 4, pp. 511-570.

10

AGRICULTURAL TYPES AND MULTILEVEL REGIONAL PLANNING IN INDIA

B.L. SHARMA

IT IS a long standing belief of many traditional societies that agriculture is a spatial distribution or organisation which serves better in national economy. But, with the changing perception of the occupation and the concept of allied activities associated with agricultural operations, the dimension of the approaches encompassing related problems in its manifestations, has changed the complexion of the whole system.

Till 1950s, the main focus of the study was based on the spatial patterns of agriculture, which means, horizontal scaling but, since then, the problem of measuring levels of agricultural development through productivity, efficiency and fertility indices, on one side, and regionalisation, regional planning and multifactor analysis techniques on the other, has changed the whole dimension. This is nothing but a shift in horizontal to vertical scaling or dimension.

To encompass the regional development in agriculture as a whole has been a problem of multivariates. While measuring development at various levels, the demarcation of regional boundary indicating development or imbalance of attributes, involves the problem of scale. Thus, there are two sets of problems while measuring the levels of agricultural development, i.e. Regional demarcation at different levels and to measure them in a suitable scale.

So far the regional demarcation of agricultural development or imbalance is concerned, there has been a lot of discussions and theoretical approaches but none of these find any amicable solution to the problems, except adopting the efficiency, productivity and fertility measures to the extent in order to encompass the levels of development. This may be referred as to the horizontal or spatial dimension of the scale. But, the task is not over here, as much criticism thwarted for not taking the taxonomic characteristics into consideration. These characteristics too, influence agricultural development to a great extent and shape agricultural economy. Thus, they play vital role in making a regional balance or imbalance.

Thus, the developed or undeveloped state of affairs on national, state and district levels is a question of scale on vertical dimension in which hierarchic orders are measured in a regional framework.

In a nut shell, if we can measure both the dimensions, i.e. vertical and horizontal of agricultural development, we may call it a multilevel regional development or imbalance in agriculture. Thus, it measures both the problems of horizontal scale or spatial dimension, i.e. National to state and state to district level planning and vertical scale in a hierarchic orders.

So far the measures of multilevel regional development in agriculture are concerned, no suitable method has been evolved so far, which may absorb types, systems and the scales simultaneously. But the out-come of the typological approach, through a IGU commission, takes the challenge of this problem in a multivariate forms. Let us examine them.

The Concept of type and System Analysis

Type forming system is essentially a system of identifying agricultural types of a region. Agricultural types and systems are the two notions used invariably when we go for locating type of enterprises in a given area. Thus, agricultural types and systems form a part of a set enterprise systems while, type denotes single set of agricultural economy. But, there may be many types in a system and many systems in a type. (Symons 1968).

Systems and type, therefore, are inter-related phenomena of functional activities carried out by farmers. Thus, if a single set of variables operating in an enterprise system, acts as an agricultural economy. While, more than one economy, well termed as enterprise, develops in a region, give rise to a system of enterprises. This enhances the operational function of agricultural activities in a multivariate dimension (Fig. 10.1).

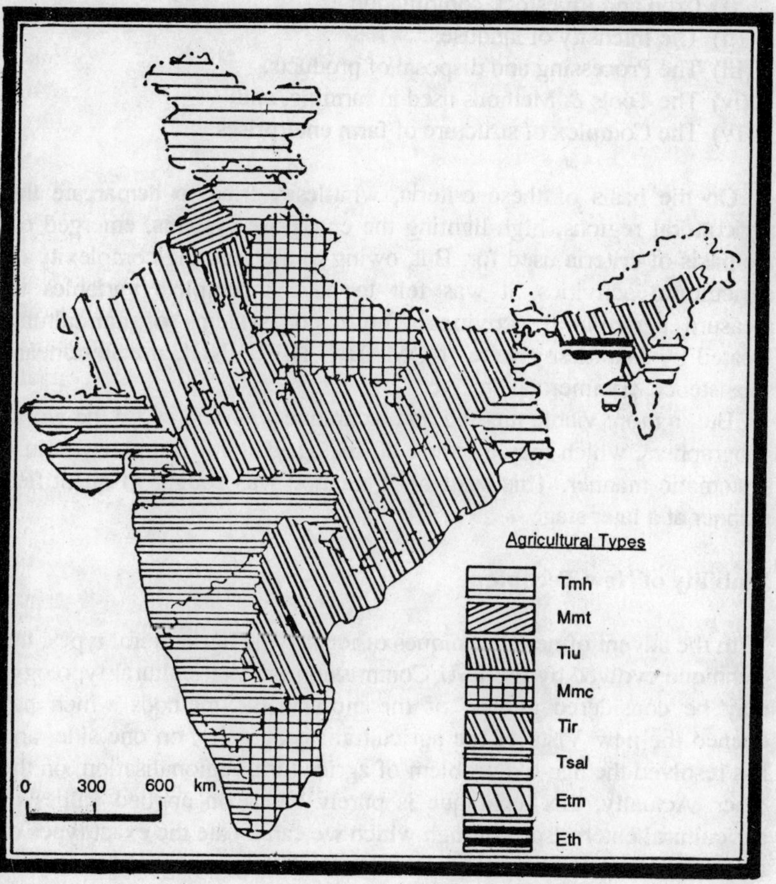

FIG. 10.1: Agricultural : Typological regions of India (1982).

Multiplicity of a multivariate activities carried out by farmers give rise to a complex system of enterprises. Thus, a set of economic patterns, based on various enterprises, form a strong agricultural infrastructure.

Whittlesey (1936), while attempting on agricultural regions of the earth, at the very outset of the criteria adopted by him, selected as such (D. Grigg, 1969) :

(i) Crop and Livestock combination,
(ii) The Intensity of landuse,
(iii) The Processing and disposal of products,
(iv) The Tools & Methods used in farming, and
(v) The Complex of structure of farm enterprises.

On the basis of these criteria, whittlesey tried to demarcate the agricultural regions, high-lighting the enterprise systems, emerged on the basis of criteria used for. But, owing to the growing complexity of agricultural activities, it was felt to have incomplete variables to measure agricultural activities. The major systems of agricultural located by Whittlesey were 'NOMADIC HERDING', mediterranean, subsistence, commercial etc. etc.

But, a more viable method and technique was desired by the world geographers, which may measure all the agricultural complexities in a systematic manner. This systematic method was sought in quantified manner at a later stage.

Viability of New Technique

With the advent of new techniques of identifying agricultural types, the technique evolved by the IGU Commission on 'Agricultural typology' may be considered as one of the most viable methods which has opened the new Vista of the agricultural geography, on one side, and has resolved the age-old problem of agricultural regionalisation, on the other. Actually, this technique is purely based on applied values of agricultural enterprises, through which we can locate the exact types of agriculture in real sense. This method of identifying agricultural types is called the "type farming system of agriculture". It is this method, through which we can separate agricultural type on one side & systems on the other.

Methodology

Typological technique helps in identifying the types of agriculture as well as the enterprises on the basis of the 27 selected variables in its coded frame-work as suggested by the I.G.U. Commission on 'World type of agriculture' led by Prof. J. Kostrowicki. The working method of the scheme helps in evolving the overall picture of agricultural attributes assigned to a region. Therefore, if one proceeds for individual attribute and comparison be drawn striking out all those prominent variables, one may reach the conclusion of a definite type of agriculture. This procedure has been adopted and applied to the districts of Rajasthan falling in desertic conditions to locate their characteristic types of agricultural economy. So that plan prospects may be highlighted for the future development of the region.

Areas of Type and Systems in Agriculture

With the advent of type farming systems, it has been possible to delineate the areas of agricultural enterprises farming types and systems. A region may have the types or systems or a combination of both in a given time. But, with development in a region, either of the two sets may lag or stranded. To confirm validity of above facts, examples may be taken up showing type on one side and systems on the other.

In support of the former case, 'Enn' type as has emerged in a desertic fold of the Wn. Rajasthan, in Jaisalmer district speaks of 'NOMADIC HERDING' may be considered as the most suitable case for the present study. While, in support of the later case, 'MMC' a type in a system, speaks of the Market oriented, mixed agriculture with crop growing prevalent in Ganganagar district of the state. Both, these cases are extremely exposite to each other cilently spell out the distinction between type and systems in agricultural economy. To have better understanding of both the systems, examples may well be illustrated from Enn and MMC-types: of Jaisalmer and Ganganagar districts respectively, as under.

Agricultural Economy

The economy destined to these units is distinctively unique in practices and ways. Environmental domain and control on the various enterprise systems is clearly reflected in animal oriented economy which determines 'NOMADIC-HERDING' as an agricultural system type in a real sense. The type forming system of agriculture is, therefore, erected on the basis of set of enterprises which are interrelated phenomenon emerged out of the environmental factors exerting influence on them.

Whittlesey writes that 'it is the simplest form of pastoralism based on sheep, cattle, goats and camels, and is primarily a subsistence form of exploitation of dry region. The length of the stay of nomades at one place and the direction of movement are generally the availability of water and natural forage' (Symons, 1968, p. 99). Thus, in search of water and forage they move out with their flocks in the neighbouring states of Gujarat and Madhya Pradesh. Thus, seasonal migration becomes a cause of tension and confrontation between people of these states which ends in a large scale killings of herds and herders at times.

$$\text{ENN : NOMADIC HERDING: } 0120141 \quad \frac{1511011}{1111110} = 151451$$

Enn is a typical type of agriculture in which a single set of enterprise system is located i.e. 'Nomadic Herding'. It is a type as has been described by whittlesey that the poorest exploitation of land economy. Thus, the taxonomy of the type is erected mainly on livestock ranching and nomadic life in search of food, forage and water. This involves large ranching fields, very low land ownership, extensive areas under grazing, very high labour input, very low capital inputs, very low land productivity, high livestock products, very low primary food production. All these characteristics lead to animal dominated economy. Whereas,

$$\text{Mmc' type: } 0140122 \quad \frac{3512433}{3333530} = 323115 \text{ (Ganganagar Distt.)}$$

On the contrary, 'Mmc' is a type of multifunctional systems of agriculture. It denotes the market oriented Mixed agriculture with crop growing prevalent type. Thus, specialisation of crops alongwith animal husbandry as developed enterprise system mainly oriented towards marketing of produced goods, are the main characteristic types of varied enterprises taking place in complex systems of agricultural economy. Different types or enterprises have their own system of working and therefore, giving a composite economy of combination of all the prevalent systems.

In general the taxonomic characteristics are high private land ownership, low, input of labour, high capital inputs, high land and labour productivity, high degree of commercialisation. This type of agriculture further may be generalised into second and third order types of its hierarchic orders, on the basis of their minor characteristics.

2. The Problem of Scaling

This, set speaks of the variable included therein relating to horizontal and vertical scales. The general distributional pattern of the organisational and production variables are well-known for their spatial distribution, whereas, social and structural characteristics are the dimensions of the verticle scale.

The aggregate value of all these variables, measured in their class intervals, brings out types of agriculture with a set of attributal characteristics in a given framework. To find an adequate expression of the type in its hierarchic order is a real and exact task of the planner to identify strong and weak links in between classes of their standings.

Thus, the identification of regional types and their hierarchic orders may be located in the following manner:-

First order-Regions

There are three main groups of 1st order regional types, if analysed on Indian Union. They are:

E = Traditional extensive agriculture.
T = Traditional intensive agriculture.
M = Market oriented agriculture.

Their main characteristics observed are:

$$E = 0240111 \ \frac{3411122}{1111110} = 124111$$

$$T = 0340111 \ \frac{3411122}{2223110} = 224111$$

$$M = 0250221 \ \frac{2233434}{4423230} = 223442$$

There are five groups of second order regional types. They are:

Et = Current follow, subsistence to semi-subsistence.
Ti = Small scale labour intensive.
Ts = Semi-commercial specialised crop.
Tm = Small scale mixed.

Mm = Market oriented mixed agriculture.

These groups are again characterised in the same manner as described in Ist order, but with the some added characteristics of particular variable, which dominantly become important.

Et = current follow, subsistence to semi-subsistence.

$$024111 \frac{3411133}{2211110} = 234111$$

Third order-Regions

In fact, all types of agriculture are identified from the 3rd order, which have less deviational differences of attributal characteristics. It is difficult to reckone and distinguish from one another, but individual variable makes it quite distinct from the rest of the types.

Thus, there are eight groups identified in the Indian Union viz. :

1. *Etm* = Current fallow mixed agriculture.
2. *Eth* = Current fallow-with migratory harding
3. *Tir* = Semi-irrigated, low productive, semi-substance.
4. *Tiu* = Irrigated, medium productive, semi-subsistence to semi-commercial.
5. *Tad* = Semi-commercial, semi-irrigated specialised.
6. *Tmh* = Semi-irrigated, Medium productive subsistence with livestock grazing.
7. *Mmn* = Small scale, mixed with industrial crops.
8. *Mmt* = Semi-commercial, mixed agriculture.

For Example:

$$\text{Rajasthan} = 0140121 \frac{3321233}{1211110} = 313231$$

$$\text{Punjab} = 0240112 \frac{4455554}{4412230} = 115231$$

Regional Imbalances

The imbalance in agriculture is reflected in different ways, i.e. on state hierarchy and on spatial patterns. The state hierarchy is a reflection of strong and weaker variables within a single set of agricultural types, while the spatial pattern is reflected in state to state or one unit to another. In both the ways, the attributal characteristics are important. Some variables are stronger at one level while others are weak. But as soon as the other variables also become stronger, the type alters abruptly and changes are evidently marked herein, as it is clear from the Punjab state.

State Level Regional Hierarchy

$$Mmt = \text{Punjab State} = 0240112 \; \frac{4455554}{4412230} = 15231 = 14$$

$$Mm = \text{Punjab State} = 0240112 \; \frac{3344334}{3422120} = 214231$$

$$M = 0120111 \; \frac{2232234}{3211420} = 114231 = 28$$

Variables (8-9-13-14-19-26)

The Planning Perspective of Regional Types of Agriculture

The agricultural types identified under various hierarchic orders, if compared and bring out the contrast between different regional groups, they speak very well of their regional development, regional imbalance and regional disparity from the planning point of view. But it is a question that how we deal with them?

If we take them for the regional development for a state alone, then we shall have to take their hierarchic orders into account. And, if we are planning for the overall regional imbalance and disparity of resources and their distribution and allocation, there, we will have to go for their 3rd order types on a national scale, i.e. state to state comparison. (State's Master Plan Rajasthan) while, dealing with both the scales, one has to go through the class values (High or low) listed in a coded framework. These values speak for the strong as well as weak points or links. Variables listed with high or very high values certainly are strong ones, in contrast to very low or low values. Thus, developmental process starts with the identification of such variables which need further treatment. Areas may thus, be located and identified and may be put for the plans. Here, there is one point to be noted, that is, we cannot locate the exact amount of the particular resources needed at a place.

Conclusion

In this context, regional planning and regional imbalance are the two apex of the study, which may be carried on simultaneously. In the present scheme of "Agricultural Typology', it is a greatest advantage of the scheme, to measure both the scales, i.e. horizontal and vertical, but, perhaps less realised at national level planning. Therefore, it needs more emphasis on application work. I hope with these words, geographers will go for its more applied values and the application aspects in the field.

REFERENCES

Grigg, D., 1969 : Agricultural Regions of the world, *Eco-geog,* Vol. 45, 2, pp. 95-132.

Hgihsmith R. M., 1966 : Now types of agriculture divided world agriculture into regions, *Foreign Agriculture.*

Kostrowicki, J. and Szozesny R., 1972 : A new approach to Typology of Polish Agriculture (in) Vanzetti Warwas; Poland.

Kostrowicki, J., 1976 : *World Types of Agriculture;* Warsaw, Poland.

Sharma, B. L., 1983 : *Agricultural Typology of Rajasthan An International Scheme;* Pankaj Prakashan, Udaipur; Rajasthan.

———, 1983 : A Typological Analysis in The Rajasthan State; *Geographia Polonica;* vol. 46, Poland, p. 71.

———, 1983 : Agricultural Typology, Case study of the Ajmer district (Raj.) *Geographia Pohonica*; Warsaw; Poland. p. 79.

Singh, Jasbir, 1983 : A typology of Agriculture. The Indian Experience; *Geographia polonica;* Vol. 46; Warsaw Poland p. 49.

Symons, L., 1968 : *Agricultural Geography*. Bell's London, p. 95.

Whittlesey, D., 1936 : Major Agricultural Regions, *Annals of this Association of American Geographer* Vol. 26, pp. 199-240.

11

TRENDS AND CONCEPTS IN AGRICULTURAL GEOGRAPHY*

B. B. SINGH

THE NAMES of Baker (1926), Jonasson (1925, 1926), Jones, (1928-30), Valkenberg (1931-36) and Taylor (1930), who had laid down the structural base to the field by contributing a number of significant articles, particularly published in *Economic Geography*, are initially associated with the origin of the discipline. They normally followed the "Inductive Method" in their studies. Agricultural regionalisation and regional description were the most common theme. A regional scheme was put forward for many countries of the world. A comprehensive scheme of agricultural regions of the world was also published in the same journal. Some of the geographers also adopted this approach at micro-regional level in studying crop variation and its causal factors. Such studies were mostly based on empirical observations. In the opinion of McCarty (1954), Buchanan, (1959) and Reeds (1964) nothing specific could be added to this field till the fifties after the publication of the agricultural region classification scheme of Baker (1926). Reds Preferred to call it a conventional approach. He advised to enrich the themes by applying statistical procedures in analysing the contents.

The study of agricultural geography in India was limited to some regional descriptions before the fifties. Geographers generally studied the regional distributions of main crops of southern India and adopted

---

* The main objectives of this research paper have been : (i) to analyse the evolution of the discipline by highlighting distinctive trends and (ii) to discuss and develop the fundamental concepts in agricultural geography.

commodity approach. Mukherjee (1942) and Dayal (1950) submitted a scheme about agricultural regions of U.P. and Bihar respectively.

There was a distinct departure in content and methodology during 1950-60. Systematic approach replaced regional ones. An overall growth of the discipline is well marked in this decade. The micro-regional model was a novel approach. Birch (1954), Helburn (1957), Scott (1957), Weaver (1954) opened a new chapter by adopting statistical approaches and thus, they may be credited to provide a scientific framework to the subject. Indeed, initiation of a quantitative and depth analysis was a new trend in agricultural geography. Zabko (1957), and others published analytical essays about the historical perspectives of the agricultural system which encouraged many geographers to follow the same approach in subsequent studies. A number of papers were published about the locational aspects of crops during this period although descriptive and regional approaches were still dominant.

Like all other countries of the world, there was a remarkable change in the theme and methodology of Indian agricultural geography in this period. The contributions were obviously different from those in the pre-1950 period in two ways : First, the study of regional distribution and variation of crops was linked with economic and socio-cultural factors. Waring and Parida (1953) evaluated the impact of land tenure and wages on agricultural operations in the Berapali area. Kulkarni (1955) analysed the impact of geographic factors on the distribution of crops of Dharwar district. Bhat (1956) highlighted the agricultural characteristics of Khumbi Valley in Kolhapur district. Mukherjee (1956) concluded that the better agro-economy of the Jats in the upper Ganga-Yamuna Doab is the result of the high percentage of the land devoted to cash crops. Lahiri (1952) established correlation between the yield of cane and soil nutrients in a sample village of West Bengal. Similarly, Ghori (1952) attempted to discuss the distribution and production of coffee with reference to physical correlates and suggested the feasibility of increase in per-hectare yield. Banerjee (1954) ascertained the impact of relief and slope on the tea plantations of West Bengal. Sinha (1954) presented a plan about the irrigation potential of the Maha Nadi Delta. Second, many valuable papers related to agricultural production, foodgrain supply and agricultural planning were also published. Chatterjee (1952) pointed out the scarcity of foodgrains in India and presented a comparative picture of the foodgrain supply of India and the world. Kurian (1952) published research papers about the food problems of India and suggested increase in the acreage under foodgrains and yield. In another research paper, Kurian (1958) submitted a substantive agricultural plan. Ghosh (1954) wrote about agricultural planning of India with special reference to soil conservation.

The post-1960 period is treated as a golden age in the field of agricultural geography. Many conceptual and thought-provoking researches were conducted. The publication of *Agricultural Atlas* by Coppock (1964) was an indication of the availability of huge agricultural data and its applicability for scientific processing. Some of the outstanding features of this period have been elaborated under the heading "Recent Trends".

Recent Trends

1. More emphasis on theoretical and systematic approach and less concern with geographical description and regional approach.
2. Retreat from deterministic interpretation of phenomenon to a probabilistic and behavioural approach.
3. Greater stress on micro-regional and representative studies.
4. Primacy of statistical analysis and synthesis.
5. Inter-disciplinary approach.
6. Regional to typological analysis.

The most significant contribution about the systematic approach was made by Burton (1963) and Bunge (1962). They are credited with laying the theoretical base. Besides these, Mc-Carty (1954), Blaut (1961), Franklin (1965) and Harvey (1966) also published valuable research papers about systematic cores.

Many justified the supremacy of cultural factors. Thus, the theory of possibilism and its applicability was strengthened. Boserup (1965) was one who treated population density as the determinant of agricultural operations. Henshall (1966) termed population as an independent factor in determining agricultural characteristics. Gould (1963) and Walpert (1964) also established the superb role of man in regional agricultural operations.

The role of sample or representative studies on micro-regional level was more stressed. Birch (1960) emphasised the role of sample farm studies in preparing a generalised mapping of farming. Brookfield (1962) deemed it easier to conduct comparative studies based on samples. Mohammad (1978) studied the agricultural geography of the Ghaghara-Rapti Doab on the sampling basis. Carter (1963) and Day (1964) also preferred to conduct sample farm studies in decision-making and resource allocation. Dean (1964), Heller (1964) and Plaxico (1963) treated sample farm studies as the basic tool for policy projection. Weaver (1956) published a model paper by treating the county as a spatial average in agricultural geography. The significance of the sample farm study was duly recognised by Blaut (1953) even before 1960. It may be safely concluded that the study of sample farm was deemed to be essential at the macro-regional level in the post-1960 period.

Statistical approach induced precision, coherence, exactness to derivative facts to this branch. It also enhanced the applicability of the results. Multivariate models were presented on the basis of statistical methods. Particularly, an application of Linear Programming was often used in determining optimum location and allocation of crops/enterprises. The contributions of Cole and King (1968), Curry (1967), Berry and Marble (1969), Haggett (1969), King L. J. (1969), Garrison (1967) and Scoff (1971) have been the most valuable in this regard. Zobler (1961), Enyedi (1967), Robinson, Lindberg, Brinkman (1961), Kostrowicki (1970) and Coppock (1962) also advocated statistical applicability in their researches.

Inter-disciplinary approach is the most significant contribution of this period. The agricultural problems were tackled by specialists of different disciplines in forecasting regional economic planning. Some of the allied disciplines and their experts like agro-economists, rural sociologists, plant ecologists, climatologists, agronomists, land economists participated together to pinpoint the problems and also to chalk out a detailed plan for agro-economic development of a region. This approach has been mainly adopted by Russian geographers.

Recently, there is an obvious shift from "Agricultural Region" to "Agricultural Typology". The concept of Agricultural Typology is the most recent which synthesises all agricultural classification connotations, i.e. agricultural system, agricultural region, agrarian structure, agricultural formation, farming type, etc. It is defined as a set or combination of similar internal socio-economic and cultural characteristics developed and shaped by historical processes in a given set of external physical (relief, climate, soil, vegetation) and other conditions (market, transport, location). Thus, it is a dynamic concept subject to transformation by evolutionary processes and hierarchical character of the spatio-temporal organisation of human activities. The concept of "Agricultural Typology" is different from "Agricultural System" as the latter is treated to be a partial Agricultural Typology. The concept of Agricultural Typology further differs from the concept of Agricultural Region as the latter is a purely spatial concept having fixed boundaries based on differences of a set of agricultural characteristics while the former is a spatio-temporal concept based on a set of similar agricultural characteristics and, thus, may not form a contiguous spatial extension. It is further pointed out that both are hierarchical in character but the concept of Agricultural Region is horizontal in character whereas the concept of Agricultural Typology is of vertical character ranging from low to high or inferior to superior (Singh, 1976), Kostrowicki (1972) may be said to be the father of the concept who devised a code for representing typology :

$$T = S\frac{O}{P}C,$$

where T = Type of agriculture, S = Social and ownership characteristics; O = Organisational and technical characteristics; P = Production characteristics; C = Structural characteristics or enterprise combinations.

Kostrowicki recognised 18 types of world agriculture based on 22 variables. Shafi (1960) used Kendall's ranking coefficient method based on yield in measuring agricultural efficiency of Uttar Pradesh. Bhatia (1967) pointed out some methodological drawbacks in former studies and suggested a better formula by taking into account the yield index divided by the area per cent under different crops. The yield of each crop of the smallest unit have been weighted to regional average. Sinha (1968) measured agricultural efficiency of India. It is based on standard score method. Singh (1972) suggested a new technique of measuring agricultural productivity of Haryana on the basis of per-acre-carrying capacity of land. Shafi (1971) in another research paper determined agricultural productivity of Great Indian Plain by applying Enyedi's formula in a modified form. Hussain (1976) recently adopted a Cash (crop-value) approach in measuring agricultural productivity in Sutlej-Ganga Plains. Singh (1977) adopted statistical approach in the measurement of agricultural productivity of U.P.

Like other countries of the world, significant researches were conducted on these aspects in India also. It may be taken as a period of rare achievement from the viewpoints of method, approach, subject matter and its integration. Conceptual researches were conducted about agricultural efficiency and productivity, cropping intensity, crop combination, crop association, crop concentration and diversification, optimum cropping and agricultural typology. Chauhan (1966) and Tyagi (1972) analysed agricultural intensity. Many Indian geographers tested Weaver's and Doi's formula of crop combination in different regions of the country (Mohammad, 1975; Mohammad and Amani (1970), Bhatia (1965) and Hussain (1970) studied crop concentration and diversification. The names of Mukerjee (1962), Gupta (1968), Ayyar (1969) and Roy (1972) are closely associated with the concepts of agricultural regions. Singh (1975) in his exhaustive critique of the Kostrowicki method of World Typology has pleaded for wide comments regarding its applicability and appropriateness for different countries of the world. He has suggested to reduce 18 types of the Kostrowicki model to nine types of the first order. He further suggested that overall, there is need for a clear perspective for agricultural geography which must retain "Ge" by concentrating mainly on area-characterising and area-differentiating features. Recently, Singh (1977) prepared a taxonomy of Indian agricultural typology. A theoretical analysis was conducted by Singh (1972) about rural settlement and landuse zoning. The publication of *Indian Agricultural Atlas* by Singh (1974) and Tiwari (1970) added much to the discipline.

Principles and Concepts

Principles, concepts and subject-matter of any discipline are mutually inter-related. Principles form the core or pivot round which concepts and subject contents revolve. In other words, subject-matter is derived from the concepts and concepts from principles. According to Morgan and Munton (1971), the fundamental concepts of agricultural geography are similar to economic geography, although it has now emerged as an independent branch. They suggested four basic principle and concepts : (a) Resource utilisation, (b) Comparative advantage, (c) Exchange of goods, (d) Optimum utilisation of the resources.

The author prefers to point out three basic principles of agricultural geography which give rise to seven fundamental concepts. Three principles may be mentioned, such as: (a) Principle of ecological adjustment, (b) Principle of functional relationships, (c) Principle of areal differentiation and organisation. The principle of ecological adjustment is related to the environmental conditions. It deals with the environmental factors as to : (i) how physico-cultural and technological factors affect the spatial distribution of crops, particularly areal differentiation and cohesion, (ii) how the factors partially and wholly determine the farming system of a region, (iii) what is the operational role of these factors in the formation of agricultural typology, and (iv) how factor-analysis has proved to be an aid in understanding the dynamic character of the landscape. The principle of functional relationship is closely linked with some of the aspects as to : (i) what is the nature of agricultural landscape, (ii) what is the level of the development of regional agricultural economy, (iii) how the functional relationship of the agricultural landscape of the region is to be established, (iv) what is the nature of mutual interaction of agricultural economies of different regions of the world.

Ultimately a question arises as to how a distinct regional organisation of agricultural economy is reflected through regional spatial-functional interactions. Thus, agricultural typologies are determined of an integrated agricultural development of the region. In brief, the principle of areal differentiation and organisation is primarily concerned with three concepts : (a) Concept of typologies, (b) Concept of equilibrium and spatial functional interactions, (c) Regional development and planning.

Seven fundamental concepts of Agricultural Geography may be expressed as such:

1. Concept of agricultural landscape.
2. Agricultural landscape is dynamic which experiences continuous change and constant evolution.
3. Agricultural landscape is a function of resource structure, process and stage.
4. Location of agricultural activities.
5. Agricultural region, areal differentiation and organisation.
6. Inter-regional equilibrium and spatial-functional interaction.
7. Regional agricultural development and planning.

Agricultural Landscape

It is a part of economic landscape. It is analysed on the basis of crops raised and farming systems adopted. It is generally marked that several crops with different rankings are grown under a distinct system in a single farming region which are being determined by the physico-cultural factors of the region. The dominance of rice followed by many other crops in subsistence system in the Monsoon countries on the one hand and wheat in commercial grain farming system of the Western countries on the other, are significant examples of prevalent environmental conditions of those areas. Thus, variation in the physico-cultural framework gives rise to distinctive regional agricultural landscape. Agriculture and its operations are visible and predominant elements of the landscape which also reveal different levels of structural frameworks depending upon the areal unit. It is, however, obviously noted that physical factors are more distinctive and operative in determining the agricultural characteristics on macro-regional levels while cultural factors like market, transport, labour, capital and technological adoptions are more operative on micro-regional levels. It is also worthwhile to mention that systems do not vary only at inter-regional but also at intra-regional levels.

Agricultural Landscape Experiences, Continuous Change and Constant Evolution

It is an ever-changing phenomenon. Its evolution is sequential in character. The rate of transformation, in fact, is dependent on the cultural and technological level of the area. The agricultural landscape changes proportionally with the rate of changing human economic activities. In the absence of positive factors, it may remain stagnant or may decline. It is true to say that present is key to past and it supports the fact that the present characteristics of any agricultural landscape are the result of past activities. The landscape bears distinctive imprints of the changing conditions. Hence, an analysis of historical perspective is a necessary prerequistic in understanding the present nature of agricultural landscape. Such analysis is, of course, basically related to transformatory processes of agricultural landscape. Thus, the study of the dynamic nature of agricultural landscape is conducted on the basis of spatio-temporal character of the area.

Landscape is a Function of Resources Structure, Process and Stage

Agricultural landscape at any moment is the product of the inter-relation and interactions of many elements. Among all these, three elements — structure, process and stage — are the most effective. The word "structure" connotes here both the resources bases, i.e. physical (land, climate, soil vegetation) and cultural (man and his economic activities). It is the basic factor in determining the nature of agricultural landscape. The term "process" implies the various ways or agencies operating in the utilisation of the resources. The evolutionary sequence of processes is closely linked with the sequential development of human society. It is the reason that the example of different technological levels ranging from traditional to the most advanced ones are visible in the resource utilisation of any area. There are many spatial examples in the world having both the extremes of technologies in a single region. Keeping in view the historical life span of a landscape, three stages may be set as: (i) Young Stage, (ii) Mature Stage, and (iii) Old Stage. These stages identify the differential level of the processes of resource utilisation. A region where man has recently started exploiting and utilising resources having high farming efficiency may be termed as resembling Young Stage. Latin America and Brazil are the representatives of this stage. Mature Stage is marked in the countries where agricultural growth has been maximised under the most advanced technologies. The U.S.A, Canada, Australia, New Zealand and the USSR are the best examples. Old Stage is observed in those countries where the history of agricultural resource utilisation is the most primitive. The rate of agricultural growth is slow under subsistence level. This stage is, however, marked in most of the Afro-Asian countries (China, India). Besides these, many West European countries also exhibit the Old Stage of the agricultural development. The concept of rejuvenation of landscape is also very much applicable. This process of rejunvention may occur owing to cultural and technological accelerations accompanied by rapid transformation of socio-economic institutional and environmental conditions.

Location of Agricultural Activities

It is based on the concept of distance as an economic factor. It particularly studies the spatial organisation of internal agricultural activities treating distance as an input cost. The concept is focused on three main aspects: (a) location of landuses around market/urban centres, (b) location of crop/enterprises in relation to farmers' settlements, and (c) location of landuses with reference to transport network. Increasing and changing utilities of land reflected by the location of new institutions/references has attracted concerned researchers to analyse the locational changes of agricultural activities. Besides formulating principles and concepts, it is also primarily concerned with the forecast of crop/enterprise locations based on past and present spatial references. Many applied linear programming approach in the optimal locations of future agricultural activities in a region.

Agricultural Region and Area Differentiation and Organisation

Since an early start of the discipline, it has always been the prime aspect of the study of agricultural geography. There is a wide variation of crops, farming systems and agricultural formations at regional and intra-regional levels. Similarities and differentiations of the systems and other associated characteristics induce the concept of agricultural regions. The regional classification schemes applied in different countries are based solely on the socio-economic conditions and technological levels of those countries. For example, any scheme of agricultural regional classification in Indian is primarily concerned with crop combination regions but it is altogether different in the case of advanced countries of the world, like the USA, Canada, Great Britain, New Zealand, etc., where agriculture has attained the form of commercial enterprise. Hence, it is also essential to examine different terms used in this regard like agricultural region, farming region, enterprises, systems, structures and formations, etc. Recently, the term agricultural typology has been used to sum up together all these connotations. It is, however, required to study and examine the agricultural typological classifications for different countries of the world for presenting a developmental planning.

Inter-regional Equilibrium and Spatial-Functional Interaction

It is a novel concept and is based on a very simple notion that all the agricultural regions of the world are functionally inter-linked and mutually interacted. Regional functional interaction operates at both levels — horizontal and vertical — which ultimately determines the nature of spatial organisation. It is, however, correct to say that the concept of inter-regional equilibrium is the result of functional interaction. It is deemed essential to analyse inter-regional similarities and dissimilarities after ascertaining the spatial location of consumers, producers and production factors. The principle of comparative advantages is useful in this regard. The main objectives of such study is to keep a balance among agricultural products keeping in view the regional food and non-food requirements. It also aims at examining inter-regional distribution (production) of crops in the changing politico-socio-economic and technological conditions of different regions. An application of linear programming has been of great utility in forecasting optimal spatial allocations. It is based on two characters: (a) input-output, and (b) spatial equilibrium analysis.

Regional Agricultural Development and Planning

It reveals the applied aspect of agricultural geography. Its ultimate objective is to analyse optimum utilisation of agricultural resources, stressing on maximum production by areal agro-economic integration. It also aims to ascertain the level of agricultural development of different regions. Here, it is necessary to determine the weightage of individual factors and factors in groups as well. The study of resource availability, abundance and scarcity and its utilisation at different cultural and technological levels is the most noteworthy feature. The planning is normally dealt with in two sub-heads : (a) Corrective — short term, (b) Prospective — long term. Recently, the concept of integrated agricultural planning has been introduced in the developmental studies of agricultural geography.

REFERENCES

Ayyar, N. P., 1969 : "Crop Regions of Madhya Pradesh : A Study in Methodology", *Geographical Review of India*, XXXI, (I) pp. 1-19.

Baker, O. E., 1926 : "Agricultural Regions of North America", *Economic Geography*, 2, pp. 459-493.

Banerjee, B., 1954 : "Relief as a Factor in the Location of the West Bengal Tea Garden". *Geographical Review of India*, 16 (2).

Berry, B. L. J. and Marble D. E. (Eds) : *Spatial Analysis : A Reader in Statistical Geography*, Englewood Cliffs, N. J., Prentice Hall.

Bhat, L. S., 1956 : "Agriculture in the Lower Khumbi Valley, Kolhapur, District", *Bombay Geographical Magazine*, 4 (1).

Bhatia, S. S., 1967 : "Patterns of Crop Concentration and Diversification in India", "*Economic Geography*, 41, pp. 39-56,

————, : "A New Measure of Agricultural Efficiency in Uttar Pradesh, India", *Economic Geography*, 43 (3), p. 248.

Birch, J. W., 1954 : "Observations on the Delimitation of Farming Type Regions with

Special Reference to the Isle of Man", *Tran. of the Inst., Brit., Geogrs,* 20, pp. 101-158.

———, 1960 : "A Note on Sample Farm Survey and Its Uses as a Basis for Generalised Mapping". *Economic Geography,* 36, p. 255.

Blaut, J. M., 1953 : "The Economic Geography of One-acre Farm in Singapore" : "A Study in Applied Micro-Geography", *Malayan Journal of Tropical Geog.,* 1, p. 38.

———, 1961 : "The Ecology of Tropical Farming Systems", *Revista Geographic,* 28 (1), pp. 47-67.

Boserup, E., 1965 : *The Conditions of Agricultural Growth : The Economics of Agrarian Change Under Population Pressure,* Chicago Aldine.

Brookfield, H. C., 1962 : "Local Study and Comparative Method : An Example from Central New Guinea", *Annals of the Assn. of Amer Geogrs.* 52 (3), pp. 242-253.

Buchanan, R. O., 1959 : "Some Reflections on Agricultural Geography". *Geography,* 44, pp. 1-13.

Bunge, W., 1962 : *Theoretical Geography,* Lund Studies in Geography, Series C, General and Mathematical Series No. 1.

Burton, L., 1963 : "The Quantitative Revolution and Theoretical Geography", *Canadian Geographer,* 7, pp. 151-162.

Carter, H. C., 1963 : "Representative Farms Guides for Decision-Making", *Journal of Farm Economics,* 45 (5), pp. 1449-55.

Chatterjee, S. P., 1952 : "Food Shortage", *Geographical Review of India,* 14 (2); (1952), India and World Food Supply, Ibid., 14 (2).

Chauhan, D. S., 1966 : *Studies in Utilisation of Agricultural Land,* p. 171.

Cole, J. P., and King C. A. M., 1968 : *Quantitative Geography,* London.

Coppock, J. T., 1962 : "Electronic Data Processing in Geographical Research", *Professional Geographer,* (14), (4), pp. 1—4.

———, 1964 : *Agricultural Atlas of England and Wales,* London.

Cury, L., 1967 : "Quantitative Geography", *Canadian Geographer,* 11, pp. 265-279.

Day, L. M., 1964 : "Use of Representative Farms in Studies of Inter-regional Competition and Production Resources", *Journal of Farm Economics,* 45 (5), pp. 1438-1444.

Dayal, P., 1950 : "Agricultural Harvest in Bihar", *Indian Geographical Journal,* 25 (1, 2, 3 and 4).

Dean, G. W. and Benedicts, D. E. M., 1964 : "A Model of Economic Development for Peasant Farms in Southern Italy", *Journal of Farm Economics,* 46(2), pp. 295-312.

Enyedi, G., 1967 : "The Changing Force of Agriculture in Eastern Europe". *Geographical Review,* 57, pp. 358-72.

Franklin, S. H., 1965 : "System of Production: System of Appropriation". *Pacific Viewpoint,* 6(2), pp. 145-166.

Garrison, W. L. and Marble D. C. (eds): *Quantitative Geography,* N. W. Uni, Studies in Geog., p. 13.

Ghori, G. K., 1952 : "Cotton in Mysore", *Indian Geographical Journal,* 28, p. 182.

Ghosh, S., 1954 : "A Critique on National Geographical Planning in India", *Indian Geographical Journal* 20 (1).

Gould, P. R., 1963 : "Man Against His Environment: A Game Theoretical Framework", *Annals of the Ass. of Amer Geogrs.,* 53, pp. 290-297.

Gupta, P. Sen and Sdayuk, G., 1968 : "Economic Regionalisation of India, Problems and Approaches", *Causes of India,* Monographs Series. 1, (9), pp. 101-138.

Haggett, P. and Charley R. J., 1969 : *Network Analysis in Geography,* London, Edward

Arnold.

Harvey, D. W., 1966 : "Theoretical Concepts and the Analysis of Agricultural Landuse Patterns in Geography", *Annals of the Assn. of Amer. Geogrs*, 56, pp. 361-374.

Helburn, N., 1957 : "The Bases for a Classification of World Agriculture". *Professional Geographer*, 9, pp. 2-7.

Heller, C. F., 1964 : "The Use of Model Farms in Agricultural Geography", *Professional Geographer*, 16 (4), pp. 20-23.

Henshall, J. D., 1966 : "The Demographic Factor in the Structure of Agriculture in Barbados", *Trans. of the Inst of Brit. Geogrs.*, 38, pp. 183-195.

Hussain, M., 1970, : "Pattern of Crop Concentration in Uttar Pradesh", *Geographical Review of India*, XXXII, (3), pp. 169-85.

Hussain, Masjid, 1976 : "A New Approach to the Agricultural Productivity Regions of the Sutlej-Ganga Plains of India", *Geographical Review .of India*, 38 (3), pp. 230-236.

Jonasson, O., 1925-26 : "Agricultural Regions of Europe", *Economic Geography*, 1, pp. 277-344 and 2, pp. 19-48.

Jones, C. F., 1928-30 : "Agricultural Regions of S. America", *Economic Geography*, 4, pp. 1-30, 159-186 and 267-294; 5, pp. 109-140, 277-307 and 390-421; 6, pp. 1-36.

King, L. J., 1969 : *Statistical Analysis in Geography*, Englewood Cliffs, N. J. Prentice-Hall.

Kostrowicki, J., 1970 : "Types of Agriculture in Poland : A Preliminary Attempt at a Typological Classification", *Geog. Polonica*, 19, pp. 99-110.

———, 1972 : "A Preliminary Attempt at a Typology of World Agriculture", *International Geography*, 22nd Intern. Geographical Congress, Montreal, pp. 1097-1100; Kostrowicki : 1964: "Geographical Typology of Agriculture : Principles of Methods", *Geogr. Polonica*, 2, pp. 1959-67.

Kulkarni, L. Y., 1955 : "Agricultural Patterns of Dharwar District", *Indian Geographical Journal*, 30 (3).

Kurian, G., 1952 : "Food Problems in India", *Indian Geographical Journal*, Subramanyam Memorial Volume.

———, 1958 : "Agriculture Planning in India", *Indian Geographical Journal*, 31 (182).

Lahiri, R., 1952 : "Soil Conditions in Sugarcane Cultivation", *Geographical Review of India*, 14 (3).

McCarty, H. H., 1954 : "Agricultural Geography", *in American Geography : Inventory and Prospects*, James, P. E. and Jones, C. F. (Ed.) Syracuse, pp. 258-77.

———, 1954 : "An Approach to Theory of Economic Geography", *Economic Geography*, W. E. 30, p. 96.

Mohammad, N., 1975 : "Crop Association Regions in Ghaghra-Rapti Doab", *The North-eastern Geographer*, Nos. 1 and 2, Vol. VII, Gauhati.

———, 1978 : *Agricultural Landuse in India*, Inter India Pub., New Delhi.

Mohammad, N. and Amani, K. C. 1970 : "Crop Combination in the Trans-Ghaghara Plain", *Geographical Review of India*, No. 1, Vol. XXXII, Calcutta.

Morgan, W. B. and Munton, R. J. C., 1971 : *Agricultural Geography*, Methuen & Co. Ltd., London.

Mukerjee, A. B., 1956 : "Agricultural Geography of Upper Ganga-Yamuna Doab", *Indian Geographer*, 11 (2).

———, 1962 : "Agricultural Regions and the Geographical Planning for Indian Agriculture", *National Geographical Journal of India*, p. 5

Mukerjee, B. N., 1942 : "Agricultural Regions of U. P. " *Geographical Review*, 4 (1), Calcutta.

Plaxico, J. S. and Tweeten, L. G., 1963 : "Representative Farm Policy and Projection Research", *Journal of Farm Economics,* 45 (5), pp. 148-65.

Reeds, L. G., 1964 : "Agricultural Geography : Progress and Prospects", *Canadian Geographer*, 8 (2), pp. 51-63.

Robinson, A. H., Lindberg J. G. and Brinkman L. W., 1961 : "A Correlation and Regression Applied to Rural Farm Population Densities in the Great Britain", *Annals of the Assn. of Amer. Geogrs*, 51, pp. 211-21.

Roy, B. K., 1972 : "An Approach to Rationalisation of Types of Farming in India", *"Economic and Socio-cultural Dimensions of Regionalisation*, Indo-USSR Collaborative Study, Census Centenary Monograph No. 7, pp. 227-51.

Scoff, A. J., 1971 : *Combinational Programming, Spatial Analysis and Planning,* Methuen & Co. Ltd., London.

Scott, P., 1957 : "The Agricultural Regions of Tasmania", *Economic Geography*, 33, pp. 109-21.

Shafi, M., 1960 : "Measurement of Agricultural Efficiency in Uttar Pradesh", : *Economic Geography*, 36 (4); Shafi, M. 1962: "Agricultural Efficiency in Relation to Landuse Survey in Uttar Pradesh", *Geographical Outlook*, 3 (1).

———, 1971 : "Measurement of Crop Productivity in India", in *Studies in Applied and Regional Geography*, Shafi, M. and Raza, M. (eds.) pp. 97-113; Shafi, M. 1972 : "Measurement of Agricultural Productivity of the Great Indian Plains", *The Geographer*, Vol. XIX, No. 1, pp. 4-13.

Singh, B. B., 1972 : "Villages Of Baraut Block : A Composite Study in Landuse and Settlement, in Rural Settlement in Monsoon Area". *"Proceedings of I. G. U. Symposium*, R. L. Singh, (Ed.), pp. 291-285.

Singh, Jasbir, 1972 : "Spatio-Temporal Developments in Landuse Efficiency in Haryana State", *Geographical Review of India*, Calcutta, 34 (4), pp. 312-316.

———, 1974 : *Agricultural Atlas of India*, Kurukshetra.

Singh, L. R., 1976 : "New Perspective" in Agricultural Geography, *National Geographer*, Vol. XI, (2), pp. 113-122.

Singh, Surendra and Chauhan, V. S. 1977 : "Measurement of Agricultural Productivity — A Case Study of Uttar Pradesh", *Geographical Review of India*, Vol. 39 (3) pp. 222-231.

Singh, V. R, 1977 : "Taxonomy of Indian Agricultural Typology", *Uttar Bharat Bhoogol Patrika*, Vol. 13, 182. pp. 17-25.

Sinha, B. N., 1954 : "Irrigation in Maha Nadi Delta", *Indian Geographical Journal,* 29 (4).

Sinha, B. N., 1968 : "Agricultural Efficiency in India" : *Geographer.*

Taylor, G., 1930 : "Agricultural Regions of Australia", *Economic Geographer*, 6, pp. 109-139 and pp. 213-242.

Tiwari, P. S., 1970 : *Agricultural Atlas of Uttar Pradesh.*

Tyagi, B. S., 1972 : "Agricultural Intensity in Chune Tehsil Distt., Mirzapur, U. P", *National Geog. Journal of India*, 18 (1), pp. 42-48.

Valkenburg, S., Van. 1931-36 : "Agricultural Regions of Asia", *Economic Geographer,* 7, pp. 217-237 and 8, pp. 109-133; 9, pp. 1-18, 109-135; 10, pp. 14-34; 11, pp. 227-246, 225-37; 12, pp. 27-44, 231-249.

Waring, P. A. and Parida, A. C., 1953 : "Agriculture of Berpali Area, Orissa, *Geographical Review of India*, 14 (4), and 16 (1).

Weaver, J. C., 1954 : "Crop-combination Regions in the Middle-West", *Geographical Review*, 44, pp. 1-13.

———, 1956 : "The County as a Spatial Average in Agricultural Geography", *Geographical Review*, 46 (4), pp. 536-565.

Worpert, J., 1964 : "The Decision Process in Spatial Context", *Annals of the Assn. of Amer. Geogrs.*, 54 (4), pp. 537-538.

Zabko-Potopowicz, A., 1957 : "The Development of Geography of Agriculture Since World War I", *Przeglad Geogrficzny* (Warsaw), 29, (1), pp. 21-46.

Zobler, L., 1961 : "A New Areal Measure of Food Production Efficiency", *Geographical Review*, 51, 549-69; Zobler, L. 1958 : "The Distinction between Relative and Absolute Frequencies in Using Chi Square for Regional Analysis", *Annals of the Assn. of Amer. Geogrs,* 48, pp. 459-67.

12

THE ECOSYSTEM CONCEPT AS A FRAMEWORK FOR STUDIES IN AGRICULTURAL GEOGRAPHY

PATRICK ARMSTRONG

An ecosystem is a segment of nature with its included organisms and their environment. Such a segment may be of any magnitude to facilitate study or understanding.

ALTHOUGH THE term ecosystem had been introduced two decades previously (Tansley, 1935, 1939), Fosberg's (1957) definition, quoted above, is perhaps the most widely used by geographers. The ecosystem idea enjoins on the inquirer to make a "whole complex of nature" his unit of study. Such a system includes animals, plants and micro-organisms within an area, as well as water, atmosphere, soils and, perhaps, man and his activities, together with the relationship that exists between these.

Although originally a biological idea, the notion has been enthusiastically adopted by geographers, particularly those interested in landuse, and by practical resource managers (Van Dyne, 1969). Stoddart (1965, 1967) was amongst the earliest geographers to stress the ecosystem concept's usefulness to geographical inquiries:

> It links geography with the mainstream of modern scientific thought, in systems analysis and related disciplines, and opens up as yet unexplored possibilities in the application to geography of the whole field of information theory and communications techniques.
>
> . . .the ecosystem is potentially capable of precise mathematical structuring within a theoretical framework, a very different matter

from the tentative and incomplete descriptions of highly complex phenomena which all too often pass for geographical "synthesis".

The idea that landuse should be seen as ecosystem modification was explored by I. G. Simmons in a paper entitled "Ecology and landuse" in 1966. Subsequently (1970) he suggested that landuse ecology might form a central theme in biogeography. In his 1970 paper, and in his important book, *The Ecology of Natural Resources*, he develops this idea : he places landuses such as wildlife conservation, water catchment, forestry, agriculture and mining on a continuum of ecological manipulation. He describes resources processes, the mechanisms of exploitation in ecological terms.

The author, with many others geographers, has attempted to develop a "landscape ecology" (Armstrong, 1970, 1971, 1973, 1975). Landscapes are seen as *mosaics* of different ecological communities. Working in the heterogeneous agricultural countryside of eastern England, he illustrated the important ecological links that exist, and have existed for many centuries, between the arable farmlands, woods, healths, wetlands and pastures of a tessellated countryside. It was these links, and the flows of matter, energy and information they represented that contribute to the long-continued stability of some types of rural environment.

Two decades earlier, Dice (1955), a zoologists, seems to have had a rather similar view of agricultural man — an organism enmeshed in a network of ecological relationships. He wrote of tribal and homestead ecosystems as well as village- town- and city-centred ecosystems, and commented:

> A human community consists not only of men, women and children, but as a minimum, it includes also certain plants that supply food to man, either directly or through animals. Every human community is actually much more complex than this: it includes food chains of numerous kinds, domestic animals and perhaps crop plants, parasites, diseases and numerous other species of organisms, some of which serve as regulatory mechanisms or perform other ecological functions.

The two weapons in the ecologist's conceptual armoury, most frequently borrowed by those interested in agricultural geography are those of the *energy-flow* and *biogeochemical circulation* models. The purpose of much of the remainder of this chapter is to illustrate how these two pieces of intellectual hardware have been utilised in attacking certain problems of land and resource use.

Energy Flows in some Natural and Agricultural Ecosystems

All ecosystems are characterised by a continuous through flow of energy. In natural communities energy from sunlight is "caught" by photosynthesis, passed along food-chains and eventually released to the environment in the degraded form of heat (Fig 12.1).

The rate at which photosynthesis leads to a built-up of organic matter in an ecosystem (or part of an ecosystem) is known as biological productivity. The term net primary productivity is sometimes conveniently used for the accumulation of dry-weight biomass per unit time by the plants in a community.

Sufficient studies have now been undertaken for reasonably accurate estimates to be made of the net primary productivity of a large number of natural and artificial ecosystems. In natural terrestrial environments plant productivity varies from a value of about 0.25 $g/m^2/day$ on the arctic tundra, through a value of about 1.4 for boreal forest to approximately 9.5 $g/m^2/day$ for tropical rain forests (Gerloff, 1963). The primary productivity of cultivated crops varies from about 0.9-1.8 for grains to an exceptional value of 25 for sugarcane plantations in exceptional circumstances (Westlake, 1963). (These "dry-weight biomass" productivity figures can be converted into values for energy fixation by multiplying by a factor for the mean energy content of plant tissue.)

More detailed inquiries are necessary to produce energy flow diagrams such as that given by Teal (1962) in his study of a Georgia salt marsh i.e. quantified versions of Fig. 12.1. He was able to evaluate the energy capital stores in plant tissues (in the *spartina* and alage), herbivores (insects and crabs), carnivores (e.g. mud-crabs and spiders) and decomposer organisms, and then to estimate the rates of energy flow between these stores. MacFadyen (1964) was amongst the first to build up a similar diagram of an artificial ecosystem. He portrayed the flow of energy through a meadow subjected to grazing by beef cattle, estimating that the bullocks consumed about one-seventh of the available forage, and of this much is excreted, only partly digested. Of the original net primary production of 5,100 $kcal/m^2$, the yield of bullock tissue was a mere 30 $kcal/m^2/year$. Another energetics study of

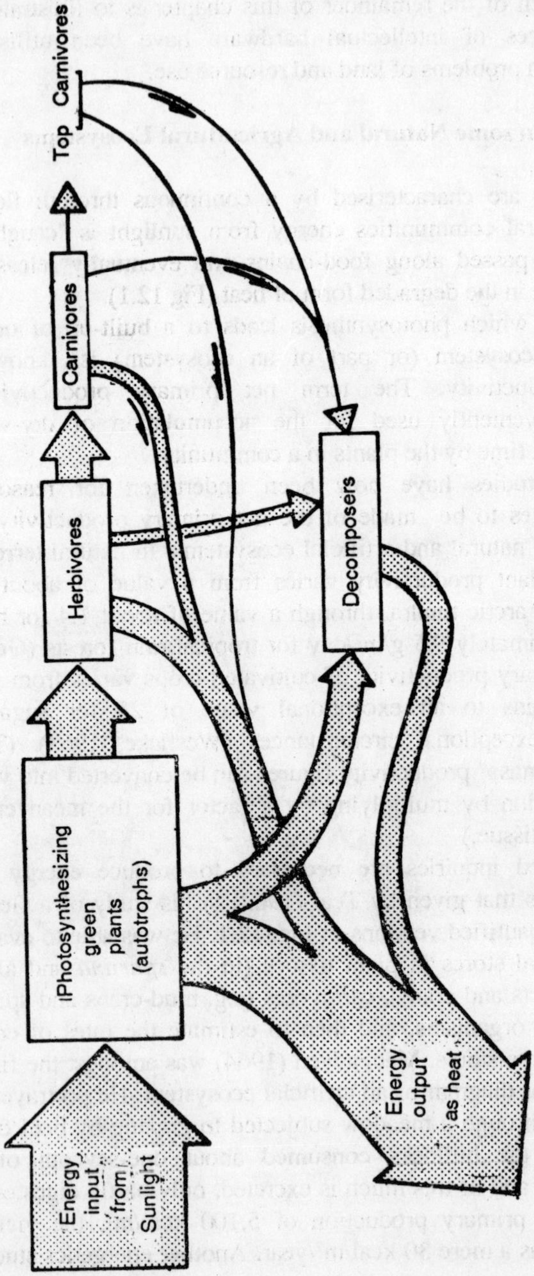

FIG. 12.1: Flow of energy through a natural ecosystem (diagrammatic).

about the same date which had clear landuse implications was Hopkins' (1965) investigation of savanna burning in Nigeria. He evaluated primary productivity, making estimates of the amounts of energy held in different parts of the ecosystem—detritus, aerial shoots and so on– and considered the flow of energy within the system and its disruption on burning.

A very useful concept in the description of the flows of energy in agricultural systems is the *energy subsidy* of H. T. Odum (1971). The highly sophisticated agricultures of western Europe, North America and Australasia (Plate 12.1) are only as productive (in terms of energy yield to man) as they are because they are heavily subsidised by energy inputs from fossil fuels (and increasingly nuclear power) via, for example, petrol and diesel oil for vehicles, tractors and combine-harvesters, electricity for farm machinery, and agricultural chemicals.

The energy subsidy concept may be used to arrive at a convenient classification of landuse and agricultural systems.

Low Energy Cultures

Here there is little non-solar energy inputs other than that of human toil. Besides the now virtually extinct hunting and collecting communities, examples include the shifting cultivators of South East Asia, and pastoral nomads.

Medium Energy Cultures

In these there is an energy input from wind and water-power for purposes such as drainage, irrigation and food-processing. There may also be considerable use of draught animals for transport and in cultivation. The peasant economy that existed in Europe and Asia in the Medieval period provides a convenient example.

High Energy Cultures

Agricultural ecosystems in many areas of, for example, the Canadian prairies (Plate 12.1), the West Australian wheat belt, or the Fenlands of East Anglia in lowlands Britain are, in ecological terms, extremely simple, yet highly productive. This combination is only maintained through vast inputs of fuel, pesticides, fertilisers and energy-using machinery in every aspect of cultivating, harvesting and marketing.

PLATE 12.1: A landscape of high energy-subsidy agriculture: Farmlands in Saskachewan, Canada. The unnatural simplicity and high yields of a wheatfield ecosystem are maintained by massive inputs of energy in the form of fuels, electricity, fertilisers and pesticides. — *Information Canada Photo.*

The distinction is not clear-cut. Fragments of all three types can be identified in many an Indian or African village. The small holder who planted and tapped the half-dozen rubber trees in the tiny clearing in the tropical forests of the West Malaysian State of Selangor shown in Plate 12.2 reached them on foot, used an absolute minimum of chemicals and equipment, but probably sent his latex to be sold by the diesel-driven truck.

A number of interesting studies have been carried out on the energetics of primitive agriculture, and other forms of subsistence (i.e. low-energy cultivating in the above terminology). These usually involve the estimation of the amount of energy used in food production and the comparison of this with the energy content of the food yielded. Examples of this type of investigation include those of Rappaport (1971), Morren (1977) and Bayliss-Smith (1977). Rappaport, for example, gives the following values for the energy required to convert an area of upland forest in New Guinea into a garden plot :

Clearing underbrush	56,628	Kilocalories/acre
Clearing trees	22,628	Kilocalories/acre
Fencing garden (to keep out pigs)	34,164	Kilocalories/acre
Weeding and burning	18,968	Kilocalories/acre
Placing soil retainers	14,476	Kilocalories/acre
Planting and weeding until end of harvest	180,336	Kilocalories/acre
Other maintenance	46,000	Kilocalories/acre
Sweet potato harvest	44,825	Kilocalories/acre
Taro harvest	5,608	Kilocalories/acre
Cassava harvest	2,184	Kilocalories/acre
Yam harvest	15,700	Kilocalories/acre
Cartage	119,764	Kilocalories/acre

One of the objectives for such studies is to obtain a value for the "efficiency" of the system in terms of energy yield, i.e., the ratio of energy output (food produced) to energy input (work involved in producing food), various terms have been suggested for this ratio. M. Harris (1971) suggested "techno-environmental efficiency", but perhaps Rappaport's "energy-yield ratio" is simpler and more easily understood. Rappaport (1971) in relating the values given above to food output in New Guinea, obtained an average energy-yield ratio of

PLATE 12.2: Rubber trees surrounded by forest in Selangor, Malaysia. The tiny holding is worked with simple technology although the product is transported by truck. — *Photo: Patrick Armstrong.*

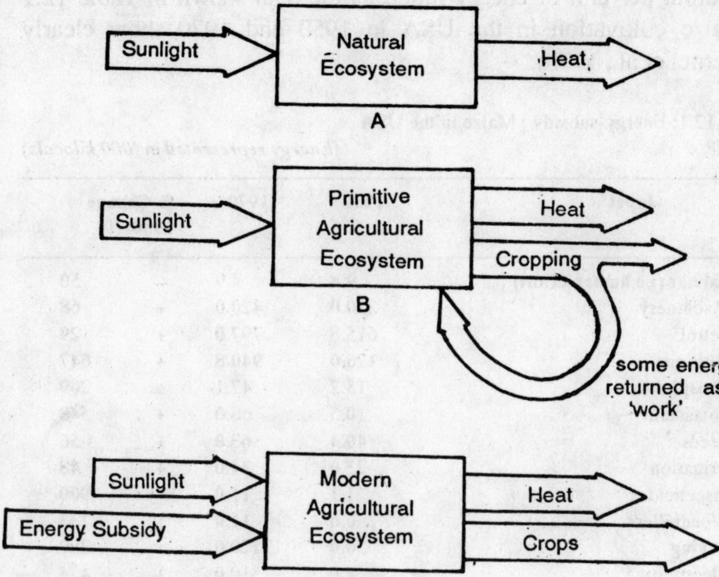

FIG. 12.2: Comparison of the energetics (Energy-flow dynamics).
A—Natural ecosystem, B—Simple cultivation system without energy subsidy, and C—Modern agricultural ecosystem with massive energy subsidies in the form of fuel, electricity, chemicals, etc.

about 16 : 1; Harris quotes 9.6 : 1 for a group of Kung bushmen (hunters and collectors) in the Kalahari, 9.8 : 1 for the Maring (another group of shifting cultivators and pig breeders in New Guinea) and 11.2: 1 for Genieri (cereal and peanut cultivators in the Gambia, West Africa).

These ratios only have meaning when compared with equivalent values for technological "high energy" agriculture, in which energy input through photosynthesis and human work is supplemented by massive infusions of energy in the form of fuel, agricultural chemicals, and so on. Of course the yield of food energy produced *per unit area* is enormously increased (as is the yield per worker), hence the relatively high biological productivity found for some artificial ecosystems (see above). But the energy-yield ratio *declines*. Moreover, the higher the energy input, the more inefficiency the system becomes in terms of food output per unit of energy input, as the data shown in Table 12.1 for maize cultivation in the USA in 1950 and 1970 show clearly (Pimpernel et al., 1973).

TABLE 12.1: Energy-subsidy : Maize in the USA

(Energy represented in '000 kilocals)

Input	1950	1970	% Change 1950-70	
1. Labour (i.e human effort)	9.8	4.9	—	50
2. Machinery	250.0	420.0	+	68
3. Petrol	615.8	797.0	+	29
4. Nitrogen	126.0	940.8	+	647
5. Phosphorus	15.2	47.1	+	209
6. Potassium	10.5	68.0	+	548
7. Seeds	40.4	63.0	+	56
8. Irrigation	23.0	34.0	+	48
9. Insecticides	1.1	11.0	+	900
10. Weedkillers	0.6	11.0	+	1,733
11. Drying	30.0	120.0	+	300
12. Electricity	54.0	310.0	+	474
13. Transportation	30.0	70.0	+	133
14. Total input	1,206.4	2,896.8	+	140
15. Maize output	3,830.4	8,164.8	+	113
16. Energy yield ratio, i.e. 15/14	3.18	2.82	—	11

To take this even further, Lawton (1973) estimated, for wheat-farming in the USA, that energy subsidy levels of 3,940 × 10^3 kcals per acre existed, giving an energy yield ratio of 1.42 : 1, and McFarlane (1974), looking at British agriculture as a whole, for 1973, found a ratio of 0.40: 1. In other words, *more than twice as much energy goes into the agricultural ecosystem along the energy subsidy routeway, as leaves it as crops!* Gifford and Millington's (1975) conclusions for Australian agriculture are in broad agreement. They included the energy costs of getting food "from the farm gate to the dining table" and quote data that suggest a ratio of 0.20 : 1.

The implication of these values, in a time of decreasing energy availability in many countries, and increasing energy costs in almost all are extremely serious.

"Cereals used to be made of sunlight, now they are made of oil"

(After Odum, 1971).

Biogeochemical Cycling

The basic facts of nutrient circulation, the movement of chemical elements between the different phases of the ecosystem — atmosphere, soil, green plants, dead organic matter, decomposer organisms — have long been established; it was Leibig in 1840 that enunciated the principles that control the nutrition of agricultural crops. The term *biogeochemical cycle* was coined by Vernardskii in 1934. (It cannot be too strongly emphasised that while there is a *one-way through-flow* of *energy* in ecological systems, chemicals — matter — *circulate*. While there are, of course, inputs to and outputs of various chemical elements from an ecosystem, the predominant movement is a circular one. The familiar nitrogen cycle diagram (Fig. 12.3) illustrates this well.

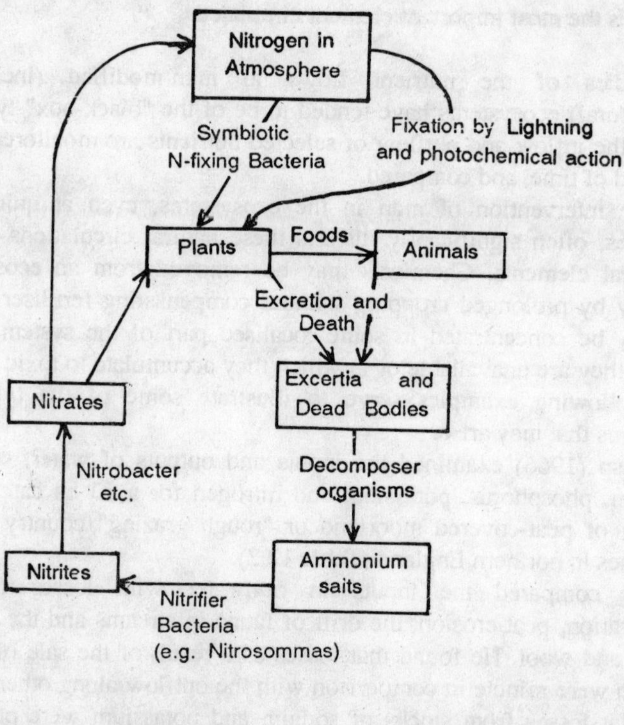

FIG 12.3: A biogeochemical cycle.

The circulation of nitrogen. Similar diagrams could be drawn for other elements such as oxygen, sulphur, carbon, phosphorus, calcium, etc.

Rodin and Basilevic (1967) have distinguished five major element circulation patterns in natural ecosystems:

1. Ecosystems in which the movement of nitrogen dominates that of all other elements, e.g. tundra areas.
2. Ecosystems in which nitrogen circulation, while important, is not quite so pre-eminent over the other nutrients, e.g. boreal (coniferous) forest, sub-tropical forest.
3. Ecosystems in which calcium is the most important element circulated, e.g. mesophytic (temperate broad-leaved) woodland.
4. Ecosystems in which silicon is the most common element in "ash" fractions (i.e. that part of the biomass that remains on burning), along with iron and aluminium, e.g. tropical rain forest, savanna, steppe, some deserts.
5. "Harsh" deserts — a few extreme ecosystems in which chlorine is the most important element circulated.

Studies of the nutrient flows in man-modified (including agricultural) ecosystems have tended to be of the "black-box" type, in which the inflow and outflow of selected nutrients are monitored over a period of time, and compared.

The intervention of man in the ecosystems, even at quite low densities, often significantly distorts these natural circulations of the chemical elements. Chemicals may be removed from an ecosystem entirely by prolonged cropping without compensating fertiliser input, or may be concentrated in some localised part of the system, from which they are unavailable or in which they accumulate to toxic levels. The following examples serve to illustrate some of the types of problems that may arise.

Crisp (1966) examined the inputs and outputs of water, sodium, calcium, phosphorus, potassium and nitrogen for a 83 ha catchment, mainly of peat-covered moorland or "rough grazing" country in the Pennines in northern England (Table 12.2).

He compared the inputs in rainwater with losses through evaporation, peat erosion, the drift of fauna in streams and the sale of sheep and wool. He found that losses as a result of the sale of sheep protein were minute in comparison with the outflow along other routes and that losses from stocks of sodium and potassium were probably largely being replaced through the weathering of underlying rock. In the case of calcium this was certainly the case. However, losses of nitrogen were more serious; about 80 per cent of the 9.48 kg/ha/year net loss was the result of peat erosion.

TABLE 12.2: **Nutrient budget for 83-hectare catchment at 760 m in Pennine Hills, Northern England**

	Na	K	Ca	P	N
Input (Kg/yr)					
Rainwater	3778	916	4864	71	1467
Output					
Streamwater	3775	744	4461	33	244
Peat erosion	23	171	401	37	1214
Sheep and wool	0.16	0.44	1.58	0.98	4.4
Net Loss					
Kg/ha/yr	20.01	7.97	49.68	0.17-0.40	9.8

The drainage basin ecosystem concept has been greatly extended by Bormann and Likens (1969) and their co-workers at the Hubbard Brook Experimental Forest in New Hampshire, USA. They compared the nutrient budgets of a forested catchment with a cover of maple, beech and birch, with a similar basin that had been clear-felled. They measured the rates of nutrient loss following clearance by monitoring the levels of calcium, magnesium, sodium, potassium and nitrogen in run-off. In the seasons following clearance the cutover system showed a net loss of 52.8 kg/ha of nitrogen as compared with a net gain (as the result of nitrogen fixation by bacteria) of 4.5 kg/ha in the case of the unaltered forest ecosystem. Increases in calcium, magnesium, sodium and potassium concentrations in stream water occurred in concert with the increase in nitrogen, and net losses of these elements were about 9, 8, 3 and 20 times greater, respectively, than those from undisturbed communities.

The clearing of forested land (for agriculture or otherwise) obviously results in the profound transformation of the hydrology and biogeochemical circulation patterns of whole regions. Some Australian examples will serve to illustrate this further.

With the removal of the eucalyptus woodland, the proportion of the water input evaporated or transpired from vegetation declines, and ground discharge increases. The lateritic soils of certain parts of south-western Australia have a high soluble-salt storage, and increased discharge results directly in increased salinity of streams. Forest clearance for farmland has thus been the immediate cause of the build-up of slats (principally, but not exclusively, sodium chloride) to levels of 600—900 parts per million (ppm) in some water bodies. In some streams of the West Australian wheat belt seasonal values of 20,000 ppm have been reported. In valley-bottom sites salt-scales form, making the soil toxic to plant growth and thus the land becomes agriculturally unproductive (Conacher and Murray, 1973; Broughton, 1976).

Studies in several parts of Australia have shown that usually about half the total phosphorus in eucalyptus forest ecosystems is in the above-ground living material. The regular removal of timber results, in the long term, in the impoverishment of the soils; Australian soils are already, naturally, amongst the most infertile in the world. One study revealed that 50 per cent of the above-ground potassium, 60 per cent of the calcium and 17 per cent of the phosphorus are in the bark. If this is not trimmed from the logs before their removal (and under modern forestry conditions it often is not), there occurs a major outflow of nutrients from the systems (Senate Standing Committee, 1977).

Burning, accidental or deliberate, has important ecological and landuse implications in many parts of the world. (The energy-flow implications of African savanna burning have already been mentioned.) Where burning occurs, the inquirer is presented with particular difficulties in the compilation of a nutrient budget. Chapman's (1967) careful study of the nutrient circulation and litter accumulation in an area of lowland heath in Dorset (southern England) was, however, able to reveal that relatively little nitrogen and phosphorus was lost from the heather (*Calluna*) litter, and also that although there was some depletion of these nutrients following burning, supplies of calcium, magnesium, potassium and sodium were capable, in the long term, of being replaced by precipitation.

The writer (Armstrong 1970, 1971, 1972, 1973, 1975), in his consideration of the mosaic of heathland, marsh and arable as it existed in east Suffolk (in eastern England) in the 17th, 18th and 19th centuries, has looked at the lowland heath ecosystem in a slightly different way. He saw the heathland and the surrounding landuse types as an elaborate network of pathways distributing calcium, phosphorus and nitrogen around an estate ecosystem. The sheep fertilised the arable lands prior to cultivation, crag and coprolites (the latter consisting mainly of calcium phosphate) were worked on the heaths and applied to the arable lands, and from the 18th century onwards clover was grown and, as a result of the symbiotic bacteria associated with its roots, served as a major input of nitrogen to the living part of the system. A typical estate in east Suffolk thus represented a remarkably stable ecosystem, many essential features persisting for several centuries.

Similarly, the human-centred ecosystems studied by Geertz (1963) in the tropics of South-East Asia — *swidden* or shifting cultivation and wet-rice culture — also represent systems in equilibrium where inputs and outputs over the long term were approximately in balance. Greetz's thesis was that while most studies of shifting cultivation have concentrated on its deleterious effects on the soil, it is most profitably viewed against the backdrop of the tropical rain forest ecosystem it replaces. In both systems matter circulates in the vegetation and upper soil-layers with great rapidity. Burning channels the nutrients locked up in the large green autotrophs of the forest — the trees — into food crops. The total biological productivity is reduced, but the proportion available to man is increased. It is only when the population density of the area increases to a high level, so that the nutrient status of the soil has insufficient time to recover between cultivations, that the symptoms of instability — soil erosion, flooding, widespread crop failure and disease — become manifest. Geertz also emphasised the role of water in the paddy fields as a vehicle for the channelling of nutrients towards man; in it mineral nutrients are carried in from outside, straw is decomposed, blue algae fix nitrogen and through the circulation of the water the soil is aerated. (Rice-field ecology is discussed in more detail below).

The conclusions drawn by Geertz on the basis of his work on shifting cultivation in Indonesia are in general supported by more recent studies by D. R. Harris (1971) in the forests of the Upper Orinoco in Venezuela. Although some anomalies appear, comparisons of the nutrient status of the soils in forests and *conuco* (cleared areas in which a variety of vegetable crops are grown) sites show that available phosphorus and exchangeable calcium and magnesium increase following clearance and burning, and that in general the soil fertility declines following cultivation.

The long-continued application of a limited range of cropping and land-management techniques may eventually result in so great a bleeding away of certain nutrients that the system becomes unstable. Countless instances could be cited : illness in sheep in southern

Scotland caused by cropper deficiency; decline in pineapple yields on plantations in Hawaii as a result of the depletion of the boron stocks in the soil. Brief details will be given of two further examples from lowlands England. Rabbits were taken from the study warrens of Breckland in west Suffolk and Norfolk for at least 600 years (Armstrong 1970, 1971, 1975). Nineteenth century evidence suggests that a typical yield was of the order of 25 rabbits/hectare/year. This represents a substantial outflow of calcium. Phosphorous and nitrogen, and a 19th century commentator (Simpson 1895) wrote of warrens that he had found.

> . . .number of examples (or rabbit warrens that). . .disregard the most ordinary rules of stock-farming, and of (the) pasturing the ground with rabbits till the pasture had become incapable of sustaining them in health or condition any longer, without any attempt even to restore the fertility by putting back an equivalent, in the shape of manure, for the flesh and bones which animals had carried away on their backs for years in succession, representing the richest ingredients of the soil.

There is evidence, from both Breckland and elsewhere, that eventually the rabbits became small and diseased.

Also in eastern England Rackham (personal communication, 1969, 1971, 1976) has investigated the history and effects of coppicing in the ancient woodlands of Cambridgeshire, and from a study of documents has shown that some of them have been regularly cut on a 10—15 year cycle basis, for many centuries. (Coppicing is the cutting of the trunks of the majority of the trees just above ground level at regular intervals; the woodland is then allowed to regenerate from the stools.) The main coppice species, ash *(Fraxinus excelsior)* and hazel *(Corylus avellana)* have a higher phosphate content than the other woodland trees, and Rackham wrote: "The coppice was not young when the proportion of twigs and bark, which would have a higher phosphate content than timber, was high; and the whole of this material was removed so that little or none of this phosphate could have found its way back to the soil". The only was in which it could have been replaced was by the long-term weathering of the chalky till (a glacial deposit) on which the woods are situated. It seems that the present shortage of phosphate in these damp woodlands and the rarity in them of phosphate-demanding plants may be the result of long continued coppicing. The total phosphate content (measured as P_2O_5) in the top 0.3 m of the soil — this being the rooting depth in the water parts of these woods — is about 4,400 kg/ha^{-1} and on the basis of analysis from elsewhere. Rackham calculated that the annual accumulation of phosphate by coppice is about 3.6 kg/ha^{-1}. Since coppicing has continued for at least 700 years considerable depletion must have occurred. There is fragmentary evidence from growth-rings in timber in ancient buildings that trees in lowland England grew more rapidly in former centuries than at present.

Coppicing in lowland Britain may be seen as part of a much broader process. One of the main ecological effects of man has been the gathering of the element phosphorus (perhaps other elements too) by his farming activities from the countryside as a whole and its storage in small areas such as rubbish tips, cemeteries, village gardens and farmyards.

In the middle distance (right-hand side) of Plate 12.3 is a small human settlement, Gilamoor Mill. It has been a continuous human settlement, at least since Saxon times. The present main building dates from 1759 but there has been a watermill on the site since before about 1050. As grain was brought for milling, through animals grazing on nearby pasture and then being brought into stables and byres at night and during the winter, and as a result of timber, food and other commodities being brought from other parts of the valley and the surrounding hillsides, there must have been a steady net inflow of some nutrients for close on 1,000 years. (No doubt there are some agricultural settlements in parts of Asia that have been continually inhabited for several times as long.)

At a very different scale we may consider the Australian farming system. The poverty of the soils of Australia's ancient unglaciated land surfaces, particularly in respect to phosphates has already been noted. The forests, with their capacity to store phosphorus, are being cleared, partly to make way for farmlands. The wheat and livestock (rich in phosphorus) produced on the farms are sent to Australia's cities (entirely on the edges of the Australian land mass) for consumption there or for export. The phosphorus removed from Australia's farmland ecosystems in this way has to be replaced at great cost (both in energy and financial terms) by inputs of phosphate fertiliser, prepared from phosphate rock mined on the Christmas Island, and

PLATE 12.3: A complex landscape ecosystem, Farndale, North Yorkshire, England. Here is a rich mosaic of communities: arable land, pasture, woodland, hedgerow, moorland and riverbank, linked by a web of ecological relationships. — *Photo: Patrick Armstrong.*

Australian territory in the Indian Ocean, Nauru, and formerly the Ocean island in the Pacific. An average of over one million tonnes of phosphate is exported from the Christmas Island to the Australian mainland each year (Department of Administrative Services, 1976.) The Significance of the current (late 1979) political and industrial disturbances on the Island, in relation to the ecological viability of Australia's agriculture, hardly needs emphasising.

How many examples of "breakdown" or of dramatic change in landuse systems in the historical past might be attributed to the depletion of supplies of a critical element or its concentration in some inaccessible part of the system ? We can only speculate, but it may be that the biogeochemical cycle model might provide a helpful framework for agricultural geography and landuse history studies elsewhere.

Conserving the Complexity of Agricultural Ecosystems

Elton (1966), from the standpoint of an animal ecologist, stressed how important relationships can exist even where there is no direct transfer of matter or energy. He coined the term *Species network* to include the totality of ecological relationship within an ecosystem. He introduced the term.

> . . .to describe any part of the interlocking system of species. It is designed to take care not only of food relationships, important as these are, but also of all sorts of other relationships that exist between species and contribute to the level of integration of communities. There may, for example, be competition for space itself, and many other kinds of competitive interferences that do not cause death of any immediate transfer of energy or materials from one species to another. Conversely there exist a multitude of ways in which one kind of animal alters a habitat so as to improve it for other species.

Relationships of this sort may be highly important in the integration of the system as a whole. Two examples, again from eastern England, may be quoted to illustrate the ways in which organisms in a simplified man-modified ecosystem "alter it as to improve it for other species" as

Elton put it, contributing to the level of integration of the community as a whole. The heaths (sheepwalks) of the Suffolk Sandlings, as well as providing grazing for large flocks of sheep, were also used as rabbit warrens. Often both sheep and rabbits grazed the same area. A document dated 1576 reveals that 100 acres (*c.* 40 ha) of heath in the manor of Benacre had then recently "been replenished with. . .sheep. . .and coneys" (*coney* is the old word for rabbit). Eighteenth and nineteenth century sources confirm that this was common practice. An interesting ecological relationship may have existed between the rabbits and the sheep; a witness before the *British House of Commons Select Committee on the Game Laws* in 1872 said of a warren in Breckland, West Suffolk :

> The sheep and rabbits would feed well together, because there is a certain portion that the sheep will take, and another portion that the rabbits will take. . . . The warren will by nature produce a certain something which the sheep will browse upon in wet weather, and a fine short grass that the rabbits will feed upon but where sheep can scarcely live.

The "certain something" was without doubt the lichen, *Cladonia*, another witness described it as "a species of white moss". Many studies have demonstrated that heavy grazing of heathland by rabbits suppresses many angiosperm species, so that eventually little remains except *Carex arenaria*, lichens and mosses. It seems that on at least some heaths grazing by rabbits encouraged the growth of the species favoured by the sheep (Armstrong, 1970, 1972).

Another linkage between the two species from the same area, where sheep grazed on heath, drained marshland pasture, arable land and the sea walls (grass-covered enclosing banks constructed to protect drained pasture and farmland) is hinted at by George E. Evans (1956), who linked the disastrous flooding of the lowland pastures in the 1950s with the disappearance of the sheep flocks from the Sandlings:

> Formerly, the sheep ate the grass on the sea-walls which have been built and kept in repair for centuries along this coast. They kept the turf on these walls firm by their constant cropping; and, by their regular tramping and treading, filled in the rat-holes and rabbit

> burrows that would naturally appear from time to time. . . . When the sheep were removed, the sea-walls gradually became weaker and less able to stand an ordinary tide, much less an exceptionally high one. . . . These turf-covered banks—although they were not spectacular to look at— were. . .effective as they were constantly renewing their own strength by their natural growth; and the sheep helped theme to do so.

The sheep here seem to have an important positions in the complex network of ecological relationships.

The species network concept may be helpful in an analysis of the colourful complexity of polycultural landuse of the Mediterranean lands and the tropics. Geertz (1963) working in Indonesia, Harris (1971) in Venezuela, Denevan (1971) in Peru and Igbozurike (1971) in Nigeria are amongst the many who have drawn attention to the heterogeneity of agriculture in tropical environments. These and other writers have stressed the slowness of spread of diseases and pests in such situations, the fact that a variety of crops of differing structure and longevity maintains a continuous cover protecting the soil from unbroken rainfall and direct sunlight, that some species (e.g. mango) provide shade for others (e.g. cocoyams, *Colacasia*), but there may be much more elaborate ecological linkages concealed in the apparent ecological disorder of the tropical forest clearing.

Heckman's (1979) study of rice field ecology in Thailand provides an excellent Asian example of an ecological approach to the investigation of an agricultural system. Research was conducted at Udorn Thani, north-eastern Thailand, in order to elucidate the effects of seasonal changes in the complex aquatic community of a rice field. An inventory was made of the organisms present — floating, creeping and submerged plants; emergent rooted plants, fungi, algae, protozoa and metazoa — "the ecosystem is a multiculture of flora and fauna". The physical and chemical changes that occurred under the strongly seasonal conditions were monitored and related to changes in biological productivity and the rates of "turnover" of certain elements. The conclusions of the study emphasise that great caution must be exercised in applying agricultural chemicals, either to improve the yield of rice or to keep in control pests e.g. the intermediate hosts of trematodes (flukes). Any distortions in the elaborate network of ecological relationship in the "community structure may break important food-chain links and reduce or eliminate the habitat's capacity for protein production".

This comes close to Elton's (1966) argument that ecological diversity in an agricultural landscape should be conserved, not only for its own sake, but also for soundly practical reasons. He saw one of modern agricultural man's main problems as:

> . . .that of fashioning over the whole extent of occupied and exploited land, a mosaic of landscape and many small habitats in it, that is as rich as possibly consistently with keeping the necessary productivity of land and its use by man for so many different purposes. There are two dangers that are now being more widely discussed than formerly. The first is that in giving priority to economic productivity, . . .the production of large cash crops from the land, the human environment itself may gradually become dull, unvaried, charmless, and treated like a factory rather than a place to live in. The second is that over-simplified communities contain within themselves, flaws of organisation that render them vulnerable to invaders of unfamiliar kinds, especially those that evolved in other continental areas.

These are probably many who would, in a world that is short of food, reverse Elton's ordering, but equally many would support him when he argues for the maintenance of:

> . . .the maximum variety of natural and semi-natural habitats and their communities in the landscape, as possible buffers against invasions and imbalance. This means preserving in as rich a form as possible all the communities that may be interspersed amongst the croplands, especially woods, scrub, roadside and field hedgerows and meadows as well as habitats traditional to water.

Plate 12.3 provide an example of the type of landscape ecosystem in which Elton rejoiced.

Bibliographic Note

Since the first publication of this work in 1981, there has been an enormous expansion of research on the dynamics of the ecosystems of many parts of the world, and a progressive inclusion of ecological ideas in the study of, and the solution to problems in, environmental and resource management. The second edition of I.G. Simmons *The Ecology of Natural Resources* (Edward Arnold, London) appeared in 1981, and is notable for the clear way in which it views resource use as ecosystem modification: agriculture, forestry fisheries and industry are described in terms of their effects on the energy flows, nutrient circulation patterns and effects upon species diversity.

Simmons took the concept of the effects of humanity on the alteration of the biosphere's energy flows even further in *Changing the Face of the Earth: Culture, Environment and History* (Basil Blackwell, Oxford) in 1989. He includes a substantial section entitled "Agroecosystems as energy traps", in which he includes many quantitative examples showing the dynamics of the pattern of the flows of energy in several types of agriculture, and also considers the ideas of ecological efficiency and stability in the context of agroecosystems.

A detailed example of the application of the energy flow concept, and other ecological ideas, to the study of an Asian city, and the agricultural land near it is given in S. Boylen, *et al.*, in *The Ecology of a City and its People: The Case of Hong Kong* (Australian National University Press, Canberra, 1981). This contains an excellent, detailed chapter entitled "Energy in the Hong Kong food system".

These attempts to understand, in ecosystem terms, human impacts on the environment, are part of a very much wider focus by geographers on the interface between human and natural systems. An expression of the enormous interest in this field over the last couple of decades, and the application of increasingly rigorous methods in its investigation, is Andrew Goudie's 1981 book, *The Human Impact: Man's Role in Environmental Change* (Basil Blackwell, Oxford), which includes sections on the human impact on vegetation, animals, soil, landforms, waters and climate and the atmosphere from an integrated, holistic, geographical perspective.

The student seeking further examples of the use of many of the ideas outlined in the preceding chapter could with advantage examine some of the above sources, several of which have had a profound effect on the development of geography over the last decade.

PLATE 12.4: A major world nutrient flow: Phosphate rock being mined on Christmas Island Ocean, to enrich the nutrient poor soils of Australia's farmlands. — *British Phosphate Commission Photo.*

REFERENCES

Armstrong, P. H., 1970 : The Heathlands of the Suffolk Sandlings in Their Setting : A Systems Approach to Landscape Study. *Ph. D. Thesis,* C.N.A.A.

———, 1971 : "The Heathlands of the East Suffolk Sandlings", *Suff. Nat. Hist.*, 15 (5), pp. 417-430.

———, 1972 : "The Rabbit in East Anglia", *East Anglian Mag*, 31 (4), pp. 162-165.

———, 1973 : "Changes in the Landuse of the Suffolk Sandlings : A Study of the Disintegration of an Ecosystem", *Geography*, 58 (1) pp. 1-8.

———, 1975 : *The Changing Landscape*, Terence Dalton Ltd., Lavenham.

Bayliss-Smith, T. P., 1977 : Energy Use and Economic Development in Pacific Communities in Bayliss-Smith, T.P. (ed.), *Subsistence and Survival : Rural Ecology in the Pacific*, Academic Press, London and New York, pp. 318-359.

Borman, F. H. and Likens, G. E., 1969 : "The Watershed-ecosystem Concept and Studies of Nutrient Cycles", in Van Dyne, G. (1969) *q.v.*, pp. 49-76.

Broughton, W. A., 1970 : *Effects of Land Management on Quantity and Quality of Available Water : A Review.* Water Research Laboratory, University of New South Wales Report No. 120.

Chapman, S. B., 1967 : Nutrient Budgets for a Dry Heath Ecosystem in the South of England, *Journ. Ecol*, 67, pp. 677-689.

Conacher, A. J. and Murray, I. D., 1973 : Implications and Causes of Salinity Problems in the Western Australian Wheatbelt : The York-Mawson Area, *Aust. Geogr. Studios*, 11, pp 40-61.

Crisp, D. T., 1966 : "Input and Output of Minerals for an Area of Pennine Moorland : The Importance of Precipitation, Drainage, Peat Erosion and Animals", *Journ. Appl. Ecol.* 3, pp. 314-327.

Denevan, W. M., 1971 : "Campa Subsistence in the Gram Pajonal, Eastern Pajonal, Eastern Peru", *Geog. Rev.*, 61, pp. 496-518.

Department of Administrative Services, 1976 : *Territory of Christmas Island*, Report for the Period July 1, 1972—December 31, 1975. Australian Government Publishing Service, Canberra.

Dice, L. R., 1955 : *Man's Nature and Nature's Man : The Ecology of Human Communities,* Ann. Arbor.

Elton, C. S., 1966 : *The Pattern of Animals Communities,* Methuen, London.

Evans, G. E., 1956 : *Ask the Fellows Who Cut the Hay*, Faber, London.

Fosberg, F. R., 1957 : *Abstracts of Papers, Ninth Pacific Science Congress*, Bangkok.

Geertz, C., 1963 : *Agricultural Ivolution : The Processes of Ecological Change in Indonesia*, Berkeley.

Gerloff, G. G., 1963 : Comparative Minerals Nutrition of Plants, *Ann. Rev. Plant Physiol.*, 14, pp. 107-124.

Gifford, R. M. and Millington, R. J., 1975 : Energetics of Agriculture and Food Production, C.S.I.R.O. *Bul.* No. 28.

Harris, D. T., 1971 : The Ecology of Swidden Cultivation in the Upper Orinoco Rain

Forest, *Geog. Rev.*. 61, pp. 475-495.

Harris, M., 1971 : *Culture, Man and Nature*, T. Crowell, New York.

Heckman, C. W., 1979 : *Rice Field Ecology on North-eastern Thailand*, Dr. W. Junk, The Hague.

Hopkins, B., 1965 : "Observations on Savannah Burning in Nigeria", *Journ. Appl. Ecol.*, 2, pp. 367-381.

Igbozurike, M. U., 1971 : "Ecological Balance in Tropical Agriculture", *Geog. Rev.* 61, pp. 514-529.

Odum, H. T., 1971 : *Environment, Power and Society*, Wiley, New York.

Lawton, J. H., 1973 : The Energy Costs of "Food-gathering", in Benjamin, B., Cox, P.R., Peel J. (eds), *Resources and Population*, Academic Press, London and New York, pp. 59-76.

Leibig, J., 1840 : *Die organische Chemie inihrer Avendung vor Agricultur and Physiologie*, Braunschweig.

MacFayden, A., 1964 : "Energy Flow in Ecosystems, and its Exploitation by Grazing", in Crisp, D.J. (ed.), *Grazing in Marine and Terrestrial Environments*, British Ecological Society and Blackwells, Oxford.

McFarlane, N. R., 1974 : "Energy in Agriculture", *Nature*, 252, 531.

Morren, G. E. B., 1977 : From Hunting to Herding : Pigs and the Control of Energy in Montane, New Guinea in Bayliss-Smith, T. (ed.) 1977, *Subsistence and Survival : Rural Ecology in the Pacific*, Academic Press, London and New York, pp. 274-315.

Pimpernel, D. et al., 1973 : "Food Production and the Energy Crisis", *Science*, 182, pp. 443-449.

Rackham, O., 1969 : "Knapwell Wood", *Nature in Cambs.* 12, pp. 25-31.

———, 1971 : "Historical Studies and Woodland Conservation", in Duffey, E. and Watt, A.S. (eds.), *The Scientific Management of Animal and Plant Communities.* Nature Conservancy, London pp. 563-580.

———, 1976 : *Trees and Woodlands in the British Landscape*. J.M. Dent and Sons Ltd., London.

Rappaport, R. A., 1971 : "The Flow of Energy in an Agricultural Society", *Scient. Am.*, 182, pp. 443-449.

Rodin, L. E. and Basilevic, N. I., 1967 : *Production and Mineral Cycling in Terrestrial Vegetation* (translated from the Russian) Edinburgh.

Senate Standing Committee on Science and the Environment, 1977 : *Woodchips and the Environment*, Australian Government Publishing Service, Canberra.

Simmons, I. G., 1966 : Ecology and Landuse, *Trans. Ins. Br. Geog.*, 38, pp. 59-72.

———, 1970 : "Landuse Ecology as a Theme in Biogeography", *Can. Geogr.* 14, p. 309.

———, 1975 : *The Ecology of Natural Resources*, Edward Arnold, London.

———, 1967 : "Organism and Ecosystem as Geographical Models", in Chorley, R.J. and Haggett, P. (eds.) : *Models in Geography*, Methuen, London.

Simson, G., 1895 : *The Wild Rabbit in a New Aspect*, Edinburgh.

Stoddart, D. R., 1965 : Geography and the Ecological Approach : The Ecosystem as a Geographical Principle and Method, *Geography*, 50, pp. 242-251.

Teal, N. M., 1962 : "Energy Flow in the Salt Marsh Ecosystem of Georgia", *Ecology*, 43, pp. 612-624.

Tansley, A. G., 1935 : "The Use and Abuse of Vegetational Concepts and Terms", *Ecology*, 16, pp. 284-309.

Tansley, A. G., 1939 : *The British Islands and their Vegetation*, C.U.P., Cambridge.

Van Dyne, G., 1969 : *The Ecosystem Concept in Natural Resource Management*, Academic Press, London and New York.

Vernardskii, V. I., 1934 : *Studies in Geochemistry*, Leningrad.

Westlake, D. F., 1963 : Comparisons of Plant Productivity, *Biol. Rev.*, 38, pp. 400-419.

13

AGRICULTURE AND ECOSYSTEM IMBALANCES

MEHDI RAZA

IN THE general eupheria about the dramatic increases in agricultural productivity and the concomitant shift in world's food supplies from deficit to surplus, the environmental consequences of this development have generally been ignored. Though there has been a surfeit of studies relating to agriculture, mostly in the productivity-surplus-deficit-regional-imbalances framework, much too little attention has been given to agricultural studies from the ecological viewpoint. Geographers, who because of their long disciplinary focus on man-environment interface, should have been in the vanguard of ecological studies, have woefully lagged behind. It is, therefore, high time that geographers shed their self-imposed pre-occupation with non-vital issue and start taking interest in issues which may be of cititical importance for human survival.

There is, perhaps, a lack of awareness about the fact that of all economic activities agriculture, in its broadest sense, has the greatest potential for ecosystem destabilisation. But it is fact that in order to tackle the twin problems of hunger and malnutrition, the agro-ecosystem round the world have adopted a typically commodity attitude towards resource exploitation, having growth as the primary concern. In this growth-oriented process of agricultural development, the natural ecological balance is increasingly being subjected to stresses beyond the tolerance limit of ecosystems. If the trend is not checked in time the function of the biosphere as a viable global life-support system may seriously be jeopardised.

What is a matter of serious concern is that in the face of an exponentially increasing demographic pressure, the agro-ecosystems of the world will be required to produce more and more food in disregard of ecological considerations (Flohn, 1980). Ever more intensive exploitation of land and water resources in order to meet the rapidly increasing food demands poses questions about the sustainability of these resources. Already there are tangible evidences of serious ecological imbalances in many parts of the world, inspiring doomsday prediction that environmental degradation will soon endanger human survival itself.

Already the agro-ecosystems of the world account for over 15 per cent of the land surface of the world. Both the intensity and extent of agro-systems are likely to increase rapidly in future to produce more and more food. According to one estimate the world's population is likely to exceed 6.35 billion by the end of the century. While the world as a whole may have the physical, economic and technological capabilities to satisfy the increasing food demand through 2000 but there is little doubt that this would be achieve at the cost of environmental quality and resource depletion.

The agro-ecosystems respond to increasing demand for food through : (a) expansion of the arable land and concomitant devegetation of the earth, and (b) massive increases in productivity-augmenting inputs such as water, fertilisers, pesticides, herbicides, energy and high-yielding varieties of seeds. Dangers to ecosystem stability are inherent in both strategies.

One of the inevitable consequences of agricultural expansion over millennia has been a dangerous devegetation of the earth. By converting forests into arable land and natural grasslands into pastures man has drastically changed the surface conditions of the earth since the Neolithic time. Flohn (1980) estimates that since the beginning of agriculture, a total of $45 \times 10^6 km^2$ or 30 per cent of the surface of the continents has been thus devegetated. Moreover, a recent forestry projection anticipates that the present net global deforestation rate of 18-20 million hectares per year will continue through the end of the century (Global Report, 1982) much of it agriculture related.

Such a massive devegetation of the earth certainly constitutes a serious anthropogenic interference in the biospheric equilibrium by removing from the system one of the most important components which may seriously affect the net primary productivity (NPP), energy flow, nutrients, water and sediment recycling in the biosphere besides an increase in albedo of the earth's surface. Of all the harmful consequences of agriculture, devegetation is, perhaps, potentially the most serious. All living things depend upon solar energy fixed in the biospheric system by photosynthesis. Human-beings, over millennia, have acquired the capability to channel and ever-increasing proportion of solar energy through themselves. It has been estimated that humans, together with their animal stocks, already consume at least 11 per cent of the net primary productivity of land (Tivy, J., and O'Hare, G., 1986). On the other hand, the NPP can safely be assumed to decrease as a result of replacement over large areas of forest and natural grasslands by sown crops and pastures. Eyre has estimated that NPP of all pasture lands has been reduced to one-third of that of the previously forested and NPP of all crop lands from one-half in mid-latitudes to one quarter in tropical areas. On this basis he calculates that the total potential NPP of the land has been reduced by about a third (Eyre, 1978).

Since NPP represents that part of the total amount of chemical energy fixed by an ecosystem per unit of area per unit of time that is actually available for moving through the system along food chains, any significant reduction in its total quantity available may have serious repercussions for the biospherical energy system, specially in view of the fact that the average biospheric photosynthetic efficiency is as low as 0.2 per cent.

It cannot, of course, be said that the dangerous devegetation of the earth is entirely due to agricultural expansion. There are many other contributory causes but there is no denying the fact since the Neolithic time agriculture has been responsible for massive devegetation. In a sense crop agriculture can be said to be anti-ecology since it cannot be practiced without removing from the ecosystem one of its most vital components i.e., natural vegetation.

Since the biosphere in essence is a vast, global ecological system whose components are mutually interactive in such a way that the functional viability of the entire system is maintained in a state of equilibrium, devegetation of the earth for agricultural purposes is bound to adversely affect the other components, specially besides energy, the biological, chemical, hydrological and sedimentation cycles.

Apart from the global ecological consequences of agriculture listed above, more perceptible regional ecological ones are manifested in terms of soil deterioration, desertification, waterlogging, salinisation and alkalinisation, general erosion, the ecological effects of the use of fertilisers and pesticides and loss of biological diversity.

One of the more obvious ecological consequences of agriculture is the accelerated loss and degradation of soil resources. It is opined by many experts that if the present farming practices continue, many now productive croplands will undergo serious soil problems. The U.S. Soil Conservation Service considers that productivity may be seriously harmed if soil loss exceeds one ton per acre for shallow soil and five tones for deep soils (Global Report, 1982). But even in as technologically an advanced a country as USA soil losses exceeding five tones per acre were reported in 1977 (Report to the Congress, 1977). On gently slopping farmlands as much as twenty tones of soil loss per acre per year was reported (Buchman and Brady, 1960).

It is a well-known fact that in most farming systems of the world there is a net loss of organic matter in the soil both as a result of low density crops, sod like root structure, burning of crop residue as fuel as well as replacement, over large areas, of organic manures by chemical fertilisers. The loss of organic matter is ecologically harmful on two counts : (a) the moisture retention capacity of the soil is impaired and the soil structure is gradually destroyed; (b) the break down of soil organic matter releases additional quantities of carbon dioxide into the atmosphere. Since there is a large amount fixed carbon in the soil its' conversion into CO_2 may add greatly to the atmospheric carbon pool which in the long term may lead to atmospheric destabilisation and undesirable climatic changes, specially the 'green-house' effect.

One of the ecologically most damaging aspects of modern agriculture is its disproportionately large amount of water requirement. Indeed, water shortage is going to emerge as the greatest constraint in increasing agricultural productivity, specially through the Green Revolution technology. Already water problems resulting from deforestation have assumed serious dimensions in at least 16 countries and in 10 countries in the form of floods. Even though water is unquestionably the most limiting factor in world agriculture, agricultural pressure on precious water resources has increased many-folds lately. Since the turn of the century, the world's irrigated acreage has increased more than four-fold. According to an United Nations estimate irrigational requirement of water, which in 1967 accounted for over 70 per cent of man's use of water, would double by 2000 (Freeman, 1974). Moreover, water used in agriculture is not immediately recyclable as much of it is lost through evapo-transpiration.

The water crunch has necessitated extensive tapping of groundwater in many parts of the world, a trend which is likely to be greatly accelerated in future. In many parts of the world — Australia, South Africa, S.W.U.S.A., India — exploitation of groundwater is proceeding at a rate greater than the natural recharge of acquifers, leading to extensive regional dents in the watertable.

Among the productivity augmenting inputs in agriculture, canalisation of water has been, over the centuries, one of the most important one. But expansion of canal irrigation, at best, has been a mixed blessing because globally more productive land is being lost due to waterlogging, salinisation and alkalinisation than is being added by irrigation. World-wide an estimated 125,000 hectares of irrigated land are becoming unproductive due to the consequences of irrigation (FAO, 1976). By the end of the century, if the likely event of this trend continuing, approximately 2.75 million acres or 1.4 per cent of the world's total irrigated land will be out of production. The problem has already become acute in countries like Pakistan where, as early as 1960, in Sind and Punjab 49 per cent of the land suffered from moderate to serve waterlogging, 50 per cent was highly saline and 27 per cent was moderately saline (U.N. Conference, 1977). Similar problems have also been reported from India, Afghanistan, Argentina, Peru and USA. It can, indeed, be said that virtually in every country with sizable irrigation facilities, a good deal of productive farmland is going out of production as a result of land degradation caused by waterlogging, salinisation and alkalinisation. Making deserts bloom is a masterpiece of human ingenuity and enterprise but in the process five time as much desert acreage has been created as it existed before by ignorance of ecological processes or by the sheer pressure of man and livestock.

In arid and semi-arid area, incompatible landuses, such as over-grazing and over-cultivation, has led to degradation of the sparce vegetation and the concomitant ecological imbalances in the cultivated steppes. The process, known as desertification, has already assumed critical dimensions specially in Africa and Middle East. When the top soil in ecologically fragile regions is regularly loosened by plough and is partly cleared of rocks, deflation of huge amounts of fine soils takes place within a short period. In the arid lands of Africa immense quantities of fine sediment, estimated at 6 million tones in 1969, are being transported into the Atlantic by the Trade winds (Mensching and Ibrahim). Desertification is an active, on going phenomena and a good deal of pasture land and cropland will be lost by the end of the century in many countries of Africa and Asia.

Modern agriculture has become increasingly dependent upon such yield enhancing inputs as fertilisers and pesticides, both having regional and global ecological implications. Fertilisers are essential for increased food production but their excessive use poses environmental threats, the dimensions of which are not as yet fully understood. Certain nutrients are integral part of the global biogeochemical cycling. All plants absorb in solution from the lithospheric nutrient pool such elements as hydrogen, nitrogen, phosphorus, potassium and calcium through the process of root osmosis. This is one reason why disruption in the hydrological cycle adversely effects nutrient cycling in the ecosystem. If soil deficiency occurs in respect of any of these elements, it can be made good by additional nutrient-loading in the form of fertiliser inputs. Generally, therefore, the effects of increased fertilisation on terrestrial ecosystems are surprisingly beneficial in terms of greatly increased output, but large scale application of the big three — phosphorus, potassium and fixed nitrogen — may affect soil properties and structure in diverse ways.

Firstly, nutrient-loading, beyond a certain limit, greatly accelerates the biological cycle and is manifested in excessive growth of undesirable species of plants and micro-organism. Secondly, excessive nitrogen-loading in course of time leaves the soil poor in organic matter. As nitrogen is often the limiting element in the decomposition of organic matter in the soil increased nitrogen usages contributes to the depletion of organic soil matter. Thirdly, in biospheric terms, if the present trend in the production of nitrogen fertilisers continues, synthetic nitrogen fixation would exceed the natural fixation by the end of the century. This would represent a significant alteration of the basic nutrient cycle. What effect this as well as a similar trend in the phosphorus nutrient cycle would have on biospheric equilibrium is not yet fully known (Soderland and Sevansson, 1976). Fourthly, according to the studies conducted by the National Academy of Sciences, USA, there is a link between anthropogenic nitrogen fixation and depletion of stratospheric ozone (NAS — 1978). The processes involved are extremely complex but it is known that they are generally the result of the release of nitrous oxide during the denitrification of fixed nitrogen. Assuming a doubling of world-wide application of nitrogen fertilisers by the turn of the century, it has been estimated that global ozone would be reduced by 3.5 per cent by 2000 (Turco, et al., 1978). Ozone depletion has extremely serious and far-reaching consequences in terms of ecosystem imbalances. By absorbing the ultra-violet solar radiation, ozone plays a crucial, protective role in the maintenance and preservation of the life-support system of the earth.

Excessive nutrient loading may have devastating effects on aquatic ecosystems. Many streams and lakes become polluted with agricultural waste. Due to heavy application of fertilisers, run-off water may also carry large amount of nitrate, potassium and phosphorus to water bodies. Apart from contamination of drinking water, the phosphorus and nitrogen components of fertilisers are known to cause eutrophication of lakes. More than 70 per cent of nitrogen entering surface water is from non-point agricultural sources.

Intensification of fertiliser input in agriculture is, therefore, fraught with serious ecological dangers. Its application should be carefully managed because excessive and improper use might lead to soil infertility and reduced yields besides leading to serious global ecological consequences.

The use of pesticides and herbicides has become a necessary evil in modern agriculture. Nutrient-loading of modern agro-ecosystem as an yield-augmenting strategy inevitably leads to higher ecosystem productivity and a concomitant excessive growth of soil micro-organism and weeds injurious to cultivated crops. That is why the use of fertilisers is generally correlated with the use of pesticides and herbicides.

The ecological consequences of large scale use of pesticides and herbicides can be summarised as follows : biological amplification and concentration of persistent pesticides in the tissues of human-beings and animals; development of increased resistance to pesticides by numerous insects and hence, possible decline in yield through increased vulnerability to pests; destruction of natural pest controls such as birds and other non-target predatory insects; emergence of new pests previously not troublesome; increased poisoning of farm workers; genotoxicological effects which may endanger the genetic inheritance; and general toxification of natural ecosystems due to the persistence and broad toxicity range of pesticides.

There are two main groups of pesticides : organochlorines (DDT, aldrine, methoxychlor, deldrin, toxaphane and other chlorinated hydrocarbons) and organophosphorus. The former is environmentally more damaging, though less expensive. The latter, though less persistent, is also apt to be harmful. Because of its relative cheapness and lesser short term toxicity to human-being, it is the environmentally more harmful organochlorines group which is the principal insecticide used in less developed countries like India (FAO, 1975), whose major concern is short term increase in food productivity regardless of ecological consequences. Thus considerable long term ecological consequences are likely to occur in view of the expected rapid increase in the application of pesticides and herbicides in the agro-ecosystems of the world.

A modern trend which is being manifested the world over is greater and greater reliance on monoculture and inbred strains. There are many ecological dangers-inherent in this trend. It is estimated that over four-fifths of the world's food supplies come from less than two dozen plant and animal species. Two crops alone (rice and wheat) provide sustenance to over four-fifth of the world's population (Harland, 1972). This represents a dangerous shrinkage of the genetic base of the agricultural system which may lead to two types of consequences : (1) modern, inbred crops are known to have a weaker defensive mechanism against pathogens and pests. The greater genetic uniformity of inbred monocultural crops, therefore, makes them highly susceptible to outbreaks of trait-specific epidemics and the consequent danger of massive crop failure. (2) The simultaneous depletion of the genetic resources available to combat trait-specific pathogens and insects also highlight the ecological vulnerability of modern monocultural agriculture.

A more basic ecological concern is that by the practice of monoculture, or using only a few crop species, man has brought about radical change in the composition of plant communities and their dependent ecosystems all over the world. In an effort to control food supplies not only innumerable species have been eliminated but many others have been transferred from their native habits to new environment and thus the prevailing homeostatis has often been drastically upset.

The extinction of biota on such a scale is always highly regrettable but what is not generally realised is that the monotonous simplicity of monoculture, with dominance of a few essential crops, will also invariably lead to monotonous short food-chains and the inherent danger to ecosystem stability. Nature's way of ensuring the stability and viability of the life-support system is diversity of life forms. As in economic and social systems diversity creates stability by acting as a cushion to any single perturbation in the system. Therefore, the more diverse and complex a system is, the more it will be able to positively respond to perturbations. Agriculture's ecological drawback is that with the extinction of species, the total global genetic pool has been depleted and the necessary biotic diversity needed for ecosystem viability reduced.

It will now be seen that agriculture has the potential of seriously undermining many of the short-term as well as long term eco-regulatory mechanisms. The great urgency to produce more food for an ever-increasing population will certainly further exacerbate the problems. What is to be done?

The basic problem of course is the exponential growth of human population. In both developed and less developed countries the continued extensification and intensification of agriculture to meet the needs of a growing population is creating major environmental problems. Growth of population must, therefore, be curbed, preferably voluntarily but by compulsion if necessary.

Assuming that effective population control and bringing down the human numbers to an ecologically acceptable level inforeseeable future to be a tall order, not likely to be achieved, we must think in terms of radically new and innovative food production strategies that would minimise, if not altogether eliminate, dangers to ecosystem stability. This may call for the changes in world's feeding habits. The vegetarian option may be a sensible alternative. Since about 90 per cent of the NPP (net primary productivity) is lost at each successive trophic level, obtaining food from trophic level is ecologically much more sensible than obtaining it from higher trophic levels.

In purely energy input-output terms also, animal food is much more expensive. Growing energy subsidy in modern agriculture requiring massive inputs of fossil energy has profoundly affected the energy budget of agro-ecosystem. The ratio between energy output (in animal feed and human food) and input has gone down from 50 to 25 to 1 or occasionally below. In this energy imbalance in modern food production as a whole animal production appears to be a greater ecological burden, giving ratio levels mostly below 0.5 (Borgstrom, 1979). Yet in the developed world animal raising has ceased to be an integral part of the cropping systems and has emerged as a major, new sector of food production. This has necessitated the transfer of an enormous quantity of proteins from human to animal consumption. Most European and North American countries are feeding their stocks with large quantities of fish proteins as well as dehydrated skim-milk and protein crops, protein that should have been channelled to human use. Since protein is the driving force in biotic systems, its misuse in this manner has great implications for human species in nutritional terms. Also, since most protein crops are acreage extensive, their cultivation for animal consumption adversely affects the landuse pattern vis-a-vis, food crops. In a sense, therefore, animal raising for food can be visualised not as supplemental to man's food requirement but as a competitor and ecological burden on vital life-support systems.

Borgstrom (1979) has propounded a new population density concept which takes into account man's total food related pressure on soil resources. The population-equivalence (PE) method has been devised to gauge the pressure which animals exert on soil resources in terms of their total intake of primary plant proteins. Computed in PE terms the total burden of living mass belonging to man or operating under his control in 1977 was not 4.2 billion human-beings but, taking into account 16.8 billion (1977) PE units of livestock, a feeding burden on food systems of 21 billion (Borgstrom, 1979).

The expansion of livestock since 1948 by 7.23 billion PE units implies a accelerating feeding pressure on the biosphere. Substituting plant food by animal production, as is happening in developed countries, not only affects land and energy requirement but also over-burdens water resources, the water requirement for a strictly vegetarian diet per person per day being 1200 litres whereas as it is 10 times more for an average western diet. In general kg. for kg. of protein, that from animal sources require 3 to 6 times as much water as from plant sources (Borgstrom, 1979). The ecological lesson to be learnt is that in view of both the energy input-output ratio and feeding pressure on the biosphere, the increasing preference for non-vegetarian diet, at least, in the affluent world, is a potential destabilising factor in ecosystem management and aught to be discouraged.

As a part of a new, ecologically viable food strategy top priority must be given to a more rational management of the protein budget. A good deal of protein which is now fed to animals (two of fifth of fish catch of the oceans, one quarter of milk protein and two-third of oil-seed protein) can and should be diverted to human consumption with suitable changes in dietary habits. It has also been suggested that recycling of the tremendous amount of organic urban waste would make available a good deal of protein for human consumption. There is, therefore, much to be gained if such non-traditional foods are meaningfully incorporated into a more rational food strategy for mankind (Borgstrom, 1979).

The foregoing analysis of the ecological implications of agriculture reveals numerous and serious environmental consequences of the high intensity, high-output nature of modern agro-systems. Numerous studies highlight the implicit dangers to the terrestrial, aquatic and atmospheric resources which are vital for the functional viability of the global life-support system. Many areas have been revealed where the capacity of the environment to provide goods and services can no longer be taken for granted. If long term man-environment symbiosis is to be maintained, there is little time to be lost for developing a new, ecologically compatible food strategy for mankind.

REFERENCES

Borgstrom, G., 1979 : "Agriculture and World's Feeding Alternatives", in Polunin, N., (ed.), *Growth Without Disaster*, p. 317.
ibid., p. 296.
ibid., p. 316.
ibid., p. 321.

Buchman Harry O., and Brady, N.C., 1960 : *The Nature and Properties of Soils*, Table 9-7.

Council on Environmental Quality and Department of State, 1982 : *Global 2000 Report to the President*, p. 318.

———, 1982 : op. cit., p. 281.

———, 1982 : op. cit., p. 285.

Eyre, S. R., 1978 : *The Real Wealth of Nations.*

F.A.O., 1975 : *Pesticide Requirements in Developing Countries.*

———, 1976 : *Production Yearbook.*

Flohn, H., 1980 : "Man's increasing impact on Climate", in Nicolas Polunin (Ed.), *Growth without Disaster*, p. 40.

Freeman, P. H., 1974 : "The Environmental Impact of Rapid Urbanisation", *Guideline for Policy and Planning*, Table 12, p. 42.

Harland, J. R., 1972 : "Genetics of Disaster", *Journal of Environmental Quality*, pp. 212-15.

Mensching, H., and Ibrahim, F., "The Problem of Desertification in and around Arid Lands", *Applied Science and Development*, Vol. 10, p. 11.

National Academy of Sciences, U.S.A., 1978 : *Nitrates: An Environmental Assessment.*

Soderland, R. and B. H. Sevansson, 1976 : "The Global Nitrogen Cycle", in Sevensson and Doderland (Ed.), *Nitrogen, Phosphorus and Sulphur-Global Cycles*, p. 66.

Tivy, J., and O'Hare, G., 1986 : *Human Impact on the Ecosystem*, p. 82.

Turco, R. P., et al., 1978 : "SST, Nitrogen Fertiliser and Stratospheric Ozone", *Nature*, 28, pp. 505-507.

U.N. Desertification Conference, 1977 : "A Case Study of Desertification", *Mona Experimental Project*, Pakistan.

U.S. General Account Office, 1977 : "To Protect Tomorrow's Food Supply and Soil Conservation Needs", *Report to the Congress.*

14

AGRICULTURAL GEOGRAPHY IN DEVELOPING COUNTRIES

J. T. COPPOCK

SINCE AGRICULTURE is the most common and widespread of the ways in which man gets his living, and since geographers are primarily concerned with his varied impact on the earth's surface, it might be supposed that a considerable proportion of geographical research would be devoted to the study of agricultural geography. Yet this is not the case, and while it is true that geographical journals contain many descriptive articles and that descriptions of agriculture are often prominent in elementary textbooks, few substantive works have appeared in this field and no adequate methodology or body of theory has yet been devised. In the developed countries some justification for such neglect might be sought in the small and declining contribution of agriculture to both national income and employment and in the urban background of most geographers, but the situation is not greatly different in developing countries where agriculture may account for two-thirds of the national income and occupy four-fifths of the population and where many of the growing number of university-trained geographers are the children of farmers or were brought up in rural communities. It seems likely that a major cause of this neglect is the difficulty which stems largely from the nature of available data, and the problems of collecting original data in the field, of undertaking worthwhile and intellectually rewarding investigations in this field by comparison with those which are possible in other branches of geography, a view which is supported by the fact that the proportion of agricultural articles in geographical journals is declining. This explanation seems even more plausible in tropical Africa where many of the investigations which have been undertaken by geographers have either been in physical geography, where the ground itself provides all necessary data, or in the study of communications, industries and settlements for which material is both more accessible and more manageable. Even in agricultural geography there has been some tendency to concentrate on those topics, such as settlement schemes, which are not representative of agricultural but which are comparatively well documented, and it is not surprising that the only monograph to appear in English on the agricultural geography of any country in tropical Africa, McMaster's (1962) study of subsistence crops in Uganda, should concern a territory with unusually rich sources of statistical data.

Despite these problems, a strong case that geographers should pay more attention to tropical agriculture can be made, on both practical and academic grounds. In the foreseeable future it is likely that agriculture will continue to occupy a leading place in the economies of developing countries, and both out of academic self-interest and as responsible citizens, geographers ought to play a part in economic development by employing their skills in describing and understanding the existing use of agricultural resources and so contributing to the planning of their more effective exploitation; for the geographers' concern with total environment, both physical and human, and his training in methods of reconnaissance survey are very appropriate in the developing countries in the tropics whose problems are often too urgent to wait detailed inquiries by many specialists. It would, of course, be wrong to allow such immediate practical objectives wholly to dictate academic priorities, but geographers may find it easier to obtain data for strictly academic investigations when they have demonstrated their capacity to make practical contributions. In any case, the distinction between academic and practical should not be overstressed; better theoretical understanding of the location of agricultural production may be expected to increase geographers' ability to predict the likely results of proposed changes and so make a more effective contribution to the planning process and, conversely, the investigation of the agricultural consequences of, say, the construction of new communications may lead to improvements in theory. In this respect, the tropics offer opportunities which are not so readily available in the developed world : the existence of large numbers of peasant farmers, operating on a small scale in an environment with few external links and a relatively simple economic climate and yet subject to radical and rapid change, offers the nearest approach to laboratory conditions that the social scientist is likely to find.

The central aim of agricultural geography as an academic discipline is an understanding of the complex variety of agricultural patterns on the earth's surface. These may be examined at all scales, from the world and continental views characteristic of general atlases and textbooks to the ultimate unit of decision-making, the individual farm. In practice, the minimum area is more likely to be a group of farms or the territory of a village or other similar rural community, since the geographer, like the economist and the sociologist, is generally more concerned with the trends and tendencies which are the resultant of many individual actions, although he recognises that individual decisions and idiosyncrasies may be important where the unit of decision-making is large and that differences in the range of variability are often as important as differences in average values. For the agricultural geographers working in developing countries in the tropics no one scale is intrinsically more appropriate than another and both large - and small-scale investigations present difficulties. The choice of the scale will depend upon the nature of the problem being investigated and upon the resources available for its solution. Where farms are very small, as in most developing countries, the territory occupied by a single rural community is likely to be the largest area which can be studied comprehensively in the field by an individual investigator, especially if he has to rely on field observation and inquiry for his data. Much larger areas can, of course, be studied more superficially on a reconnaissance basis, by sampling, by reliance on secondary sources, by team work, or by limiting the aspects to be investigated, and all these approaches have a part to play.

Ideally, each investigation in agricultural geography should involve four stages : the identification of a problem; the formulation of a hypothesis or hypotheses; the collection of the relevant data; and the testing and modification of the hypothesis to provide an adequate explanation. Such a sequence presupposes that the problem can be readily identified, that sufficient is known to permit the formulation of hypotheses and that appropriate data can be obtained, but it will rarely be possible for all these of conditions be to satisfied in the present state of knowledge of the tropics and objectives must consequently be more limited. Like the fly on the newspaper photograph, the earthbound geographer can observe at any one time only the details of the picture and one of his first tasks must frequently to be construct a reconnaissance agricultural geography, giving a general conspectus of the main agricultural features of the country in which he is working from such data as are available, perhaps in the form of an agricultural atlas. Such a reconnaissance study is unlikely to be intellectually rewarding, but it will help to define problems and to provide the context in which specific investigations can be set. It will also serve a useful purpose by influencing the mental maps which all administrators and politicians have of the territories for which they are responsible and whose form is indicated by Gould's (1966) pioneer study in Ghana. Such maps must play an important part in decisions about where agricultural and other developments will occur and the nearer they approach reality, the better the prospect that decisions will match the actual needs of the country.

Such a reconnaissance study will, of course, be less necessary in territories where the World Bank, FAO or similar bodies have made general surveys, although the published reports often provide only a very sketchy geographical appraisal. It can be argued that to attempt such a broad survey is to reverse priorities, in that it ought to await the completion of many detailed investigations, yet as Buchanan and Pugh (1958) have convincingly shown, it is possible to provide a surprisingly full picture of the agricultural geography of a territory, for there is often much information, whether statistical data, reports of administrators, agricultural officers, soil surveyors and visiting experts, or investigations by research workers in a wide range of disciplines, which, however unsatisfactory or meagre its contribution may be in isolation, can be usefully brought together in such a synthesis. A valuable first step would thus be a thorough bibliographic survey, without regard for subject boundaries, of all books, papers and published and unpublished reports relating to the agriculture of each territory.

To complement this work there is also an urgent need for detailed studies of individual agricultural communities; in many areas too little is known about the organisation of agriculture, especially in relation to local environment, for adequate hypotheses to be made about such topics as the choice of particular crops or particular agricultural methods. Consequently, in advance of, or parallel to, more problem-oriented investigations, studies of the agricultural organisation of selected communities covering a range of environments should be undertaken in-depth. Anthropologists have long realised the importance of such investigations, which presuppose a willingness to undergo long residence in a rural community, and have accumulated a wealth of information, much of it germane to an understanding of the use of agricultural resources. Agricultural economists and sociologists, too, have come to appreciate the necessity for such studies in depth and there is increasing need for geographers to follow the example of Hunter (1961), Prothero (1957) and others in undertaking surveys of small areas. Such tasks are particularly appropriate for geographers in developing countries, unhampered by linguistic barriers or cultural differences.

Between these two extremes of broad appraisal of the whole agricultural economy and the detailed study of small areas, the range of possible problems which might be investigated is very large, although most will involve either the study of a single agricultural enterprise, such as cocoa-growing or yam production, throughout a territory, or the investigation of a more limited problem area, particularly from the viewpoint of agricultural development. In practice, the choice will tend to be limited by the information that is available or that can be collected and by the skills of individual geographers; a major obstacle here is the lack of any formal training in agriculture provided in most courses in geography.

The nature of available data is of critical importance. Written records, of the kind found in the journals of learned societies, in government archives and libraries, and in the publications of official and semi-official bodies, can be valuable in suggesting possible hypotheses and giving an insight into information gleaned from other sources; but, while they may sometimes contain the results of special surveys, they will not usually yield sufficient specifically located data to form the basis of a satisfactory geographical analysis. Even so, in the absence of suitable statistical data, written records, especially reports on local administrative areas like the agricultural notebooks prepared by district agricultural officers in Northern Nigeria (Nigeria, 1957) or the agricultural surveys in Uganda (Uganda, 1960), can sometimes provide sufficient qualitative data about the occurrence of particular crops and perhaps even their relative importance to allow the preparation of distribution maps showing the approximate limits of their cultivation and the areas of greatest importance. Although such maps must be treated with great care, they will be particularly useful where a distribution has changed markedly over the course of time and they can be used also as a check on statistical data.

Agricultural geographers in developed countries frequently

complain about the inadequacy of the statistical data available to them, and it is true that the more interesting the data, the more likely they are to be regarded as confidential and hence to be highly generalised and difficult to locate exactly; but for may aspects there is an abundance of material rather than a lack and the principal difficulty, now more readily soluble with the advent of digital computers, has been how to handle it. In Africa the situation is very different, for the major problem is the paucity of statistics of any kind and confidentiality is much less likely to be a major obstacle. In many ways the problems of the statistician in tropical counties resemble those of the agricultural geographer in that he must attempt to describe, with quite inadequate resources, the characteristics, of large areas about which comparatively little is known. For some territories there are still no available statistics which can be used to establish the regional character of agriculture, and for others there are only rough estimates collected by administrative officers and other officials. In an increasing number of States, statistics are derived from small samples, designed to produce acceptable national totals rather than the detailed breakdown by areas desired by the geographer.

Where statistical material exists the agricultural geographer can use it to provide, at least in outline, the main spatial characteristics of his problems, but his first task must be to evaluate its accuracy, both by inquiry into the way in which it has been collected, and where possible, by comparing it with other sources, e.g. the records of purchases by marketing boards or by commercial companies. Statistics derived from the informed guesses of district officers or collected as a by-product of taxation returns are not likely to have a high degree of accuracy, but, unless there is reason to suppose that there are marked variations in the scale and direction of error between one administrative area and the next, even these can be used for the production of qualitative maps of distribution and perhaps of relative importance by arraying crops or livestock in rank order (International Bank, 1955; Coppock, 1965). It may also be possible to use indirect measures of the distribution of different crops or livestock by plotting features such as cotton-buying stations, rice mills or veterinary inspection posts whose location is often accurately known and can be presumed to bear some relation to the particular agricultural enterprise being investigated (Johnsrud, 1960).

Where statistics are derived from samples they are potentially more useful, provided that the sample has been properly drawn and the sample areas accurately measured and recorded (Holleman, 1964). While the basis of sampling is usually made explicit in official reports it must be admitted that in view of the scarcity of skilled staff, the absence of large-scale maps, the small size of most farms and the complexities of inter-cropping and successional cropping, these are big assumptions. Nevertheless, the use of sampling does conserve scarce resources of skill and makes it possible for survey teams to spend some considerable time in each sample area and to measure plots by chain and compass, so that the results are likely to be more accurate than any other available data (Sierra Leone, 1967). If it is possible to have access to the sample data for each village or other enumeration area, if these can be correctly located and if the crop or livestock being considered is of major importance and not too highly localised, it is probably better to treat each as a point sample rather than to use the estimates which have been computed for the administrative areas in which they lie for not only are these areas often very unsatisfactory areal units, but the factors used in raising the sample e.g. tax lists, are likely to be less reliable than the sample data themselves. An increasing number of statistical procedures is available for the analysis of sample data of this kind such as trend-surface analysis (Chorley and Haggett, 1965) and they are also amenable to systems of computer mapping such as Symap, in which isopleths may be interpolated between sample points. The use of sampling also offers other potential benefits, for if the records of the sample villages are accessible (and this is likely to be a less serious problem than in the developed countries where rules on disclosure are very restrictive), they can provide a wealth of detail about the agricultural structure of individual communities. Such data have already been used to good effect by Floyd (1965) in his study of terrace agriculture, but their potential has hardly begun to be exploited.

Data on yield are sometimes available, and where based on actual measurements of sample plots, as in Sierra Leone where produce from randomly chosen plots 22 ft. square is weighed, are likely to be more reliable than the generality of estimates of yields (Sierra Leone, 1967). Economic data, on the other hand, are very scarce, although the use of indirect measures, such as standard labour requirements, is probably more justifiable than in temperate latitudes where livestock play a more important role and there are wide variations in the use made of machinery. Nor is it generally possible to obtain satisfactory data about the seasonal aspects of agriculture which are of such importance throughout the tropics or about the seasonal movements of crops, livestock and people which link the agricultural economies of widely separated areas, although there are sample ergographs and records of veterinary checkposts have proved valuable in analysing movements of trade cattle (Fricke, 1964, Morgan, 1959a).

Little if any of the statistical information available will permit detailed comparisons over time, although some indications of major change can be obtained by a subjective visual comparison of maps of different periods (Mabogunje, 1959). No direct comparisons will be possible because the bases of collection have generally changed and, whatever the accuracy of recent statistics, it is probably much higher than that of any earlier estimates. Plotting indirect indices, such as the location of buying stations for different periods, can also provide a useful indication of changes in distribution, although the data are essentially qualitative and must be handled with caution (Coppock, 1965).

Potentially one of the most valuable sources of data is remote-sensing. Air photographs are increasingly available in many parts of the tropics, but inter-cropping makes crop interpretation difficult and the large-scale photographs necessary would demand resources beyond the capabilities of the individual research worker if any considerable area was to be covered. Where photographs are already available their most immediate application is in the identification and measurement of the major categories of landuse, especially the extent of land actually under cultivation and available for cultivation, as in the maps of Gambia prepared by the Directorate of Overseas Surveys, and in the preparation of accurate base maps for the study of sample areas (Gambia, 1958-59). Even colour photographs taken from air with a hand-held 35-mm camera can throw useful light on landuse. For broad categories of landuse and for seasonal variations, e.g. in the extent of flooding, satellite imagery from the LANDSAT programme provides an increasingly important source, the significance of which will grow as resolution improves.

For the investigation of any individual country or of any large part of it the agricultural geographer will have to depend on data such as those collected by others, often for quite different purposes. At best, he can hope to supplement the information they provide by field inquiries of his own, and given the resources he can command, this too will generally involve sampling. The value of the impressionistic reconnaissance survey must not, of course, be discounted, the experienced geographer with a good eye for country can glean much valuable information by a carefully planned reconnaissance survey of an area, recording absence or presence and relative frequency of different crops and classes of livestock.

The possibility of using the rapidly growing number of secondary school pupils to undertake qualitative surveys should also be borne in mind, for while the nature of landuse in many parts of the tropics and lack of large-scale maps makes landuse surveys similar to those carried out by schoolchildren in Great Britain too slow and too difficult, it is quite possible to record the number and relative abundance of crops, the nature and timing of events in the agricultural calendar and the like.

Nevertheless, these impressions will have to be checked by sample surveys, for which the tropical world is well suited, with its relatively small variation in size of farm and traditional agricultural practices over large areas. There is much scope for field experimentation in different methods of sampling, but the primary rural community is probably the most useful sampling unit, partly because the universe from which the sample is to be drawn can be readily identified from census records and often from medium-scale maps and partly because the community frequently plays an important part in decision-making and is, therefore, a meaningful agricultural unit in a way that the parish or village in a developed country is not. Resources of manpower are likely to require some sub-sampling of farmers within the sample areas and some form of areal stratification will also be desirable, although the theoretical advantages of random sampling will thereby be lost. Line samples have been used elsewhere in sample work, as in the Soil Survey of the Nigerian Cocoa Belt, although difficulties of access are likely to be a major obstacle in wooded country (Vine et. al., 1954). Traces cut for overhead electricity cables offer one possibility for fairly rapid traverses across wooded country and, while they have the disadvantage of being neither random nor parallel, they are not tied to existing tracks and roads which are themselves likely to modify the pattern of landuse. Traverses by air have obvious advantages in this respect and have been successfully used in counts of cattle in Northern Nigeria (Fricke, 1965).

The geographer's contribution to agricultural development can be made in four ways: by the provision of broad studies of the major crops and livestock and of the physical parameters of soil and climate on which they depend, which will facilitate the evaluation of existing resources; by reconnaissance studies to identify and clarify the nature of particular problems and problem areas; by focussing attention on and evaluating the spatial links between farmers and their markets; and by service as a member of a team on particular development projects. Contributions of the first type, which serve both practical and academic ends, can be made in advance of any specific need, although most national development plans in fact call for extensions of the acreages of the leading crops, both to provide additional foreign exchange through increased production of export crops and to feed rapidly growing populations. The second type of contribution can also be made either as a by-product of academic inquiry or for specific purposes. The geographer's training well qualifies him for the task of identifying the main features of a problem area; he is accustomed to working with data of doubtful accuracy on problems involving many variables and his tradition of exploration (in the widest sense of the term) helps to avoid some of the pitfalls which arise from excessive specialisation. A much neglected but vitally important aspect of the geography of agriculture has been the study of linkages between farmers and their domestic markets, whether these be processing plant or the ultimate consumer, especially of food products. Such neglect is equally true of developed and developing countries, but it is of critical importance in the latter as urbanisation gathers pace and towns depend increasingly on more distant hinterlands for their food supplies or else rely on imports that require valuable foreign exchange. There has been a general tendency to focus attention solely on production and to assume that food is consumed either where it is produced or in the immediate locality; but, as Mortimore (1979) and others have shown highly complex linkages have developed. The nature and efficiency of these is a topic demanding great attention from geographers.

Geographic skills can also be valuable in development projects which can originate either independently or as a result of such reconnaissances; for once the problems has been identified in outline it must be the subject of more detailed surveys, so that solutions can be devised and plans prepared to put them into effect. Resources of time and money will generally be inadequate for prolonged investigations and the geographer's familiarity with both physical and human aspects can make him a valuable member of a small team. As yet, there has been comparatively little opportunity to test this assertion, although geographers such as N. Hilton have been employed by FAO on work of this kind, sometimes under designations other than geographer. It is a matter of regret that valuable experience of such team work in

Ashanti should not have led to many similar investigations elsewhere (Fortes et al., 1948).

Although such work is important, the attraction of the tropics for the agricultural geographer lies also in its possible contribution to the study of agricultural geography as an academic discipline. The spatial arrangement of agriculture around settlements is of particular interest, but it is difficult to evaluate in developed countries where holdings are often so large that individual preferences can greatly modify the zonation of landuses, where both the availability of mechanical transport and pricing policy have made location of declining importance and where the hurdle of disclosure of confidential data has to be surmounted successfully before any satisfactory analysis can be undertaken (Chisholm, 1962; Prothero, 1957). None of these problems is a serious obstacle in the tropics. Holdings are small, so that individual decisions matter less, the great bulk of local transport and travel is undertaken on foot, so that effort is directly proportional to distance, and the question of disclosure of confidential information does not generally arise. It should be possible to discover settlements in which the surrounding land was sufficiently homogeneous for land quality to be discounted as a variable so that the effects of distance alone could more easily be measured and, although it is true that the motivation of farmers in developing countries is only to a limited extent economic and that systems of land tenure and traditional ways of working the land are likely to modify the theoretical arrangement of landuse, some of the more serious distortions can be avoided by the careful selection of examples. Indeed, it ought to be possible to find a sequence of settlements which illustrates the effects of a progressive relaxation of the idealised conditions assumed in theories of landuse zonation. Alternatively, the rapidity of change in some areas makes it possible to achieve the same end by studying landuses in selected settlements over a period of years.

There are other situations in which practice and theory might be studied with equal profit, such as the areal diffusion of new crops and new techniques, changes in the values of a single variable and the introduction of some new factor. The tropics have certain advantages for the study of the spread of new agricultural enterprises and techniques, partly because channels of communication are relatively few, partly because innovations are generally so different from existing practices that they can be fairly easily mapped by observation, and partly because it may frequently be possible to determine fairly accurately the date and place of more recent introduction since these have usually been made through official channels. The effects of price changes are also more easily studied in areas where a single cash crop is produced and marketed through official buying points and where marketing boards and governments can cause prices to very widely and uniformly. Thus a study of the distribution of export crops since the Second World War might be rewarding in this respect, especially where, as in Nigeria, marketing arrangements have changed considerably. The introduction of a quite new factor is also more likely than in developed countries. There are numerous instances of situations which have been radically transformed by the building of new roads and even of new railway lines, or by the erection of a new processing plant. Where firm proposals for new communications are known sufficiently far in advance, as with the Bauchi-Maiduguri railway, it should be possible to study selective aspects of the agricultural geography before and after the opening of the new link: the effects of the creation of a new rubber factory or palm-oil mill could be studied in a similar way. Such situations would have the great merit of allowing the construction of hypotheses which could then be tested by the actual course of events (Barbour, 1967).

Other topics within the field of agricultural geography will also repay study. Many of these are primarily of intellectual interest, such as the investigation of a highly localised and distinctive type of agriculture or the study of regional variations in methods of grain storage or in the plants used to surround enclosures, although each of these has practical applications in countries where soil erosion is a hazard, where as much as quarter of harvested crops may be lost in storage and where the integration of crops and livestock and the individualisation of land tenure are likely to require the use of some cheap and effective hedge material such as the euphorbia used on the Jos Plateau in Nigeria or the Mauritius thorn used by the Kipsigis in Kenya (Morgan, 1959b).

In all these investigations the geographer's main contribution must be to identify the spatial characteristics of the topic he is investigating and to place them firmly in their human and physical context. Admittedly this will not be easy in countries where large-scale maps are rare and where so much of the work of environmental stock-taking has yet to be done. It is for this latter reason that much of the work done by geographers in tropical countries has been concerned with inventories of physical resources rather than with their agricultural utilisation. The geographer's contribution here is likely to be especially valuable in those fields where geography and the systematic sciences overlap, notably in biogeography, climatology, hydrology and geomorphology, and it is no accident that some of the most noteworthy practical contributions have been made in these branches, such as the reconnaissance studies of soil erosion in Nigeria undertaken by Grove (1951). Attempts to assess agricultural potential are particularly important in this connection.

The context in which agricultural features are placed must not only be local. Comparative studies have an important place in academic geography and there is much to be gained by comparing homologous situations existing in other countries in tropical Africa, Asia and Latin America. Conditions will never be identical, but comparison will focus attention on both similarities and differences and the identification of these is likely to yield fresh insights and to draw attention to features which might otherwise escape notice; the appearance of a crop atlas of West Africa is especially welcome in this connection (Papadakis, 1965). This comparative approach will be particularly valuable in the study of individual crops, agricultural systems and agricultural practices; but it will also be appropriate in investigations of land settlement, land consolidation and the like.

In the short run, at least, the most valuable service that the few agricultural geographers working in tropical countries can perform is essentially educational bringing together a range of material from a wide variety of sources to provide a new perspective and possibly even a new dimension for the technical expert and for the administrator. The volume of relevant material which can be used to construct at least an outline geography of agriculture for any developing country is very considerable and new sources are continually coming to light. There is also increasing awareness of the need for agricultural surveys as a basis for planning research and development and hence an opportunity for the geographer with some training in agriculture both to provide what is currently available in an appropriate team and to employ his skills in the filling of major gaps. The intellectual harvest from work in agricultural geography in tropical countries also offers exciting prospects but it, too, will require much patient recording and investigation, for as Weaver (1958) has noted identification and description may not be our ultimate aim, but they do constitute the first order of geographic business.

REFERANCES

Agriculture and Food Supply in Developing Countries, Deptt. of Geography, University of Edinburgh.

Barbour, K. M., 1967 : "A Survey of the Bornu Railway Extension in Nigeria: A Geographical Audit", *Nigerian Geographical Journal*, 10, pp. 11-28.

Buchanan, K. M. and Pugh, J. C., 1957 : *Land and People in Nigeria*, London.

Chisholm, M., 1962 : *Landuse and Rural Settlement*, Hutchinson, London.

Chorley, R. J. and Haggett, P., 1965 : "Trend Surface Mapping in Geographical Research", *Transactions of the Institute of British Geographers,* 37, pp. 47-67.

Coppock, J. T., 1965 : "Agricultural Geography in Nigeria", *Nigeria Geographical Journal*, 7, pp. 67-90.

Floyd, B., 1965 : "Terrace Agriculture in Eastern Nigeria : The Case of Maku", *Nigerian Geographical Journal*, 7, pp. 91-108.

Fortes, M., Steel., R. W. and Ady, P., 1948 : "Ashanti Survey 1945-46 : An Experiment in Social Research", *Geographical Journal*, 110, pp. 149-79.

Fricke, W., 1964 : *The Cattle and Meat Industry in Northern Nigeria*, Chapter A, Unpublished Report, Frankfurt-am-Main.

———, 1965 : Herdenzahlung mit Hilfe von Luftbildern. *Die Erde* 96, 206-23.

Gambia (1958-59) : *Landuse, 1 : 25,000*, Directorate of Overseas Surveys.

Gould, P., 1966 : *On Mental Maps*, Discussion Paper No. 9, Michigan Inter-University Community of Mathematical Geographers.

Grove, A. T., 1951: *Landuse and Soil Conservation in Parts of Onitsha and Owerri Provinces*, Bulletin 21, Geological Survey of Nigeria.

Holleman, J. F. (ed.), 1964 : *Experiment in Swaziland*, Cape Town.

Hunter, J. M., 1961 : "Akotuakrom : A Devastated Coca Village in Ghana", *Transactions of the Institute of British Geographers,* 29, pp. 161-86.

International Bank, 1955 : *Report on the Economic Development of Nigeria*, Map 2.

Johnsrud, R. O., 1960 : "A Decade of Nigerian Cotton", *Nigerian Geographical Journal*, 3, Fig. 3.

Mabogunje, A. L., 1959 : "Rice Cultivation in Southern Nigeria", *Nigerian Geographical Journal*, 2, pp. 59-69.

McMaster, D. M., 1962 : *A Subsistence Crop Geography of Uganda,* World Landuse Survey, Occasional Papers, No. 2.

Morgan, W. B., 1959a : "Agriculture in Southern Nigeria (excluding the Cameroons)", *Economic Geography*, 35, Fig. 2.

———, 1959b : "The Distribution of Food Crop Storage Methods in Nigeria", *Journal of Tropical Geography*, 13, pp. 58-64.

Mortimore, M. J., 1979 : "The Supply of Urban Foodstuffs in Nigeria," in Coppock J. T. (ed.) *Agriculture and Food Supply in Developing Countries*, Deptt. of Geography, University of Edinburgh.

Nigeria, 1957 : Onwards, *Ministry of Agriculture,* Kaduna.

Papadakis, J., 1965 : *Crop Ecologic Survey in West Africa*, 2 Vols., F.A.O.

Prothero, R. M., 1957 : "Landuse at Soba, Zaria Province, Northern Nigeria", *Economic Geography*, 33, pp. 72-86.

Sierra Leone, 1967 : *Agricultural Statistical Survey of Sierra Leone, 1965-66,* Appendix, Form AS9, Freetown.

Uganda, 1960 : *The Systems of Agriculture Practised in Uganda,* Memoirs of the Research Division, 5 Vols. Kampala.

Vine, H., Weston, V. J., Montgomery, R. F., Smyth, A. J. and Moss, R. P., 1954 : "Progress of Soil Surveys in South Western Nigeria", *Proceedings of 2nd International African Soils Conference*, 1, pp. 211-36.

Weaver, J. C., 1958 : "A Design for Research in the Geography of Agriculture", *Professional Geographers,* 10, p. 5.

15

AGRICULTURAL LOCATION THEORY

A Review

EDISON DAYAL

AGRICULTURE USES more space than any other human activity, and is still the most important provider of sustenance for the human society. As geography is mainly concerned with the spatial organisation of human activities, agriculture deserves special attention of geographers. For a long time geographers felt satisfied with unique explanations of real agricultural patterns mainly in terms of physical environment and rarely a passing reference was made to economic and social variables as determinants of agricultural location (Baker, 1926; Jonasson, 1925; Whittlesey, 1936). Nevertheless, this was a significant development and led to identification and establishment of agricultural patterns at different scales — comparable to the identification of elements and compounds by chemists before they entered the formulation of laws and principles (Weaver, 1958). The agricultural organisation in particular regions was explained in specific terms based on empirical work and, therefore, highly coloured by the point of view and the training of observers. Thus, the need arises for an objective and general explanation for the spatial organisation of agriculture unbiased by individual viewpoints. The last two-and-a-half decades have seen among geographers a growing interest in the theory of knowledge. Bunge particularly emphasised the need for a more theoretical approach to the analysis of geographic distribution (Bunge, 1962). The geographers in the 1950s had perhaps caught the spirit of scientific thought and had begun to divert attention from the particular to the general. This intellectual awakening of geographers revived much interest in several classical theories of location, and also led to a re-examination and elaboration of Von Thunen's theory of agricultural location.

Agricultural location is affected by several factors, such as soils, climate, slope of land, areal extent of plant and animal diseases, and cultural and emotional characteristics of people. Besides these physical, biological and social variables, agricultural organisation depends, perhaps more strongly, on relative location expressed in terms of distance.

The relative location is an economic variable, which influences the operational costs and disposal of farm products, affecting the net return. Because of the extensive nature of agriculture, distance acts as an important input. The influence of relative location on agriculture was first demonstrated by Johann Henrich Von Thunen in 1826.

Ricardo and Von Thunen

In order to explain the spatial variation of net income per unit of land both Von Thunen and Ricardo introduced the concept of economic rent independently. Ricardo describes economic rent as income arising from the use of land of superior fertility, in comparison with the income that comes from the use of marginal land. Ricardo said that the most fertile land (first grade) would be brought under cultivation first, and cultivation would extend to less productive land (second grade) as demand for agricultural products increased due to population growth. The increased demand will cause scarcity of products in the market, resulting in higher prices. This will enable farmers to bring less productive land (second grade) under cultivation for the welfare of society. When less favourable land or second-grade land is brought under cultivation, the farmers working on a more productive land will have an advantage over those cultivating second-grade land, in that they will earn additional income per unit of land because of the "original and indestructible power of the soil", that is, the difference in soil productivity. Thus, the additional income that the farmers get from the first-grade land becomes the rent of first-grade land.

McCarty and Lindberg have presented a diagrammatic version of Ricardo's theory of rent. In the centre of the diagram they indicate an area as having "optimum" conditions for the production of a certain crop (Fig. 15.1). The optimality is measured in terms of two variables — moisture and temperature. Towards the periphery from the centre the conditions become less and less favourable, and a limit is finally reached beyond which production would be impossible, owing to severe limitations of moisture and temperature. The cost of production will be the lowest in the central "optimum" zone and will increase towards the periphery. The rent of land will change the opposite direction and will be highest in the optimum zone. Beyond the zone of production feasibility there will be no rent. In Fig. 15.1 it has been assumed that a price of "7" will meet the cost of production up to the end of the fourth zone, and beyond it the production cost will be too high. The land beyond the fourth zone will be cultivated only if there is a rise in market prices, which will increase the marginal revenue productivity of land.

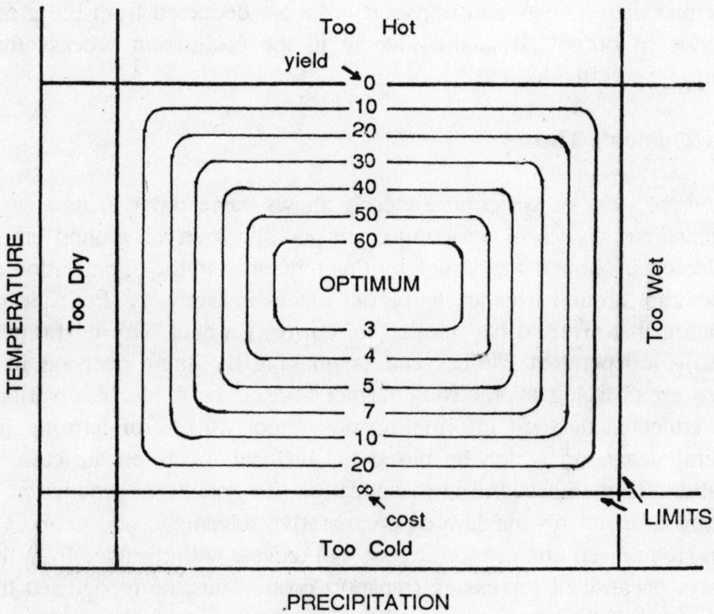

FIG. 15.1: Optima and limits schema.

Von Thunen explained the variation of economic rent in terms of locational advantage. He said that the economic rent of land would decrease with increasing distance from the market due to transport cost differences. The farmers close to market would pay less on transport costs and would make higher net income on output per unit of land which is a result of distance-decay function. The economic rent in Von Thunen's theory is often referred to as location rent — rent that arises from the advantage of location in relation to market.

It has long been recognised that the two theories of rent are very similar to each other and D.W. Jones have recently proved mathematically that the two theories are economically equivalent (Jones, 1978; Stevens, 1961). However, Thunen's theory is more known to geographers because of its more explicit concern with the spatial organisation of agriculture.

At this point a distinction must be made between ordinary rent and economic rent. The ordinary rent is the amount of money which a tenant pays to the landlord for the use of property for a specific period. Economic rent, on the other hand, is the surplus net income from land after production costs and transport costs are deducted from the gross revenue of output. It is the income in the production process that should be credited to land.

Von Thunen's Theory

The areal view of agriculture clearly shows some order in its spatial organisation. Types of agriculture are nodally arranged around urban settlements. Thunen was struck by this regularity in the organisation of agriculture around Rostock and other places in Germany. Perhaps the question that aroused his intellectual curiosity was : Why do farmers making independent choices end up making the same decision in a given area? Being a practising farmer himself he made observations and collected detailed information on various aspects of farming for several years, on which he based his brilliant thesis on agricultural location. Thunen recognised that location of agricultural production is mainly affected by the law of comparative advantage and economic rent. He realised that economic rent will decline with distance from the market because of increasing transport costs. Thus, he recognised the influence of distance on the arrangement of landuse patterns around markets. To prove his ideas he created a controlled environment with the help of a series of assumptions.

The Isolated State

At the centre of a fertile and extensive plain there is a large city. Throughout this vast plain the soil fertility, labour wages and all other costs of production are uniform. The plain turns into uncultivated wilderness at some great distance from the market and is not connected to the outside world. There is no transport route passing through the State. All parts of the State are, therefore, equally accessible to the central market through horse-drawn wagons, and transport cost is a direct function of distance. As there are no other towns in the State the central town must supply the manufactured goods to the entire rural community surrounding the town and, in return, get all its food and raw material requirements from the surrounding area. In this isolated State of uniform surface the only variable that will affect the economic rent is the distance from market. After creating this isolated State he expressed : "The Problem : what pattern of cultivation will shape under these conditions? and how will the farming systems of different districts be affected by their distance from the town (market)"? This statement of Thunen clearly indicates that he was concerned with a normative theory (describing how agriculture ought to be organised under certain assumptions), and that he was quite aware that the location of agriculture is affected by several factors. He made no claim that his theory was capable of explaining arrangement of landuse in the real world under all conditions. However, he realised that distance through its influence on transport cost determines the economic rent per unit of land, and affects the arrangement of landuse patterns around a central market.

The economic rent is a function of distance and may be expressed algebraically.

where $R = E(p-a)-E_{fk}$

R = economic rent

E = yield per acre

P = market price per unit weight of commodity

a = production cost per unit weight of commodity

f = transport cost per unit of distance per unit weight of commodity

k = distance

The above equation expresses a linear relationship between rent and distance, and can be graphically presented, as in Fig. 15.2.

At this point it must be mentioned that Thunen introduced two distinctly different ideas in his *Isolated State* — an intensity theory and a crop theory (a theory of agricultural location). These two theories are often confused with each other in the subsequent literature on agricultural location. Although the crop theory is indirectly concerned with intensity, the main emphasis is on something else (Hall, 1966).

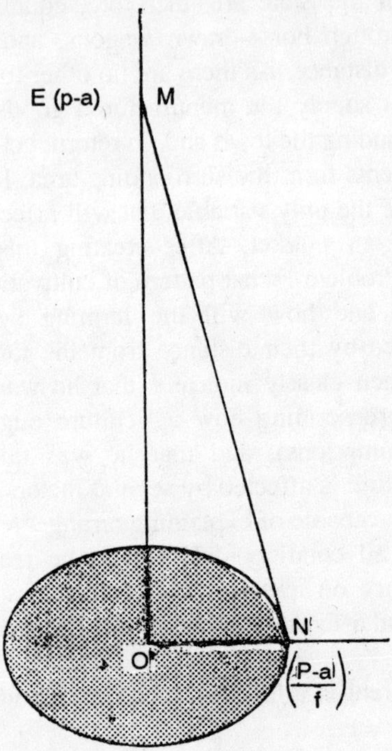

FIG. 15.2: Economic rent and location of market.

The intensity theory states that agricultural intensity decreases with distance from the market. In a situation where only one crop is produced around a market, its cultivation will be intensive near the market and extensive at some distance from it. This happens because the economic rent near the market is high due to savings on transport costs, which leads to high land values and occupancy rent. Thus, land is more valuable relative to labour and other inputs. The farmers who own land near the market strive to maximise output per unit of land by substituting other inputs for land, i.e. they try to maximise the profit on the most valuable factor of production. Furthermore, it pays to produce more per unit area, because the transport cost is lower near the market. Away from the market the economic rent declines, owing to high transport cost, which leads to low land values and occupancy rent.

Land becomes a cheaper input relative to labour and other inputs. The farmers maximise the utility of more expensive inputs. They use extensive cultivation and apply scarce labour and other inputs over larger areas. Hence, they produce more per unit of input but less per unit of land. In this way they also minimise the transport cost per unit to land.

Thunen's crop theory is concerned with agricultural location, i.e. the organisation of agricultural systems around a central market. The rationale, which determines the location of agricultural production, is the maximisation of profit — the economic rent — as in any other economic venture. Since economic rent is dependent on the value of outputs and the cost of inputs, a higher economic rent will be derived through intensive farming near the market for reasons explained earlier. However, in comparing the economic rent derived from diverse agricultural products or systems, we are concerned with different patterns of costs and returns. Different crops will have different yields per acre, costs of production per acre and a range of market prices, depending on the transportability of the crops. The products that have a high yield per acre, but have a low value per unit weight, will be produced near the market. Also, products that have a high yield per acre, high value per unit weight, but low transportability, like vegetables and milk, will be produced near the market. The products that have low yield per acre and low value per unit weight, but good transportability, will be produced far from the market. Thus, it is quite evident that the economic rent in the crop theory is a complex function of several variables. It is determined by the yield per acre, multiplied by the market price, minus production cost and transport cost.

When only one product is produced around the central market, the maximum economic rent per unit of land will be derived near the market *O* (Fig. 15.2), where is will be equal to $E(p-a)$. Its maximum extension outward from the market will depend on the slope of the rent function *MN*, and will be reached at the point *N*, where it will be $p-a/f$. At the point *N* the outermost limit of economically viable production will be reached where marginal revenue is equal to marginal cost. The maximum area of viable production will be the area enclosed by radius *ON*.

In a more complex situation, where several products are competing for land around the central market, the products that experience maximum cost reduction near the market will make the highest bid to occupy the land near the market.

In Fig. 15.3 product-I (whose economic rent function is expressed by the line *PQ*) can be produced profitably up to the point *Q* from *O*, but beyond the point *E′* product-II (whose economic rent function is *RS*) will be more profitable than product-I as it will yield more economic rent. Therefore, product-I will be produced over the distance *OE′*, and product-II over *E′S*. In a similar manner the outer and inner boundaries of several products can be determined. Dunn (1954) has presented an excellent extension of Thunen's theory and has mathematically determined the conditions for the formation of zones, entry of crops in the competition and market equilibrium.

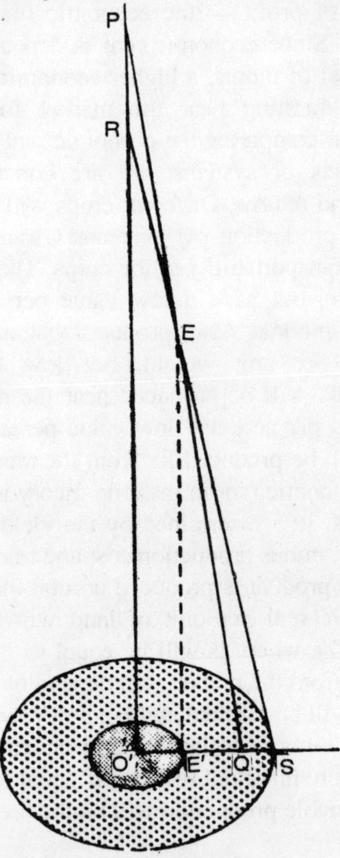

FIG. 15.3: Cost of production and location of market.

Von Thunen and the Real World

Von Thunen sought to identify the major principles that determine the diversity of agricultural systems around market centres. He first developed a pure theory of agricultural location in a restricted environment, knowing fully well that the relaxation of limiting assumption will distort the shape of the agricultural zones. By introducing a navigable river and multiple markets in the model, he demonstrated that despite distortions in the shape of agricultural zones their order of arrangement around the market would hold and would be determined by economic rent. His model is a partial equilibrium model as it does not take into account the variable character of demand, supply and prices. Day and Tinney (1969) introduced dynamic demand, farming practices and regions, and found that the theory in essence will be the same, but the number of possible equilibrium solutions will increase. For example, if we insert variable costs into the model, then we find that economies of scale will be more around larger markets relative to small markets. If savings on cost near large markets more than offset the transport cost of reaching smaller towns, then the influence of smaller towns on the organisation of agriculture will be negligible and the pattern will strengthen around large markets.

Thunen's theory has stood the test of time. It has been tested in a variety of environmental contexts and the spatial expression of the theory is well supported in a large majority of cases. It has been applied at the international, national, regional and micro-scales and has been found to hold its validity. Richard Peet (1969) applied it at the international scale to explain the expansion of commercial agriculture in the 19th century. He considered Western Europe and Eastern America as the world-urban industrial nucleus and demonstrated the existence of Thunen type spatial organisation of agriculture around the "world city". He examined the distance from which various agricultural products were imported into Britain and found that vegetables, dairy products and live animals were imported from much shorter distances compared to foodgrains and wool. As the transport technology improved the supply areas for agricultural products expanded, but vegetables, dairy products and live animals still came from a much shorter distance compared to foodgrains and wool. Crossley (1976) in a recent study demonstrated the existence of Thunen-type supply rings for the beef processing industry in England.

Muller (1973) and Griffin (1973) applied Thunen's model at the national scale for the USA and Uruguay, respectively, and both found that the spatial organisation of agriculture was in conformity with his ideas. Horvath and Dayal tested the model at the regional scale around individual urban centres and found strong evidence of the spatial expression of Thunen's model in Ethiopia and Australia. Blakie (1971) examined the spatial organisation of agriculture at the micro-scale, and found Thunen-type zones of agriculture in several villages in North-west India.

Although Thunen made no claim to the universality of his model in his time, or in the future, because it was based on a specific transport technology and local conditions in Mecklenburg, there is mounting evidence that his model is still valid in explaining spatial organisation of agriculture. "Modernisation forced changes in the universe and assumptions of the original model, but in spite of some inevitable problems stemming from the structure and from the theory, it is still applicable to urban areas of different scales than originally intended" (Kellerman, 1978). Nevertheless, Thunen's work has been subjected to criticism, and a variety of views have been expressed about his theory. Some critics have stated that distance from the market should not be the only criterion on which to base the analysis of agricultural location. Thunen was well aware of this fact and, therefore, he introduced other variables in the model, such as land quality, differentiated transport surface and multiple market centres. One important contribution of Thunen is that he introduced a method of analysis, which enables one to isolate an important factor from those considered important for examining its influence on agricultural location.

Some others have overexaggerated the influence of transport development on Thunen's theory. Although it is true that transport cost is no longer directly related to distance, it still continues to be an important variable in determining transport cost. For example, if the transport cost for 100 km for a given consignment is $ 10, for 200 km the cost may not be $ 20, but certainly a lot more than $ 10, and may well be only slightly less than $ 20. Also, Agricultural products in many countries are more commonly transported by road, in which cost is more dependent on distance. Furthermore, the disproportionately rising fuel prices will greatly enhance the importance of distance in locational decisions in the near future, and will provide greater support to Thunen's logic than ever before. The other advantages, such as land quality, are very likely to become much less important relative to distance.

Thunen's major concern was the analysis of the mechanism that operates in a commercial farming environment. Under such environment the differentiated land rent has been found to be an important regulator of commercial agriculture. It is apparent that intensification of landuse will be more economical where produce prices are higher. This concept is also supported by Marx who wrote :

> By its very nature, intensive agriculture, meaning systematic investment of 'capital in a given piece of land, is found predominantly, or to a greater extent, on better land.

Yet another view is that the influence of urban sprawl on the spatial organisation of agriculture is such that it tends to reverse the formation of Thunen type agricultural zones. The argument given is that Thunen assumed a static city in his postulates, but cities in today's world are rapidly expanding over the surrounding areas, leading to a competition between urban and rural landuses in the urban fringe. According to Sinclair (1967), land near the expanding city has a high potential for urban use and, therefore, a farmer is reluctant to make investment in it because he knows that in the near future he may have to sell the land to the urban developers, so that farming intensity in terms of inputs is less near the city and increases with distance from the city. This leads to a reversal of farming intensity zones suggested in Thunen's intensity theory. Sinclair's views are, however, not based on any measured attributes of agriculture, but are purely conjectural. Only one study, that by Mattingly, has provided any empirical evidence to support Sinclair. But, Mattingly (1972) measured agricultural intensity only in terms of labour input, which for a technologically advanced nation, such as the USA, cannot give a proper idea of farming intensity, because of a whole range of inputs including enormous capital used in agriculture. Recently, Dayal (1974) tested the spatial expression of Thunen's theory around two rapidly growing towns over space and time in Australia. His results seem to refute Sinclair's views. Around both centres the intensity of cultivation was found to be declining with distance. The test of variation of farming intensity over time on sampled farms, within a zone where farmers may anticipate urban expansion in the near future, also did not provide any support for Sinclair (Fig. 15.4). On a majority of farms the farming intensity was either constant or was slightly increasing. A few farms that gave some indication of declining intensity had a rate of decline so small that it was statistically not significant. Interviews with farmers within this zone revealed that a majority of them had made no cuts in the inputs used. Some had, in fact, increased the level of fertilisers and labour inputs. The tendency of increasing farming intensity on farms within the urban fringe appears to be understandable. Knowing that they will shortly have to sell the land and move out, farmers try to maximise short-term returns. Investments yielding long-term benefits may have already been made by the farmers, perhaps long before they began to fear urban encroachment, as for example, the investment in farm machinery, farm buildings, fences, which have a long life. The investments that yield short-term gains, such as investments in seeds, fertilisers and insecticides do not involve large sums and often repay the same year. Hence, urban anticipation does not appear to be a strong reason that would discourage a farmer from making short-term investments. Also, if a farmer is a rational decision-maker, as assumed by Thunen and maintained by Sinclair, he will want to maximise his profit. He will, therefore, use his land productively almost up to the time he actually sells it. This may also be made necessary by increased taxes. By stopping cultivation or by reducing intensity a farmer gains in no way. Once the land is sold for urban use it becomes urban land and it is not fair to include it in calculating farming intensity. Thunen based his theory on the land actually used for agricultural production and so it should be tested accordingly.

Conclusion

Von Thunen's theory of agricultural location has been tested in a variety of environmental contexts and has proved a useful tool for explaining the spatial organisation of agriculture. The basic pattern of agriculture in Europe and Anglo-America is in keeping with the order found in Thunen's theory (Muller, 1973; Peet, 1969; Crossley, 1976). Chisholm (1962) cited Thunen-type zones around cities in Spain, Italy, Finland, the Netherlands, Bulgaria and Pakistan. Similar zonation of agriculture around cities has been noted in Africa, Asia and Latin America (Horvath, 1969; Eyre, 1959; Griffin, 1973; Blakie, 1971). Whereas there is mounting evidence, drawn from a variety of physical, economic and social environment, that supports Thunen's theory there is as yet no solid empirical evidence to show that it is no longer valid. The energy crisis, which the world is now facing, has led to rapidly escalating transport costs. Thus, in future, the spatial organisation of agriculture will be more in keeping with the arrangement found in Thunen's theory.

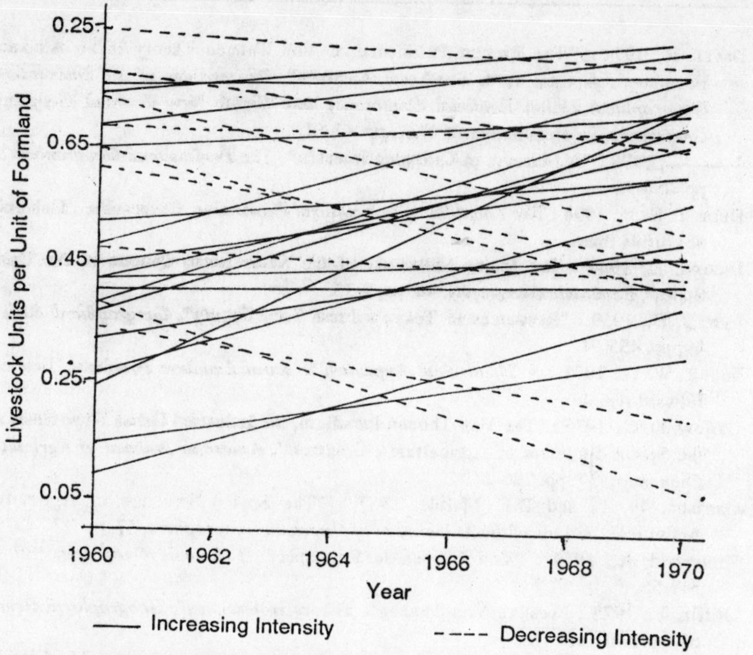

FIG 15.4: Trend of farming intensity on selected farms.

Thunen's theory of agricultural location is based on a solid empirical foundation and logical reasoning that few would dispute.

REFERENCES

Alonso, W., 1965 : *Location and Landuse*, Cambridge : Harvard University Press.

Baker, O. E., 1926 : "Agricultural Regions of North America", *Economic Geography*, 2, pp. 459-93.

Blakie, P. M., 1971 : "Spatial Organisation of Some North Indian Villages", *Transaction of the Institute of British Geographers*, 52, pp. 1-40.

Bunge, W., 1962 : *Theoretical Geography*, Lund : Lund Studies in Geography.

Chisholm, M., 1966 : *Rural Settlement and Landuse*, London : Hutchinson University Library.

———, 1969 : "The Relevance of Von Thunen", *Annals of the Association of American Geographers*, 59, p. 401.

Crossley, J. C., 1976 : "Location of Beef Processing Industry", *Annals of the Association of American Geographers*, 66, pp. 60-75.

Day, R. H. and E. H. Tinnery, 1959 : "A Dynamic Von Thunen Model", *Geographical Analysis*, 2, pp. 137-51.

Dayal, E., 1974 : "The Present Pertinence of Von Thunen Theory in an Advanced Economy : A Case from Southeast Australia", *Proceedings of the International Geographical Union Regional Conference and Eighth New Zealand Geography Conference*, Palmerston North, N.Z., pp. 43-51.

———, 1978 : "A Measure of Cropping Intensity", *The Professional Geographer*, 30, pp. 289-96.

Dunn, E. S. Jr., 1954 : *The Location of Agricultural Production*, Gainesville : University of Florida Press.

Durand, L., 1964 : The Major Milksheds of the Northeastern Quarter of the United States", *Economic Geography*, 40, pp. 9-33.

Eyre, J. D., 1959 : "Resources of Tokyo's Fresh Food Supply", *Geographical Review*, 49, pp. 455-74.

Found, W. C., 1971 : *A Theoretical Approach to Rural Landuse Problems*, London : Edward Arnold.

Garlow, D. C., 1975 : "The Von Thunen Paradigm, the Industrial Urban Hypothesis and the Spatial Structure of Agriculture : Comment", *American Journal of Agriculture Economics*, 57, pp. 726-27.

Garrison, W. L. and D.F. Marble, 1957 : "The Spatial Structure of Agricultural Activities", *Annals of the Association of American Geographers*, 57, pp. 137-44.

Grotewald, A., 1959 : "Von Thunen in Retrospect", *Economic Geography*, 35, pp. 346-55.

Griffin, E., 1973 : "Testing Von Thunen's Theory in Uruguay", *Geographical Review*, 63, pp. 500-16.

Gregor, H., 1970 : "Large Industrialised American Crop Farms, *Geographical Review*, 60, pp. 151-75.

Hall, P. (ed), 1966 : *Von Thunen's Isolated State*, An English Edition of *Der Isolierte Staat, London* : Pergamon Press.

Harvey, D. W., 1966 : "Theoretical Concepts on the Analysis of Agricultural Landuse Patterns in Geography", *Annals of the Association of American Geographers*, 56, pp. 361-74.

Horvath, R. J., 1967 : "Von Thunen and Urban Sprawl" *Annals of the Association of American Geographers*, 57, pp. 811-12.

———, 1969 : "Von Thunen's Isolated State and the Area Around Addis Ababa, Ethiopia", *Annals of the Association of American Geographers*, 59, pp. 308-23.

Hoover, E. M., 1971 : *An Introduction to Regional Economics*, New York : Alfred A. Knof.

Isard, W., 1956 : *Location and Space Economy*, New York : John Wiley.

Johnson, H. B., 1962 : "A Note on Thunen's Circles", *Annals of the Association of American Geographers*, 52, pp. 213-20.

Jonasson, O., 1925 : "Agricultural Regions of Europe", *Economic Geography*, 1, pp. 277-314.

Jones, D. W., 1978 : "Rent in Equilibrium Model of Landuse, *Annals of the Association of American Geographers*, 68, pp. 205-13.

Jumper, S. R., 1969 : "The Fresh Vegetable Industry in the U.S.A. : An Example of Dynamic Inter-regional Dependency", *Tijdschrift Voor Economische en Sociale Geografie*, 60, pp. 308-18.

Katzman, M., 1974 : "The Von Thunen Paradigm, the Industrial Urban Hypothesis and the Spatial Structure of Agriculture", *American Journal of Agricultural Economics*, 56, pp. 683-96.

Kellarman, A., 1978 : "Determinants of Rent from Agricultural Land Around Metropolitan Areas", *Geographical Analysis*, 10, pp. 1-12.

Krymoneski, R., 1928 : "Graphical Presentation of Thunen's Theory of Intensity", *Journal of Farm Economics*, 10, pp. 461-82.

Losh, A., 1954 : *The Economics of Location*, New Haven : Yale University Press.

Mattingly, P. F., 1972 : "Intensity of Agriculture Landuse Near Cities", *The Professional Geographer*, 24, pp. 7-10.

McCarty, H. H. and J. B. Lindberg, 1966 : A *Preface of Economic Geography*, Englewood Cliffs, N. J. : Prentice Hall.

Muller, P. O., 1973 : "Trend Surface Analysis of American Agricultural Patterns : A Macro-Thunian Analysis", *Economic Geography*, 49, pp. 228-42.

Muller P. O. and G. J. Diaz., 1973 : "Von Thunen and Population Density", *The Professional Geographer*, 25, pp. 239-41.

Norton, W. and E. C. Conkling, 1974 : "Landuse Theory and the Pioneering Economy", *Geografiska Annaler*, 56B, pp. 44-56.

Peet, J. R., 1969 : "The Spatial Expansion of Commercial Agriculture in the Nineteenth Century : A Von Thunen Interpretation", *Economic Geography*, 45, pp. 283-301.

Pred, A., 1967 : Behaviour and Location : Foundations for a Geographic and Dynamic Location Theory, Part I, Lund Studies in Geography, Series B, *Human Geography*, 27.

Rakitnikov, A. N., 1978 : "Von Thunen and the Significance of His Work for the Geography of Agriculture" *Soviet Geography*, 19, pp. 567-71.

Sinclair, R., 1967 : "Von Thunen and Urban Sprawl", *Annals of the Association of American Geographers*, 57, pp. 72-87.

Stevens, B. H., 1961 : "Linear Programming and Location Rent", *Journal of Regional Sciences*, 3, pp. 21-25.

Stryker, J. D., 1976 : "Population Density, Agricultural Technology and Land Utilisation in a Village Economy", *American Economic Review*, 66, pp. 347-58.

Whittlesey, D., 1936 : "Major Agricultural Regions of the Earth", *Annals of the Association of American Geographers*, 26, pp. 199-240.

Weaver, J. C., 1958 : "A Design for Research in the Geography of Agriculture", *The Professional Geographer*, 10, pp. 1-5.

16

QUANTITATIVE METHODS IN SPATIAL ANALYSIS OF AGRICULTURE

K. BIELECKA, M. PAPRZYCKI AND
Z. PIASECKI

THE NEED for the application of quantitative methods has been rising steadily. This is due to two circumstances : first the problems researchers face now have been increasingly complex, and they can often be solved only by handling observation materials that are huge and hard-to-grasp and interpret; secondly, there is the old drive in science toward employing methods as objective and producing results as accurate as possible. Quantitative methods safeguard objectiveness of research procedures, they lead to unequivocal and precise results, secure the repeatability of definite research procedures with various empirical materials, the identity of results regardless of the researcher's idiosyncrasies and of the place and time at which they are employed, and they make possible the use of techniques suitable for long calculation series as well as EDP techniques that save a lot of effort and shorten the time necessary for arriving at the results.

The utilisation of quantitative methods in the non-technological research domains, in which they have never been employed before, is usually a hard and complex task. Want of experience may jeopardise any new attempt at applying quantitative methods. The dilettantish "mathematisation" of researches, which is rooted in the belief that the above-mentioned advantages of quantitative methods by themselves guarantee the substantial correctness of analytical results, involves the greatest dangers in this respect. It is tantamount to attributing to mathematical methods some magic power by which they are supposed to unravel all methodological problems any given research project may involve. A researcher, who is uncritical enough of the power of mathematics and who may not have fully mastered the morphology of the phenomenon he is to analyse, will venture headlong into an interpretation of his calculations heedlessly ignoring the possibility of their being theoretically erroneous. On the other hand, a critical approach on the part of a branch specialist well acquainted with the morphology of the given phenomenon to the results obtained by means of quantitative methods and his rejection of them when there is doubt as to their adequate representation of reality will prevent him from endorsing a false interpretation but — and this is what most often happens — it may also discourage him from taking recourse to the research apparatus of quantitative methods. Research experience and intuition are doubtless more reliable as foundations for the exploration of phenomena than inadequate uses of even the most sophisticated and accurate methods. Dubious results are the main reason for the rarer than necessary application of quantitative methods in research projects in the non-technological domains.

We must realise that the use of quantitative methods is in many cases simply a necessity rather than a whim. Hence it is imperative to be fully aware of certain pre-conditions for the application of such methods to be effective enough.

Quantitative methods can be employed effectively only if definite requirements are fulfilled. The inter-disciplinary approach, that is, the collaboration — of the given branch specialist (geographer, economist, planner, etc.) intending to use quantitative methods in his study with the mathematician — in formulating and solving the research problem, is of fundamental significance in this respect. The specialist has his task in defining the subject and the aim of the research and in formulating the questions the study is to answer. The mathematician, on the other hand, must match the quantitative method — or sometimes even construct a new one — that is likely to provide adequate answers to the questions posed by the specialist. Thus the mathematician must scrutinise the subject-matter and the research aim as they have been defined by the specialist and to check the information furnished by the latter for its adequacy to the subject and the aim of the given research project.

Experience has shown that branch specialists find it very hard to formulate adequately their questions. In such cases this methodologically important link in the cognitive process is bypassed and the research instead undertaken on the grounds of intuition and conjecture alone.

The consequences of conducting research without properly

formulated questions, without a precise enough definition of the subject-matter and aim of the research and without correct information matching come to light in doubtful results. It is then that the quantitative methods are often blamed in the sense that "the mathematics has failed"; mathematicians rejoin by insisting that "mathematics does not lie and is pitiless". This reply implies that substantially (cognitively) dubious results are a specific test of both the adequacy of the quantitative method employed and of the methodological aspect of the study undertaken. Dubious results point to the need of submitting the entire research programme to a verification. The guiding rule in this respect says that the more precise the given quantitative method is the more profound must be the knowledge of the morphology of the research problem.

The inter-disciplinary work method requires a definite amount of knowledge of the properties of quantitative methods on the part of the branch specialist whereas the mathematician must have some knowledge of the problem to be studied. It is moreover very important to realise that the conditions for the application of quantitative methods are of a double aspect, and hence that their effectiveness is double too : first, their formal effectiveness depends on the match of the method — the most efficient one of all adequate ones — for solving the given task, and then there is their substantial (cognitive) effectiveness — which depends on whether the matched most efficient quantitative method has been applied to the collected observational materials that are adequately chosen for the subject and the aim of the given study. Hence follows that practical indication that a knowledge of the efficiency of quantitative methods is indispensable. The possibilities in this respect have so far been fairly limited because of the relatively modest knowledge available as compared to the current needs in this respect. Whenever we are unable to draw on other researchers' experiences we are then compelled to precede the special study proper (in which the quantitative methods are to be applied) by an assessment of the effectiveness of those methods that are practicable in the given case of research problem. It is quite possible that none of the methods known so far will fulfil all the conditions put to it. Then one has either adapted some of those most suitable for our purpose or else taken recourse to mathematical ideas to provide the foundation for a new efficient method.

The above praxiological remarks are based on the experience gained in studies in agricultural geography intended mainly to assess the possibilities of applying quantitative methods in the typology of agriculture : to the mathematician, the task following from the research aim thus conceived was to answer the question — which of the quantitative methods might be most suitable for developing a method of agricultural typology.

The present study discusses the results obtained so far by our research team in this line of study.

Scope of Studies

To solve the task the mathematician had to pick the methods suitable for the given typology and next to diagnose the selected methods with a view to matching the one that was most efficient of all.

As the theoretical starting point, the following definition of the notions of typology and type were taken:

— a typology is a systematising operation which involves the identification and grouping of similar objects by virtue of definite properties in mathematical language. Such an operation is called the classification of a set of multi-feature objects, that is, objects described by several variables;

— the notion of type is used in classificatory description; in the case of empirical material a type is the set of all objects from different populations and displaying a definite similar combination of features (characteristics);

— type is conceived of as an object or a notional construct in the nature of a pattern for systematisation and the scientific description of the given domain of reality;

— type of agriculture is a taxonomic, hierarchical and dynamic notion, it is determined on the basis of mutual relations between all significant ownership, organisational-technical and production features of agriculture;

— the combination of features determining the given type is recurrent, which implies that the given empirical type of agriculture may occur in various geographically not adjacent areas at the same time or at different intervals;

— the notion of type of agriculture is not identical with that of agricultural region; in contrast to type, which is a taxonomic notion, a region is a territorial notion, one which is non-recurrent, unique, and geographically continuous (spatially compact or cohesive); a really existing empirical type may be connected with a region in that the spatial pattern of types of agriculture may be used as the foundation for delimiting agricultural regions; the procedure of passing from the pattern of types of agriculture to that of regions consists in generalising the types occurring in the given area; with the boundaries of the regions being determined in virtue of a definite structure of types, that is, of the dominance of one or the codominance of two or more types in the given area.

The kind of quantitative methods that are to be submitted to appraisal for the prospective research project is determined by the main methodological problems of the typology of agriculture. These include:

1. The selection and adequate expression of the diagnostic features of the types of agriculture, and
2. The method of grouping the set of objects (spatial units) that safeguard the comparability of results in time and space.

The following are the possibly adequate quantitative methods of solving the problems involved in the above-mentioned methodological problems of agricultural typology:

1. The methods of factor analysis, which are used to:
   (i) pass from the system of mutually uncomparable features to a system of statistically comparable factors,
   (ii) analyse the dependences between the diagnostic features.
   (iii) reduce the set of diagnostic features to a smallest possible collection of factors, without any significant loss of information contained in the original collection of features,
   (iv) analyse the similarities between objects.
2. The methods of numerical taxonomy, which are applied to:
   (i) order the manufacture objects, that is, to represent in a formal way the structure of similarities between the objects of the analysed set,
   (ii) classify, that is, divide the non-homogeneous set of multi-feature objects into groups mutually similar in the features studied.

The scope of assessment of the selected quantitative methods follows in our researches from two possible situátions in set classification:

First, when there are certain patterns to which we can compare the objects analysed; then the task consists in their discrimination or identification,

Second, when no patterns are available, then the task consists in dividing the set into groups of similar objects.

The scope of analysis of the methods is adequate to their specific structure; this latter has its elements in the measure of similarity of objects, the criterion of division of the set, and the algorithm of division of the set (the plan for indispensable operations).

There are many similarity measures, division criteria and algorithms. Various taxonomic methods differ in their selection which accounts for the different divisions of the same set that each of those methods yields on application.

As there are no mathematical recommendations in this respect, the choices of the similarity measure, of the division criterion as well as of the form of the algorithm are arbitrary. The only condition for the similarity measure is that it should represent as truly as possible the dependences between the objects analysed and be adequate in its substance to that of features describing the object (in our case, to the substance of the diagnostic features of agriculture). Due to the integral connection between the diagnostic features and the şimilarity measure of objects, the choice of that measure should be preceded by the question: which similarity measure will be most adequate to the collection of features to be studied? In the case of "distance" type measures it is also necessary to seek an answer to the reverse question: which features and weights should be adopted for the calculated distance (similarity) to be as close to reality as possible?

Every grouping procedure contains an implicit or explicit criterion of division of set. The selection of such a criterion is a troublesome matter while its adoption is arbitrary, for we have not yet identified any super-criterion by which we might appraise the criteria for the optimality of division pertinent to the various grouping procedures at the phase of constructing the method and its algorith. Hence, the validity of any given division criterion is checked indirectly, on the ground of the results of grouping the given set by various methods. The solution we propose is to adopt a criterion minimising intra-group and maximising inter-group differences within the set to be divided; this is done in respect to the variation in the descriptive features. Such a criterion could be, for example, the maximum of a function that is the ratio of the multi-dimensional inter-group variance to the multi-dimensional intra-group variance. Unfortunately, it has not been possible so far to solve numerically the task thus formulated.

The many similarity measures and criteria for set division as well as the arbitrary nature of their choice in constructing grouping methods on the one hand and the researchers' engrossment in quantitative classification methods on the other have produced a huge variety of grouping procedures. Yet as no satisfactory theory of classification was at hand at the time when we were undertaking our inquiry into the possible applications of such methods we had no clear idea of which of them might be formally superior to others. The prevailing view then was that, for practical reasons, the most reasonable way is to experiment with methods picked for the specific needs of the given discipline until it has become clear that one of them is most suitable for the solution of the given problem.

The selection of the diagnostic features is a problem in itself. This is by no means a simple task, as here, again, no theory of feature selection has been developed yet. This accounts for the arbitrary nature of such selections; it further explains why the substantial adequacy of the collection of diagnostic features is so dependent on the knowledge of the morphology of the research problem the selecting specialist commands and on his experience and intuition. Mathematically, there are no recommendations regarding the selection of the features for the description of the object. The one possibility there is, is to make a formal appraisal of the given collection of features (which can be made, for example, by factor analysis) for their statistical properties to gain thereby some idea as to which of the features should be adopted and which ruled out. Let us emphasise here that by doing so one gets no assessment of the completeness of the collection of features: in other words, a formal appraisal of some such collection will by itself not indicate that certain features not included in that collection would characterise the given problem more adequately. These limitations notwithstanding, the formal appraisal of future collections is of fundamental importance to the results of set grouping, for by ruling out the dispensable features one gets a clearer picture of the set's division.

The next methodological problems to be solved relate to the phase of statistical preparation of the observation materials the branch specialist has collected; that phase precedes the classification proper.

The statistical materials collected are, as a rule, expressed in natural measurement units (m, kg 1 ha, piece, etc.) which are obviously mutually non-summable, whereas the collection of features often contains mutually dependent ones. The task the mathematician faces here consists in finding some way of transformation of the statistical material into a system of summable and mutually independent features. Theoretically, there are several ways to do that, yet the want of any clear indications makes their choice a matter of intuition for the practical researcher and leaves it an open problem to theory.

Our researches into the possibilities of application of the apparatus of quantitative methods for the purpose of agricultural typology comprise issues relating both to the phase preceding the classification and the classification proper. What are to be explained include the following:

1. The use of definite quantitative methods for assessing how suitable the object-describing features are for diagnosis;
2. The effect of different ways of passing to the system of summable features on the procedure and the results of grouping the set;
3. The suitability of available recommended grouping methods for agricultural typology; and
4. Whenever necessary, the possibility, of constructing a method more efficient than the one available on the ground of mathematics.

This report presents the results of studies under way. These latter have been carried to various points of advance in solving the specific substantial problems, but on the whole we are still far from a solution to the entire problem of a quantitative method of agricultural typology. Nonetheless even in that fragmentary form the results of our studies do explain quite a few significant issues thus bringing us closer to a global solution.

We have based our assessments on the results of many tests Bielecka, Paprzycki and Piasecki, 1975, 1979a, 1980, made on empirical materials that constitute different purposefully selected sets of sizes big enough for the efficiency of the given method to be tested experimentally.

More Significant Results so far Obtained

Possible Applications of Factor Analysis in Agricultural Typology

In our researches we have considered the following problems:

(i) The suitability of factor analysis in diagnosing the object-describing variables;
(ii) The effect of extremum units on the results of factor analysis; and
(iii) The possibilities of using factor analysis for the classification of sets.

1. It has turned out that the R-technique of factor analysis as a statistical method is suitable for a formal assessment of the collection of the diagnostic features of agricultural types. The method indicates which of the features in the collection should be retained and which discarded, but says nothing about the completeness of the collection of the features with respect to the studied problem, that is, the results of factor analysis alone do not suffice to state that the statistically analysed collection of features is complete or whether or not other features would perhaps characterise the analysed object (phenomenon) better.

2. Our studies have further shown the suitability of the R-technique of factor analysis for the preparation of statistical materials in the phase preceding the classification of set proper. The passage from a system of diagnostic features to the system of common features (Hotelling components) is purposeful and justified. As we could gather, factor analysis largely rules out all information that is of little significance for the differentiation of units as well as dispensable (or apparent) information, referred to as information noise, whereas it reinforces the significance of information differentiating the studied material. Factor analysis enables the researcher to handle set objects described by component scores, a task much easier than handling the original values of the diagnostic features.

Our attempts at grouping sets by Berry's method, which relies on the original values of diagnostic features, have stumbled over considerable difficulties in the identification of groups in the studied sets; the minimum distance values determined by that method tend to increase almost identically, which prevents the statement of existence of groups in the analysed set on this ground. Groups can be stated to occur during dendrite analysis only, though there too group identification is difficult. In an attempt at grouping by Berry's method based on the values of Hotelling components group identification turned out to be much easier, for significant changes (jumps) can be observed in the analysis of the minimum distance values already; this vindicates the confirmation of the existence of groups in the set which are then relatively easy to distinguish by means of the dendrite.

3. Extremum units occurring in a set can strongly affect the results of factor analysis. How great that effect actually is cannot, however, be said without certain relevant tests. For this reason, factor analysis should be preceded by and analysis of the original material as regards the occurrence in it of extremum values and their possible effect on the structure of relations between the diagnostic features.

4. In considering the possibilities for utilising factor analysis in carrying out a proper set classification we examined the efficiency of the "popular" graphical method of division of the set in the system of two most significant common factors determined by the R-technique of factor analysis, and the efficiency of the Q-technique.

In the graphical method, each object in the set is described by the values of only two common factors and is represented as a point on a plane (in a rectangular coordinate system), with the values of the first and second factors being its coordinates. The representation of all objects of the set in the system of rectangular coordinates as points on a plane yields a picture of the mutual situation of objects and the similarity structure between objects in a system of the two factors adopted. The distance between points indicates the similarity of the set's objects. On the ground of a visual appraisal of the distance between the points one can draw the lines of division of the set of points into their subsets, which is here equivalent to a division of the set into groups of similar objects. The results obtained by that method are the more reliable the greater the scope of common variation (part of the information contained in the original material) that is explained by the first common factors. The division of the set is "coarse" because the method of visual analysis is rather imprecise. Whenever a less accurate reading is sufficient for research purposes the simplicity of the method together with the possibilities of its easy graphical presentation (Fig. 16.1) speak in favour of its application. But when a more accurate division of the analysed set is necessary all common factors have to be taken into account, and in such cases the graphical method is impracticable.

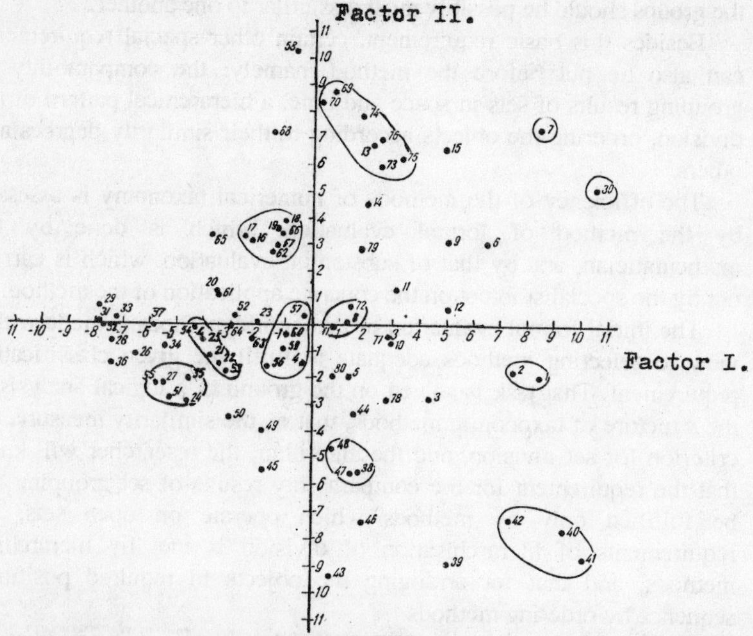

FIG. 16.1: Division of the set in the system of rectangular coordinates (example of the division the set composed of 80 spatial units of Poland).

When applying the Q-technique of factor analysis to the division of a set one must bear in mind that it may yield a division into classes greatly differing as regards the degree of their self-homogeneity. Hence it seems that every division of a set by that technique ought to be submitted to a verification on the ground of an assessment of the degree of self-homogeneity of the classes it yields. That means that the measure of suitability of the method is determined by the concrete result of the classification made which depends on the material to which the method is applied, and it is the individual researcher who decides whether the result obtained is sufficient or not for his research aims and whether it can be accepted or not.

Techniques of Evaluating the Efficiency of Classification Methods

The basic requirement of a method of classification of a set of multi-feature objects is that the singled out groups should include objects as similar to one another as possible, and, at the same time, that the groups should be possibly most dissimilar to one another.

Besides this basic requirement, certain other special requirements can also be put before the method, namely: the comparability of grouping results of sets in space and time, a hierarchical pattern of the division, ordering the objects according to their similarity degrees and others.

The efficiency of the methods of numerical taxonomy is assessed by the method of formal evaluation, which is done by the mathematician, and by that of substantial evaluation, which is carried out by the specialist intent on the concrete application of the method.

The initial formal evaluation of taxonomic methods is made with a view to selecting methods adequate to fulfil the given classification requirement. That task is solved on the ground of a logical analysis of the structure of taxonomic methods, that is, the similarity measure, the criterion for set division, and the algorithm; the researcher will know that the requirement for the comparability results of set grouping can be fulfilled only by methods which operate on open sets, the requirements of hierarchisation of division is met by hierarchical methods, and that for arranging the objects in required positional sequence by ordering methods.

The choice of the most efficient method, that is, that one likely to yield the best division of the set, from among the given group of adequate methods, can with the present state of things be made by comparing the divisions of the given set by several such methods.

One method of formal evaluation of goodness of a set's division which is frequently applied is the logical analysis of the observation material and of the divisions obtained as well as personal knowledge and intuition. That method enables the researcher to evaluate the following:

1. The sharpness of the divisions, that is, he can find out at which procedure groups tend to be distinguished most clearly;
2. The inclination of the method, that is, he can disclose the structure of the given division, for example, whether there is an inclination to form small numbers of large groups, or to join the objects into chains, etc.;
3. The minuteness (level of generality) of the divisions; and
4. The internal homogeneity of the groups in a highly generalised degree.

This potentially possible scope of recognition and formal evaluation of goodness of the set's division, by applying logical analysis in combination with personal knowledge and intuition, is insufficient and, therefore, the substantial evaluation by a branch specialist is of great significance. The latter using all his knowledge of the morphology of the classified object should decide whether the obtained division of the analysed set is, in his opinion, acceptable or not. The knowledge of the object classified plays here a fundamental role and one can rely on it. Yet there is the justifiable need for caution in judging because the branch specialist evaluates the set division obtained by thought comparing it to his own idea of the structure of a multi-feature phenomenon, which he had up to the moment analysed with regard to the single features and hence it may not coincide fully with the complex reality it is to represent. Moreover, one should also consider that the classification concerns always only the scope of information contained in the adopted collection of descriptive features and hence the results of the grouping of the set are approximate to the reality in such a degree in which the collection of features adopted and the manner of their measurement are complete with respect to the object classified.

The substantial evaluation of the goodness of the set's division may be correct only in the case of a good knowledge of the morphology of the studied subject. However, it often happens that the subject classified is not well-known yet and then the substantial evaluation of the goodness of the set's division can be based only on intuition, which is not a satisfactory enough criterion.

Thus, both the "newness of the subject" and the researcher's "uncertainty" as to which of the divisions of the set should be chosen to channel the search for solutions toward quantitative criteria, and which would be objective, and very sharp tools for diagnosting divisions of sets.

Formal Criteria for Evaluating the Goodness of Divisions of Set

Science has not, as yet, worked out the formal criterion of the evaluation of goodness of division of a set of multi-feature objects. Elaboration of such a univocal criterion — minimising the intra-group differences and maximising the inter-group differences, on account of the variability of descriptive features — is a very involved problem and numerically a hard-to-solve task. We attempt to solve it in the way of successive approximations. In doing so we propose to employ collections of formal purposefully matched partial criteria.

The first proposal in this respect was made by F.A. Szczotka, who adopted the following collection of criteria for a formal evaluation of division of set:*

I. Indexes characterised the frequency of groups :
   1. Number of one-object groups.
   2. Number of objects in the most numerous group.
   3. Variance of the number of the objects in groups.

II. Indexes characterised the homogeneity of the groups
   4. Sum of squares of distances inter-groups.
   5. Average square of distances inter-groups.
   6. Relative sum of squares of distances inter-groups.
   7. Sum of squares of distances from the centre of gravity inter-groups.
   8. Sum of squares of distances between groups.
   9. Ratio of the sum of squares of distances between groups and sum of squares of distances from the centre of gravity inter-groups.

Using F. A. Szczotka's results M. Paprzycki suggested a "System of 14 Criteria" in working out which he was guided by the rule that various aspects of the intuitive taxonomic premiss of minimising intra- and maximising inter-group differences ought to be taken into account. These are the criteria he has proposed:

1. Average inter-group distance $K_{10} = d_{sm}$

$$(1) \quad d_{sm} = \frac{2}{n(n-1)} \sum_{i=1}^{n} \sum_{j=i+1}^{n} \min_{\substack{k \in N_i \\ l \in N_j}} (d_{kl})$$

---

* F. A. Szczotka analysed the goodness of divisions of a set of spatial units of Poland, classified by the following method: the single-linkage method, the complete linkage method, the paired-group method, the Hubert divisive method A, the Hubert divisive method B [23].

2. Maximal inter-group distance $K_{11} = d_{mx}$

$$(2)\quad d_{mx} = \max_{\left\{\substack{i,j \in N \\ i = j}\right\}} \min_{\left\{\substack{k \in N_i \\ l \in N_j}\right\}} (d_{kl})$$

3. Minimal inter-group distance $K_{12} = d_{mx}$

$$(3)\quad d_{mx} = \min_{\left\{\substack{i,j \in N \\ i = j}\right\}} \min_{\left\{\substack{k \in N_i \\ l \in N_j}\right\}} (d_{kl})$$

4. Proportion of average inter-group distance of average of averages inter-group distances $K_{13} = \frac{d_{sm}}{d_{ws}}$

$$(4)\quad d_{ws} = \frac{1}{n} \sum_{i=1}^{n} \sum_{\left\{\substack{j \in N_i \\ j \neq k}\right\}} \sum_{k \in N_i} d_{kj} \frac{1}{(n_i - 1) n_i}$$

where: N—is the set of numbers of groups

n—is the number of groups

d_{kj}—is the euclidean distance between the k-th and j-th objects (spatial units)

N_i—is the set of numbers of objects belonging to the i-th group

n_i—is the frequency of the i-th group.

5. Product of maximal inter-group distance and average diameter of groups $K_{14} = d_{mx} \cdot d_{wm}$

$$(5)\quad d_{wm} = \frac{1}{n} \sum_{i=1}^{n} \max_{\left\{\substack{j \in N_i \\ k \in N_i \\ j \neq k}\right\}} (d_{kj})$$

6. Product of maximal inter-group distance and average of averages intra-group distances $K_{15} = d_{mx} \cdot d_{ws}$
7. Proportion of average diameter of groups to minimal inter-group distance

$$K_{16} = \frac{d_{wm}}{d_{mx}}$$

8. Proportion of average of averages intra-group distances to minimal inter-group distance

$$K_{17} = \frac{d_{ws}}{d_{mx}}$$

9. Average inter-group ratio $K_{20} = d_s$

$$(6) \quad d_s = \frac{2}{n(n-1)} \sum_{i=1}^{n} \sum_{j=i+1}^{n} D_{ij}$$

10. Maximal inter-group ratio $K_{21} = d_m$

$$(7) \quad d_m = \max_{\begin{bmatrix} i,\, j \in N \\ i \neq j \end{bmatrix}} (D)_{ij}$$

11. Minimal inter-group ratio $K_{22} = d_m$

$$(8) \quad d_m = \min_{\begin{bmatrix} i,\, j \in N \\ i \neq j \end{bmatrix}} (D)_{ij}$$

12. Product of average inter-group ratio and average of averages intra-group distances $K_{23} = d_s \cdot d_{ws}$
13. Product of maximal inter-group ratio and average diameter of groups $K_{24} = d_m \cdot d_{wm}$
14. Product of maximal inter-group ratio and average of averages intra-group distances $K_{25} = d_m \cdot d_{ws}$

Criteria K_{10}, K_{11}, K_{12}, K_{20}, K_{21} and K_{22} are based exclusively on the inter-group and intra-group variance evaluated by means of diverse, contradictory manner.

Criteria K_{13}, K_{14}, K_{15}, K_{16}, K_{17}, K_{23}, K_{24} and K_{25} are based on the inter-group and intra-group variance. This means that these are some of many analytically possible manners of expressing the primary criterion of taxonomy, i.e. they make it possible to obtain as a result of taxonomy (classification) such a division that simultaneously fulfils the following two conditions:

1. The groups are internally homogeneous, and
2. The groups substantially differ from one another.

It should, however, be emphasised that all selected criteria are constructed of intra-object distances and are not directly influenced by the values acquired by the objects (spatial units) in their features which cause that they are not directly associated with the mutual positioning of objects in the space of the features.

In accordance with the given idea of the criterion of goodness of the division of a set, which implies that the most profitable situation is when the intra-group distance is minimal and the inter-group distance is maximal — the division of a set will get better with smaller values of criteria K_{11}, K_{14}, K_{15}, K_{16}, K_{17}, K_{20}, K_{21}, K_{23}, K_{24} and K_{25} and higher values of criteria K_{10}, K_{12}, K_{13} and K_{22}.

The Application

A System of 14 Quantitative Criteria was used in the formal evaluation dividing results of a set of 61 spatial units of the world, described by means of the codes worked out by J. Kostrowicki (1976) — classified by the five following taxonomic methods: WARD'S, Z Piasecki's ORLINE, M. Paprzycki's FARELL-mod, M. Paprzycki's GRAVITY and Z. Piasecki's IDWER; which were selected from among 16 taxonomic methods analysed earlier and matched in such a way that they represented various kinds of taxonomic methods and various areas of potential applications.

Ward's method (Vanzetti, 1975) is one of the dendritic methods. It is based on the principle of division of the set by using the criterion of the minimal growth of intra-group variance.

The ORLINE method (Bielecka, Paprzycki and Piasecki, 1980, 1979a, 1979b) represents diagraphical methods. This method, which is a formalisation of the idea of Czekanowski's diagraphic method (Czekanowski, 1913, 1909) utilises distances and orders the objects linearly.

The FARELL-mod method (Bielecka, Paprzycki and Piasecki, 1979b) represents the methods of pattern recognition. The idea of the FARELL method, worked out at Novosybirsk University, is retained in the modification. The FARELL method is an attempt at a simulation of methods which would be used by man if he visually divided a set of points in the two-dimensional space into internally homogeneous groups. Its modification by M. Paprzycki consists of introducing the requirement of the affiliation of the object to a group with the centre of gravity lying closer to the geometrical centre.

The division of the analysed set was obtained for the radius $r = 6.5$.

The GRAVITY method (Bielecka, Paprzycki and Piasecki, 1975, 1979a, 1979b) represents centroidal methods. This method is an adaptation of the physical phenomenon of gravity to numeric taxonomy, combined with the use of statistical evaluation of significance of changes in the internal structure of groups in the process of grouping.

By means of this method a hierarchical division of the analysed set was obtained, while the following three levels of significance: α_1 = 0.01, α_2 = 0.05 and α_3 = 0.10 were used. As the GRAVITY method — contrary to other methods applied here — produces a hierarchical division, it had to be modified to ensure comparison. Therefore, the division by the GRAVITY method was replaced by two non-hierarchical divisions: into eight groups, and into 26 groups corresponding with the division into sub-groups of the first rank.

The IDWER method (Bielecka, Paprzycki and Piasecki, 1979b, Piasecki, 1975a, 1975b) represents methods which apply a frame of reference. Its elements are as follows:

I. The manner of transformation of features describing the object leading to the acquisition of the so-called angular values depicting structural dependence between features;
II. The frame of reference, called in the IDWER method the "basic set" which in theoretical assumption is a set of *differentiable representants*, i.e. which can be distinguished, of types of agriculture;
III. The classification of the set, based on the principle of identifying the given set with a frame of reference (a basic set), which can be referred to sets with a very big, theoretically infinite, number of objects, i.e. on open sets; this makes the IDWER method different from the previous four taxonomic methods, operating on closed sets. The conception of the IDWER method implies that comparability in time is equiponderant with comparability in space because every object described in two different time cross-sections may be treated as two spatially differing objects.

When the given set of the spatial units of the world was divided by means of the IDWER method, the following basic set, composed of the pairs of units: 42-41, 6-2, 60-55, 10-11 and 26-27, was used.

Results of the Applications

Dividing the results of the set of 61 world types of agriculture by 5 chosen taxonomic methods are as follows:

1. *Grouping by the WARD's Method*

Group I. (50 53 54 56)
Group II. (55 57 60)
Group III. (48 58 59)
Group IV. (46 47 49 51)
Group V. (31 33 34 35 36 37 38)
Group VI. (39 40 41 42)
Group VII. (28 30 32 45)
Group VIII. (16 17 18 27 29)
Group IX. (22 23 24 25)
Group X. (8 9 13 15 19 20 21)
Group XI. (11 12 14 26)
Group XII. (3 6 7 10)
Group XIII. (1 2 4 5)
Group XIV. (44 61)
Group XV. (43 52)

2. *Grouping by the Z. Piasecki's ORLINE Method*

Group I. (5)
Group II. (1 2 3 4)
Group III. (9 12)
Group IV. (7 17)
Group V. (6 8 10 11 13 14 15 16 17 18 19 20 21)
Group VI. (23 25)
Group VII. (26 27)
Group VIII. (22 24)
Group IX. (28 29 30 31 32 33 34 35 36 37 39)
Group X. (38 45)
Group XI. (40 41 42)
Group XII. (46 47 48 58)
Group XIII. (49 50 55 56)
Group XIV. (59)
Group XV. (51 53 54 57)

Group XVI. (60)
Group XVII. (43 44 52 61)

3. *Grouping by the M. Paprzycki's FARELL-mod Method*

Group I. (3 6 7 8 9 10 11 12 13 15 18 19 20 21 23)
Group II. (1 2)
Group III. (3 5)
Group IV. (16 17)
Group V. (22 29 30 31 32 33 34 35 36 37 38 39 40)
Group VI. (24 43)
Group VII. (25)
Group VIII. (14 26 28 45)
Group IX. (27)
Group X. (41 42)
Group XI. (46 47 48 49 50 51 53 54 56 57)
Group XII. (52)
Group XIII. (58 59)
Group XIV. (55 60)
Group XV. (44)
Group XVI. (61)

4. *Grouping by the M. Paprzycki's GRAVITY Method*

Group I. (1 2 3 4 5 6 7)
I. 1. (1 2)
I. 2. (3 6)
I. 3. (4 5)
I. 4. (7)
Group II. (8 10 11 12 13 14 15 19 20 21 26) + (9 extreme)
II. 1. (13 19 20 21)
II. 1.1. (13 21)
II. 1.2 (19)
II. 1.3. (20)
II. 2. (10 11 12)
II. 3. (8 15) + (9 extreme)
II. 4. (14 26)
Group III. (16 17 18 25 27)
III. 1. (16 17 18)
III. 2. (25)

	III. 3.	(27)
Group	IV.	(30 31 32 33 34 35 36 37 38 39 40 41 42)+ (22 extreme)
	IV. 1.	(31 33 34 35 36 37)
	IV. 1.1.	(31 35)
	IV. 1.2.	(33 37)
	IV. 1.3.	(34 36)
	IV. 2.	(30 32)
	IV. 3.	(39 40) + (38 extreme)
	IV. 4.	(41 42) extreme sub-group
	IV. 5.	(22) extreme sub-group
Group	V.	(46 47 48 51 53 54 55 56 57 58 59 60)+ (50 extreme)
	V. 1.	(46 47 48 49 51 58) + (59 extreme)
	V. 1.1.	(46 47)
	V. 1.2.	(48 58)
	V. 1.3.	(49)
	V. 1.4.	(51)
	V. 1.5.	(59) extreme
	V. 2.	(53 54)
	V. 3.	(55 56 60)
	V. 3.1.	(55 60)
	V. 3.2.	(56)
	V. 4.	(57)
	V. 5.	(50) extreme
Group	VI.	(28 29 45)
	VI. 1.	(28 45)
	VI. 2.	(29)
Group	VII.	(24 43 44 52 61)
	VII. 1.	(24 43 52)
	VII. 2.	(44 61) extreme sub-group
Group	VIII.	(23)

5. *Grouping by the Z. Piasecki's IDWER Method*

Group	I.	(5)
Group	II.	(43 44 52 61)
Group	III.	(3 4 33)
Group	IV.	(22 24 35 36 37 47 49 51 57)
Group	V.	(53 54)

Group VI. (34 38 46 48 50 55 56 58 59 60)
Group VII. (19 21)
Group VIII. (23 26 28 29 30 31 32 39 40 41 42 45)
Group IX. (27)
Group X. (2 6 7 8 10 11 12 13 14 15 16 20 25)
Group XI. (9 18)
Group XII. (17)
Group XIII. (1)

Numeral symbols applied here correspond with those used by J. Kostrowicki (1980) in his paper, *A Hierarchy of World Types of Agriculture*.

Table 16.1 contains the results obtained by using the "*System of 14 Criteria*" for evaluation of goodness of the divisions of the analysed set and the selected taxonomic method.

TABLE 16.1: Results of various methods

	Criteria	Ward's method	Orline method	Farell-mod method	Gravity method I	Gravity method II	Idwer method
	d_{wm}	7.28	6.16	4.12	8.16	3.70	6.20
	d_{ws}	5.63	5.22	3.41	5.54	3.18	4.54
K_{10}	d_{sm}	8.10	8.82	9.49	6.48	8.63	8.15
K_{11}	d_{mx}	13.53	19.03	15.97	10.05	14.76	15.23
K_{12}	d_{mx}	3.87	3.46	3.87	4.24	3.87	3.16
K_{13}	d_{sm}/d_{ws}	1.44	1.69	2.78	1.17	2.71	1.79
K_{14}	$d_{mx} \cdot d_{wm}$	98.47	117.25	65.79	82.00	54.58	94.48
K_{15}	$d_{mx} \cdot d_{ws}$	76.15	99.30	54.41	55.70	46.99	69.16
K_{16}	$d_{wm} \cdot d_{mx}$	1.88	1.78	1.06	1.92	0.95	1.96
K_{17}	$d_{ws} \cdot d_{mx}$	1.45	1.51	0.88	1.31	0.82	1.44
K_{20}	d_s	1.68	1.49	1.44	2.18	1.32	1.80
K_{21}	d_m	3.00	3.15	3.46	3.50	2.53	4.05
K_{22}	d_m	1.14	1.00	1.00	1.25	1.00	1.00
K_{23}	$d_s\ d_{ws}$	9.47	7.78	4.90	12.09	4.22	8.15
K_{24}	$d_m\ d_{wm}$	21.84	19.42	14.27	28.57	9.36	25.13
K_{25}	$d_m\ d_{ws}$	16.89	16.45	11.80	19.41	8.06	18.40
	n	15	17	16	8	26	13

The results obtained reveal that:

(i) No division is best if all evaluation criteria are taken into consideration;

(ii) Each division possesses a criterion which makes it good;

(iii) The biggest number of best ratings was yielded by the GRAVITY method 26 groups;

(iv) In the light of 14 criteria adopted in our research the order in which the methods are arranged is as follows: starting with GRAVITY II, GRAVITY I, FARELL-mod, WARD'S, ORLINE, IDWER.

Conclusions

1. As an outcome of the application of a number of criteria of the formal evaluation of classification methods a table of numerical results of evaluation of the analysed quantitative methods is obtained (Table 16.1). This table should, in turn, be formally evaluated be means of a method which, to a great extent, is conditioned by the subject of research. Various formulas are used to reduce the series of figures describing every method to an unequivocal criterion, for example:

(i) The maximum of the number of the highest ratings,

(ii) The maximum of the number of the lowest ratings,

(iii) The maximum of the sum of ratings normalised in the criteria of the formal evaluation,

(iv) The maximum of the weighted sum of ratings normalised in the criteria of the formal evaluation (weights depending on the significance of the criterion),

(v) The maximum sum of "places" obtained by the ratings in the criteria of the formal evaluation,

(vi) The maximum sum of the ranks of ratings in the criteria of the formal evaluation.

The above-described procedure makes the evaluation of goodness of the classification methods an easier task; however, it does not provide an answer to the question whether the method found the best from the formal point of view is the most efficient from the cognitive point of view.

The "chosen" method must be tested by:

(i) a formal evaluation;

(ii) a cognitive evaluation of results obtained by its application;
(iii) the verification of conclusions drawn from the application of the method to research object.

And only then the method can be recognised as the most efficient.

The methods should be tested in the given order 1, 2, 3, and at every stage ineffective methods should be eliminated. This means, for example, as regards the formal evaluation, that those methods out of the analysed collection should be eliminated which received the biggest number of bad ratings in the criteria of the formal evaluation.

2. When evaluating the results obtained through division of the set of 61 spatial units of the world be means of the six selected methods the following should be taken into account:

(i) Differences in the character of divisions (hierarchical division obtained by means of the GRAVITY method, and other divisions) are caused by various properties of the methods,
(ii) The analysed methods differ in the scope of their application: the IDWER method can be used not for grouping but for ordering and identifying the objects with a model structure,
(iii) If all the criteria of the formal evaluation of division of the set are based on the distances between the objects, a certain one-sidedness would occur,
(iv) A possible enrichment of the body of the formal evaluation criteria would only widen the gap between the ratings given to every classification method and bring nothing to the assessment of the method itself as far as its goodness is concerned.

3. The results of the grouping carried out in our research prove that the classified set of the 61 agricultural types of the third rank would be a perfect "basic set" for the IDWER method and, therefore, the result of grouping by the IDWER method cannot be positive in this case.

4. The System of 14 Criteria worked out by M. Paprzycki for the formal evaluation dividing results is a diagnostic tool enabling the researcher to obtain an answer — on the basis of the formal evaluation of the results of the grouping of a set — which of the two classification methods is more efficient from the formal point of view.

REFERENCES

Aitchison, J. W., 1972 : "The Farming Systems of Wales", in Vanzetti, C. (ed), *Agricultural Typology and Land Utilisation*, Verona, pp. 109-146.

Berry, B. J. L., 1958: "A Note Concerning Methods of Classification", *Annals of the Association of American Geographers*, 48, pp. 300-303.

———, 1961 : "A Method for Deriving Multi-Factor Uniform Regions", *Przeglad Geograpficzny*, T. XXXIII, z. 2, pp. 263-282.

Bielecka, K., Paprzycki, M. and Piasecki, Z., 1975 : "An Evaluation of the Applicability of Selected Mathematical Methods for the Typology of Agriculture", in Vanzetti, C. (ed), *Agricultural Typology and Land Utilisation*, Verona, pp. 55-70.

———, 1977 : "Badania nad stosowalnoscia metod ilosciowych w typologii rolnictwa — Refleksje metodologiczne" (Sum. Research on the Applicability of Quantitative Methods in Agricultural Typology — Methodological Reflections), *Przeglad Geograficzny*, T. XLIX, z. 4, pp. 807-817.

———, 1978 : "On Methods Applied in the Evaluation of the Effectiveness of Numeric Taxonomy Methods", (in Regional Conference of International Geographical Union, Lagos, Nigeria) *Resources and Development in Africa*, Lagos, Vol. I.

Bielecka, K. and Paprzycki, M., 1979 : "Evaluation of Taxonomic Methods from the Point of View of Comparability of Results in Space and Time — in Optimisation Aspect", *Geographia Polonica*, 40, pp. 187-190.

Bielecka, K., Paprzycki, M. and Piasecki, Z., 1979a : "Ocena staosowalnosci wybranych metod ilosciowych w typologii rolnictwa" (Sum. Evaluation of the Applicability of Selected Quantitative Methods for Agricultural Typology) *Studia*, Vol. LXVIII, p. 93, Warszawa, Polska Akademia Nauk, Komitet Przestrzennego Zagospodarowania Kraju.

———, 1979b . "Proposal of New Taxonomic Methods for Agricultural Typology", *Geographia Polonica*, 40, pp. 191-200.

———, 1980 : "Applicability of Numeric Taxonomy Methods in Agricultural Typology — Problems, Criteria and Methods of Evaluation", *Geographia Polonica*, 43.

Czekanowski, J., 1909 : Zur Differential diagnose der Neander-talgruppe, *Koresspondez-Blatt der Deutschen Geselschaft fur Antropologie, Ethnographic und Urgeschichte*, Vol. XL, pp. 44-47.

Zekanowski, J., 1913 : "Zarys metod statystycznych w zastosowaniu do antropologii" (Sum : An Outline of Statistical Methods Applicable in Anthropology), *Prace Towarzystwa Naukowego Warszawskiego*, III Wydzial Nauk Matematycznych i Przyrod-niczych, No. 5, pp. 167-172.

Harman, H. H., 1960: *Modern Factor Analysis*, The University of Chicago Press.

Holzinger, K. J. and Harman, H. H., 1941 : "*Factor Analysis a Synthesis of Factoral Methods*", The University of Chicago Press.

Kaiser, H. F. 1958 : "The Varimax Criterion for Analitic Rotation in Factor Analysis", *Psychometrica*, Vol. 23, No. 3.

———, : "Formulae for Component Scores", *Psychometrica*, Vol. 27, No. 1, March.

Kostrowicki, J., 1974 : *The Typology of World Agriculture. Principles, Methods and Model Types*, Warszawa, p. 74 International Geographical Union, Commission on Agricultural Typology (*Mimeographed*).

———, 1976 : *World Types of Agriculture*, Warsaw, 1976, 49. p.+fig., 1 International

Geographical Union, Commission on Agricultural Typology (*Mimeographed*).

———, 1980 : "A Hierarchy of World Types of Agriculture", *Geographia Polonica*, 43.

Munton, R. J. C., 1972 : "Farm Systems Classification : A Use of Multivariate Analysis", in Vanzetti, C. (ed), *Agricultural Typology and Land Utilisation*, Verona, pp. 89-106.

Piasecki, Z., 1975a : "Proposal of a New Identification-Verification Method to be Used in Agricultural Typology", in Venzetti, C. (ed), *Agricultural Typology and Land Utilisation*, Verona, pp. 269-276.

———, 1975b : "Typology of Poland's Agriculture — A Study of Application of the IDWER Method", A Paper, read at the Paris Conference, September, 1975, Fontenay-aux-Roses, duplicated typescript, pp. 8+1 Map.

Szczotka, F. A., 1976 : "Analiza porownawcza wybranych metod taksonomicznych na materiale empirycznym" (Sum. Comparable Analysis of Selected Taxonomic Methods on the Empirical Material) (*Mimeographed.*), Institute of Geography and Spatial Organisation, Polish Akademy of Sciences, Warsaw, Sygn. A. 835/4.

Vanzetti, C. (ed), 1975 : *Agricultural Typology and Land Utilisation*, Centre of Agricultural Geography, Verona, p. 498.

Ward, J. H. : "Hierarchical Grouping to Optimise An Objective Function", *Journal of the American Statistical Association*, V. 58, pp. 236-244.

17

LANDUSE ANALYSIS THROUGH DIGITAL IMAGE PROCESSING OF LANDSAT THEMATIC MAPPER DATA

S. P. GARG AND INDERJIT SINGH

LANDUSE PLANNING is an essence for the determination of the optimum use of every area of available land. Clearly optimum or optimal use is not the same, therefore, it must necessarily change from decade to decade with preventing economic conditions. Any landuse planning must accordingly be dynamic and not static, flexible not rigid, capable of being adopted to changing conditions, not for getting changing with the habits of the people. The land, as a whole, must be used as to satisfy many of the possible needs and legitimate desires of the people, the nation, as a whole.

The importance of landuse mapping has gained momentum in the recent past, with the advancement in remote sensing technology, because of increase in the awareness of the planners and decision-makers of optimal landuse planning. As the landuse is a dynamic phenomenon, it is very essential to have reliable and up-to-date data acquisition system. Information required for monitoring the changes in the landuse for the purpose of preparation of optimal landuse plan, is prerequisite. Therefore, it is necessary to have up-to-date data of existing landuse types of the region. The main advantage of using remote sensing technique over the conventional is that the remote sensing data provide the synoptic view of the large area, which is not possible by conventional ground surveys (Fig. 17.1). The data recorded by the remote sensors of the satellite permanent unbiased. The data are amenable to multi-disciplinary use. Digital Image Processing is numerical manipulation of digital images and it includes pre-processing enhancement and classification (Table 17.1).

TABLE 17.1: The various bands of the Landsat Thematic Mapper are as follows:

Band	Wave-length (μm)	Resolution of pixel	Remarks
1.	0.45-0.52	30 mts.	For increased penetration of water bodies as well as supporting analysis to land, soil & vegetation characteristics.
2.	0.52-0.60	"	Records reflectance peak of vegetation, vegetation discrimination and vigorous assessment possible.
3.	0.63-0.69	"	Chlorophyll absorption region emphasis contrast between vegetation and non-vegetation.
4.	0.76-0.90	"	Responsive to amount of vegetation biomass, land and water contrast.
5.	1.55-1.75	"	Records soil moisture and crop water content.
6.	10.4-12.5	120 mts.	For analysing vegetation stress, soil moisture and thermal related phenomenon.
7.	2.08-2.35	30 mts.	For discrimination of rock formation.

Source: Lilis & Thomson M. & Ralph W. Kiefer (1987) *Remote Sensing and Image Interpretation*, p. 567, 2nd John Wiley & Sons, New York.

Location and Extent

The study area (79° 14′00" E to 79° 31′30" E longitude and 29° 32′00"N to 29° 49′30"N latitude) falls in the Ranikhet Tehsil of district Almora. The total area is 1406.25 sq. km. Entire area lies in the hilly terrain of the Lesser Himalaya. The slopes are gentle to steep (5°-35°) with altitudinal variation from 500 metres near Gagas to 2242 metres near Rinchi village with sub-tropical type of climate (Mean Summer Temperature 21.5° C, Mean Winter Temperature 5°C.) the monsoon is active from mid June to mid September (mean annual rainfall 1293 mm). The general pattern of rock distribution forms the following tectonic units calcareous formations-fault-quartzite formation-thrust-crystalline formation. The crystalline formations consist of schists, gneiss, mica schist, thin bands of quartzite and basic intrusions. On the northern parts, the depth of soil is more than on the southern slopes, as they are thickly vegetated. The forests are classified into following types:

(i)	Type 3C/C29	-	Moist Siwalik Sal
(ii)	Type 3C/C3	-	Moist mixed deciduous forests
(iii)	Type 12C/C 1C	-	Ban Oak forests

(Champion & Seth Classification, 1968).

The economic conditions of the people of the study area is poor. They are mainly dependent upon Agro-forestry pursuits (Fig. 17.1).

Data and Material Used

The following data were used to carryout the present study :

1. Landsat Thematic Mapper (TM), False Colour Composite (FCC) of band 2, 3 & 4 (Blue, Green & Red) combination of 9th March, 1986.
2. The Computer Compatible Tape (CCT) of Landsat TM Data of 23rd January, 1987 of 1st quadrant (1st & IInd Volumes).
3. Survey of India maps on 1 : 50,000 scale.

Methodology

The Landsat TM Digital Data (CCT) was used for Digital Image Processing on RRSSC/SEP System having VIP-32 software. The process was carried out into following steps:

1. Pre-processing of the data — The digital data obtained in the form of CCT was first prepared for pre-processing, in which enhancement with different methods was carried out in the Computer System.

(a) Linear contrast stretching — this contrast enhancement technique was used to expand the range of brightness value in the image to video display, in an identifiable manner for each band and by superimposing the FCC of different band combinations were analysed for their utility to make different type of landuse. The output was obtained in the form of photo prints, and the results were compared with original Landsat TM(FCC) scene of the study area.

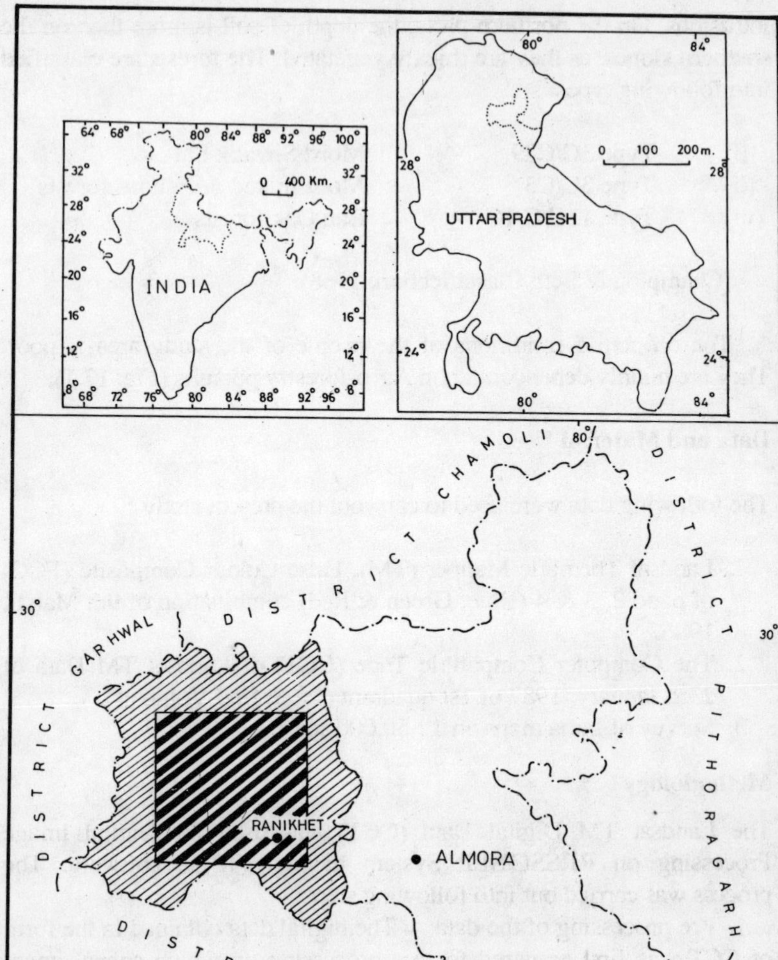

FIG. 17.1: Ranikhet tahsil location.

(b) Band Ratioing — Only two bands of Landsat TM were considered for the study important landuse types of the region, which was done by two methods.

- Single Vegetation Index/Simple Ratio

Band 4, 3 i.e.

$$\frac{\text{Band 4}}{\text{Band 3}} \text{ or } \frac{\text{Ir}}{\text{R}} \text{ was analysed.}$$

Normalised Vegetation Index (NVI) — here also bands 4 and 3 were used with the formula

$$\text{NVI} = \frac{\text{Band 4} - \text{Band 3}}{\text{Band 4} + \text{Band 3}} \quad \text{and}$$

and the results were studied for the differentiation of vegetation and other non-vegetated cover types.

(c) *n-Space Combination*: In this technique, the following parameters were considered i.e. greeness of vegetation, brightness index for soil and wetness index for highlighting the vegetation cover. The hybrid colour composite of normalised vegetation index (NVI), Greeness Vegetation Index (GVI) and Wetness Index (WI) were superimposed in the form of picture product on the video display unit. The following Landsat TM bands and transformation were used in statistical analysis (Crist 1983).

Brightness Index (TM B) = 0.3037 (TM 1) + 0.2793 (TM 2) + 0.4743 (TM 3) + 0.5585 (TM 4) + 0.5082 (TM 5) + 0.1863 (TM 7)

Greeness Index (TMG)

- 0.2848 (TM 1) - 0.2435 (TM 2) - 0.5436 (TM 3) + 0.7243 (TM 4) + 0.0840 (TM 5) - 0.1800 (TM 7)

Wetness Index (TMW)

0.1509 (TM 1) + 0.1973 (TM 2) + 0.3279 (TM 3) + 0.3406 (TM 4) - 0.7 112 (TM 5) - 0.457 2 (TM 7)

Digital Image Processing

To extract the useful information about landuse types from the remotely sensed data, the process of multispectral classification was performed by supervised classification using maximum livelihood algorithm on VIPS32 Computer System of RRSSC, Dehra Dun. The ground truth collected in the form of training sets, which was fed to the Pericolour 2000. The separability of individual landuse types were tested by bivariate distribution in 4 band combination i.e. 2, 3, 4 & 5. These were used for final classification for the following landuse types:

I. *Land Not Available for Cultivation*

- River bed and Water bodies
- Barren Land and Uncultivable land
- Forest land : Pine Dense, Pine Medium Dense, Pine Low Dense Pine Regeneration, Oak and Sal.
- Habitation

II. *Cultivable Land*

- Grass land
- Scrub/Shrub land

III. *Land Under Cultivation*

- Irrigated
- Unirrigated

and shadowed areas (unilluminated area).

The final classification image was smoothed using 3×3 moving window to make the classified image more readable (Table 17.2).

Results

Important landuse types are as follows:

1. *Land Not Available for Cultivation*:

(a) Forest Land

- Pine forest (*Pinus rooxburghii*)
  Forest dominated with Pine trees
- Dense : Pine forest having more than 60% density
- Medium dense : Pine forest having 60-40% density

- Low densely : Pine forest having less than 40% density
- Regeneration : Young pine trees on hill slopes.
- Oak forest : Forest dominated (*Overcua Incana*) with Oak trees
- Sal forest (*Shorea robusta*) : Forest dominated with Sal trees in Shiwaliks.

(b) Barren Land and Uncultivated Land

- Completely denuded or eroded hill slopes and exposed rocks

(c) River bed : Rocky and sandy river beds
(d) Water body : River course with water
(e) Habitation : Human settlements.

II. *Cultivable Land*

(a) Grass (Pasture) Land:
Hill slopes covered with thin grass cover
(b) Scrub/Shrub Land:
Hill slopes covered with shrubs and bushes.

III. *Land Under Cultivation*

(a) Irrigated : Agriculture done in valley and alongwith river beds with irrigation facilities
(b) Unirrigated : Agriculture done on slope with no irrigation facilities.

Discussion

The enhancement techniques used i.e. linear contrast, enhancement for the purpose of brightening the value in the image which gave good results and the land features including forest land became clear. In this technique, the contrast density values in the 512 × 512 scene are pulled further apart, so as to use full range (0-255) of grey values, resulting in the increase of visual contrast between the two areas of different density. This type of enhancement is appropriate if the area of interest is limited (Jenson, 1986). For band ratioing, due to hilly terrain, the illumination was different and shadow did not allow the visual analysis of forest and non-forest categories, therefore, the simple ratio i.e. Band

4 divided band 3 (IR/R) was used, which gave good contrast between different landuse types. In this technique, two spectral images are divided pixel by pixel and the resultant data are displayed on the video display unit (Jenson 1986) were studied. When normalised, vegetation index of IR-R/IR+R was carried out, the boundaries between the forest and non-forest became sharp, also reducing the partial shadow. For the study of n-space combination to enhance the vegetation and soil separately for delineation purpose, one of the most important vegetation index developed by Kauth & Thomas (1976) called Tessled Cap Transformation, where soil brightness index, vegetation and atmosphere information are taken care. The hybrid composite (FCC) had shown good result by enhancing the suitable contrast which were difficult to detect on simple FCC or ratioed image in the study area, however, the shadow effect was not fully removed.

TABLE 17.2: Landuse types

Sl.No.	Landuse Types	Sub-class	Area (in sq.km)	Percentage
1.	Land not available for cultivation	(a) River bed & water body	21.87	1.56
		(b) Barren land & uncultivated land	361.50	25.71
		(c) Forest land		
		-Pine dense	84.56	6.01
		-Pine medium dense	216.39	15.39
		-Pine regeneration	45.11	3.21
		-Oak	15.62	1.11
		-Pine low dense	107.81	7.67
		-Sal	80.27	5.71
		(d) Habitation	20.02	1.42
2.	Cultivable land	(a) Grass land	117.57	8.36
		(b) Scrub/shrub land	132.61	9.43
3.	Land under cultivation	(a) Irrigated	54.49	3.87
		(b) Unirrigated	100.58	7.15
		Shadowed Area	47.85	3.40
		Total :	1406.25	100.00

To extract the useful thematic informations from remotely sensed data. The process of multispectral classification gave good results in which 10 landuse classes were identified on the basis of training sets fed to the computer system and results were compared with the ground checks which were found to be almost 90 per cent correct.

REFERENCES

Anderson, J. R., E. E. Hardy and Roach, J. T., 1972 : *A Landuse Classification System with Remote Sensing Data*, M.S. Geological Survey Circular 671.

A User Guide using Remote Sensing Data for Landuse Mapping and Inventory (July, 1974), United States, Deptt. of Interior, Geological Survey Inter Agency Report, USGS-253.

B. M. Singh et. al., 1983 : Use of Remote Sensing Techniques for Agricultural Landuse Survey Classification and Planning, *PHOTONIRVACHAK Journal, Ind. Soc. Photo. Int. and Remoter Sensing*, Vol. II, No.3, pp. 1-13.

Champion H. G. and S. K. Seth, 1968 : *A Revised Classification of Forest Type of India*, Govt. of India Publication, Delhi.

Crist E. P., 1983 : *The Thematic Mapper Tasseled Cap — A Preliminary Formulation*. PWC 9th Int. Symp. on Machine Proceeding Remotely Sensed Data, Oyrde Univ., W. Lafayette Ind., pp. 357-363.

Domoghue Daniel N. M. and Ian Shennan, 1987 : A Preliminary Assessment of Landsat TM Imagery for Mapping Vegetation and Sediment Distribution in Wash Estuary. *Inc. Jour. Remote Sensing* Vol. 8 (7), pp. 1101-1008.

Ellefsen, R. A., 1978 : Computer Aided Mapping of Landuse and Mapping of Landuse and Landcover using Landsat Multi-special Scanner Data. *ESEMS*, Vol 4, Ed. by J. R. Anderson, U.S.G.S. Reston.

Gautam, N. C. and L. R. A. Narayan, 1982 : *Tentative Landuse/Landcover Classification System in India using Remote Sensing Techniques*, NRSA Pub. Hyderabad.

———, 1981 : Landsat : A Complete Data Source for Geographer. *Proc. of the IInd Annual Conference for Geographers* (NAGI), Geography deptt., S.V. University, Tirupathi.

———, 1983 : Landsat MSS Data for Landuse Inventory and Mapping. A Case Study of Andhra Pradesh. *PHOTONIRVACHAK Journal Ind. Soc. Photo. Int. and Remote Sensing*, Vol. II, No. 3, pp. 15-27.

Jenson, John R., 1986 : *Introductory Digital Image Processing*, Prentice Hall, Englewood, New Jersey, pp. 117-229.

Lillesana, Thomas M. and Kiefer Relphi W., 1987 : *Remote Sensing and Image Interpretation*, 2nd Ed., John Wiley and Sons Inc., New York.

Manchanda, M. L. et. al., 1983 : Multi-level Landuse Surveys using Remote Sensing Techniques, *PHOTONIRVACHAK Jour, Ind. Soc. Photo Int. and R.S.*, Vol. II, No.3, pp. 67-75.

Pofali, R. M., 1981 : 'Landuse and Planning', in N. Mohammed (ed) *Perspective Agriculture Geography*, Vol. III Concept Publishing Co., New Delhi, pp. 185-208.

Raghavaswami, V. and N. C. Gautam, 1982 : Application of Remote Sensing for Optimal Landuse Planning and Environmental Conservation in Different Physical Environments. Abstract. *Ind. Soc. of Photo Int. and Remote Sensing*, pp. 8-9.

R. K. Saxena, R. M. Pofali and L. R. Hire Kerur, 1983 : Preparation of Optimal Landuse Maps for Agricultural Development using Remote Sensing Techniques. *PHOTONIRVACHAK Jour. Ind. Soc. of Photo. Int. and R. S.*, Vol. II, No.3, pp. 29-38.

Roy, P. S., R. N. Kaul and S. S. Gabriyal, 1985 : Forest Cover Stratification in Tropical Forests and Delineation of Shifting Cultivation in Eastern Part of Arunachal Pradesh using Landsat MSS Data, *Ind. Jour. of R.S., Vol.* 6 (3 & 4), pp. 411-418.

Roy, P. S., 1987 : *Mountain Vegetation Stratification through Digital Processing of Landsat MSS Data*, Geo. Cart. Int., pp. 19-26.

Strahler N. Arthur, 1976 : *Remote Sensing Techniques in Geographical Research — Physical Geography*, 4th Ed., John Wiley and Sons Inc., pp. 632-643.

Singh, V. P., 1967 : *Working Plan of West Almora Forest Division*, Kumaon Circle, U.P. 1966-67 and 1975-76, Govt. Publication.

Sahai, Baldev, 1983 : Remote Sensing in Landuse Land Cover Mapping, *Proc. Natural Seminar NNRMS*, Hyderabad.

Singh Inderjit, 1989 : Evaluation of Landsat TM Data for forest cover type and Landuse Classification in subtropical forests of Kumaon Himalaya (U.P.) *PHOTO-NIRVACHAK Jour. of Ind. Soc. of R. S.* Vol.16, No.4, pp. 43-51.

Unni, N. V. Madhavan, 1983 : *Forest Survey and Management using Remote Sensing*, Remote Sensing Ed., B. L. Deekshatulu and V. S. Rajan, Ind. Academy of Science, Bangalore, pp. 115-133.

18

QUANTITATIVE ASPECTS OF REMOTE SENSING

A Tool for Agricultural Mapping

SANJIB K. GHOSH

THE HISTORY of man is laced with chaos, contention and suffering. It is equally saturated with the search for order, peace and enjoyment. One aspect of this search in the modern world is "planning". We do what we can today with the hope that our lives will be better tomorrow.

Food being the most essential element for man's survival, agriculture plays a great role. Agricultural planning, therefore, remains one substantial necessity in man's life. For this, aerial and space imageries provide the greatest tools available to the modern man (Munshi, 1983).

For providing an overall view of the components with which he must work when concentrating on local, regional or global problems, the planner's design tools par excellence are the maps. Recent studies (United Nations, 1989) indicate the practically all topographic mapping in the world is done with aerial (and to a small extent, space) photography. For such work on a regional or country-wide basis, use of satellite imagery is unsurpassable. All these are generally known a "Remote Sensing".

Remote sensing (i.e. to acquire information about an area without coming into direct contact with it) can be done in various ways. There are, however, two distinct aspects of Remote Sensing:

1. *Qualitative* (or interpretative), which deals with the recognition and interpretation of details in the area; and 2. *Quantitative* (or mensural) which involves precision measurements to obtain direct information related to the size, shape and other dimensional features as well as derived information such as change (e.g., area change, etc.),

statistical (e.g., area distribution, time variation, etc.) or associated parameters of inference. Within the scope of the latter (Quantitative Remote Sensing), modern technology with various sensing and measuring devices and the computers, deals with the solution of problems by mathematical computations, by using measurements (made directly on the imagery or otherwise digitally obtained) as input data.

The best known remote sensor is the human eye which can perceive only a narrow band of electromagnetic radiation from 0.4 to 0.7 μm. The eye has, in a way, three different types of detectors, sensitive to blue, green and red lights. Different wave lengths stimulate these sensors to different degrees, transmitting pulses via the optic nerves which the brain interprets as single colours. All remote sensing devices function in similar fashions. Incoming electromagnetic radiation stimulates a photosensitive device either photographically (i.e. chemically) or electrically.

Apart from the direct photographic imaging (black and white, color, false color, etc.), during the last three decades several Electro-optical imaging as well as Radar imaging and Passive Microwave Radiometric imaging systems have been developed. Ideas on them may be obtained from the Manual of Remote Sensing (1983, Published by the American Society of Photogrammetry).

All these can produce data in the form of a film copy, a computer lineprinter character map or digitally on a CRT (Cathode Ray Tube) display screen. The display may be chosen according to the available equipment and the user's requirements. The analyses and handling of the data, however, is usually complex — requiring the use of a high speed computer, or sophisticated restitution equipment (analogical or analytical).

Sensor Systems

Quantitative and cartographic potentials exist in the following systems:

1. Landsat Imageries (USA)
2. Metric Camera Photography (Spacelab Imagery, ESA)
3. Large Format Camera (LFC) Photography (USA)
4. SPOT Imagery (France)
5. MKF-6 and KFA-1000 Photography (both USSR.)

In understanding their cartographic potentials it would be pertinent to note the technical details of each imaging systems and their capabilities as have been evaluated through recent research in various organisations in the world.

1. *Landsat Imageries*, launched by the US National Aeronautics and Space Administration (NASA) are multispectral systems. In this series, Landsat 1, 2 and 3 with a satellite altitude of 900 km gave a full coverage of the earth's surface in 800,000 images (frames) having a ground resolution of around 80 m. The imaging process terminated in 1983. Landsat 4 and 5 (also known as Thematic Mapper, TM), launched in 1983 with the satellite altitude of 700 km yielded a ground resolution of ≃ 30 m. Research studies made at the US Geological Survey indicate that, in single stereo-models with given ground control, a planimetric (X, Y) accuracy of 10 to 20 m and an elevation (Z) accuracy of 25 m (at mid-latitude) with a detectability limit of 30 ≃ m are obtainable. The ground coverage per image is 34,000 km². However, there is no full ground coverage with Landsat 4 and 5 imageries. US President Bush approved in June 1989 the recommendations of the US National Space Policy Council whereby the Landsat System would continue to be operational through the fiscal year 1990 and the development and launching of Landsat 6 would be assured.
2. *Spacelab Imageries* were taken with the Metric Camera (Zeiss RMK-A 30/23) on board the Shuttle Mission of NASA-USA for the European Space Agency (ESA). Only one mission was launched in 1983. It was an experimental system where a satellite altitude of 250 km was used yielding a ground resolution of ≃ 20 m with photo scale 1 : 800,000. The system produced 1000 photographs over fragmentary portions of the Americas, Europe, Africa and Asia. The ground stereo-coverage per photo is 36,000 km². Research studies made at the ESA and University of Hannover, FR Germany by using analytical stereo-restitutors indicate a planimetric (X, Y) accuracy of 15 m and an elevation (Z) accuracy of 20 m with a detectability limit of ≃ 20 m. In tests for direct mapping in scale 1 : 100,000, some minor roads and buildings were found to be missing.
3. *Large Format Camera* (LFC) *Photography* was taken with the

LFC (manufactured by Zeiss) on board the Shuttle Mission of NASA-USA launched in 1984 with a satellite altitude of 240 km, the system yielded a ground resolution of ≃ 13 m with photo scale 1 : 800,000. Upto 1985, the system produced 2,200 photos giving partial coverage of the earth with stereocopic overlap variable due to the large format (23 × 46 cm^2). The ground coverage per photo is 20,000 km^2. Research studies made at Laval University, Canada by using analytical plotters indicated a planimetric (X, Y) accuracy of 9 m and an elevation (Z) accuracy of 17 m. Photo-triangulation of a strip of 9 photos with ground control at both ends and the middle yielded an accuracy of 15 m in all directions. The detectability limit was found to be ≃ 10 m. The system was found to be generally adequate for 1 : 50,000 scale topographic mapping with 50 m contour interval.

4. *Systeme probatoire d'observation de la terre* (*SPOT*) launched on board the Ariane Satellite Mission by Centre national d'etudes spatiales (CNES) of France in 1986 with a satellite altitude of 820 km, yielded a ground resolution of ≃ 10m with image scale of 1 : 400,000. Stereoscopic ground coverage per photograph is 3,600 km^2. Research studies made at the French Institut geographique national (IGN) and elsewhere by using processed data at analytical plotters indicate a planimetric (X, Y) accuracy of 7 m and an elevation (Z) accuracy of 5 m. With a detectability limit of 10 m, the system is generally found to be good for 1 : 50,000 scale topographic mapping with 20 m contour interval. The sponsors claim a cost savings of 50 per cent when compared with conventional mapping with aerial photography.
5. *Messkammer F-6 (MKF-6) photography* taken with a metric camera manufactured in the German Democratic Republic and launched on board the USSR Salyut 6 and 7 Intercosmos Space Missions at a satellite altitude of 250 km yielded a ground resolution of ≃ 10 m. Stereoscopic ground coverage of the earth surface of the order of 50 million km^2 has been obtained. This also offers good cartographic potential in scale 1 : 50,000 or smaller. Similarly taken, the KFA-1000 Camera (USSR) photography with ≃ 5m resolution can be purchased from Sojuskarta, Moscow, USSR. It appears that the space missions of the USSR are at the moment the only source for continuous

delivery of space photographs of the earth. However, potentials exist in the following systems which are mostly in the process of development:

6. Japanese Earth Resources Satellite (JERS) Imagery, scheduled to be launched in 1992 with continuous stereo images by means of two linear array sensors.
7. Other Satellite Imaging Systems, operational or in development (Chinese, Indian and European), with rather limited potentials.
8. Radarsat Imagery (NASA — USA), scheduled for 1994, being a joint venture of Canada, USA and UK. Full earth coverage would be a possibility. Radar imaging would be able to penetrate through clouds and haze, which often limit the full potentials of other imaging systems.

Mapping involves, beyond the human element, three items: (1) *Sensor data* (or the photograph) with the inherent problems, especially because of its dynamic mode of data acquisition; (2) *Ground* with its diversities; and (3) *Map* with its various possibilities. In order to comprehend the complexity involved in the process of mapping, one must recognise the sensor data acquisition procedures and understand the mutual relationships between these elements.

The general sensor data handling concept in Remote Sensing is presented in Fig. 18.1. For specific systems one should refer to the particular handbook or manual.

With regard to the mutual relationships, firstly the relationship between the sensor and the ground can be established through the principles of *photogrammetry*, at the backbone of which lies the condition of "collinearity".

Secondly, the relationship between the ground and the map can be established through the principles of cartographic map projection. The specific projection should, however, be considered in view of the desired type of information.

Sensor — Ground Relationship

The basic transformation equation describing the relationship between two location vertors $\vec{R}$ and $\vec{r}$ defining the ground point (P) and the corresponding photo point (P) respectively with regard to the perspective centre (O) may be written as (Fig. 18.2):

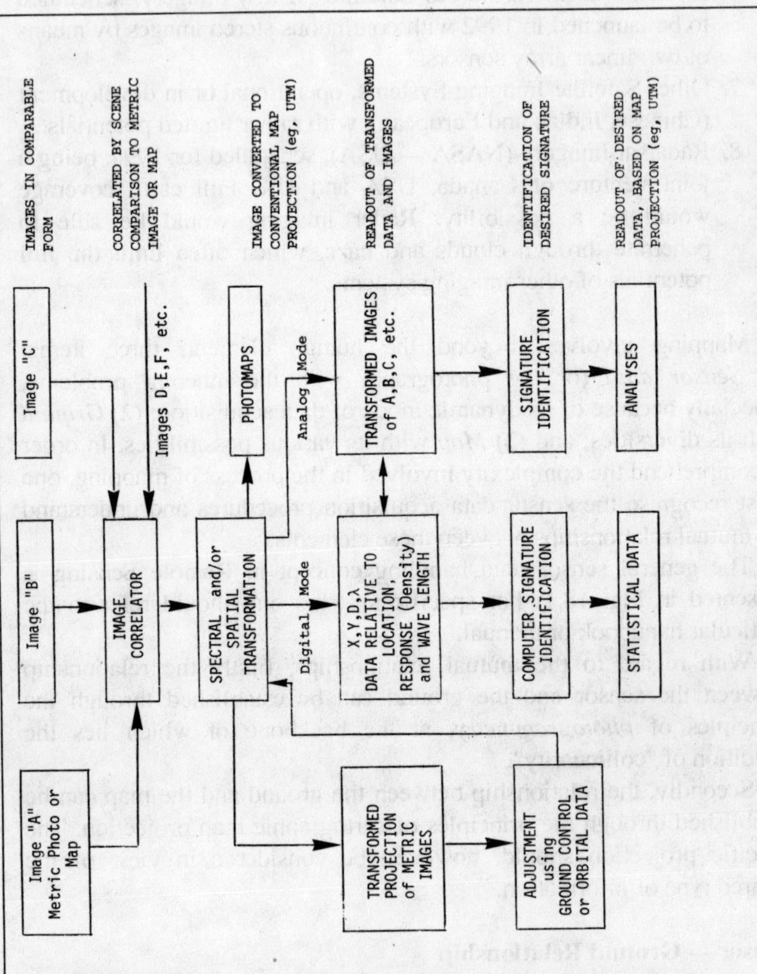

FIG. 18.1: Data handling concept in quantitative remote sensing.

$$\vec{r} = \begin{bmatrix} x_p & - & x_o \\ y_p & - & y_o \\ & - & f \end{bmatrix} = k\,[M] \begin{bmatrix} X_p & - & X_o \\ Y_p & - & Y_o \\ Z_p & - & Z_o \end{bmatrix} = k\,[M]\,\vec{R} \qquad [1]$$

where x_p, y_p are photo-coordinates of point p;
x_o, y_o are the coordinates of the photo principal point;
f is the focal length;
k is the scale factor;
M is the "Rotation" matrix;
X_p, Y_p, Z_p are the ground coordinates of point P;
and X_o, Y_o, Z_o are the coordinates of the origin (perspective centre) in the X, Y, Z (ground) system.

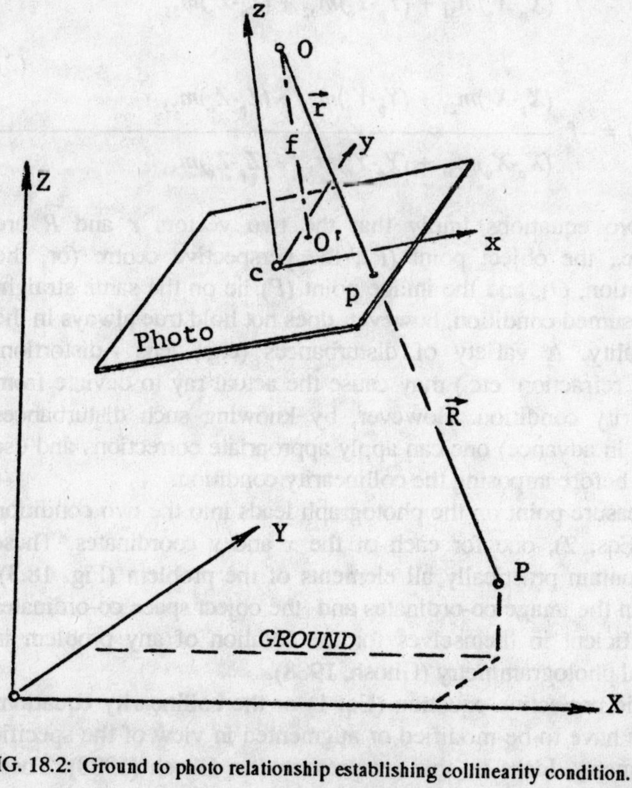

FIG. 18.2: Ground to photo relationship establishing collinearity condition.

The rotation matrix is a 3 x 3 orthonormal matrix made up of 3 independent rotations (ω around *X*, φ around *Y* and κ around *Z* axes):

$$M = M_\kappa \cdot M_\phi \cdot M_\omega = \begin{bmatrix} m_{11} & m_{12} & m_{13} \\ m_{21} & m_{22} & m_{23} \\ m_{31} & m_{32} & m_{33} \end{bmatrix}$$

The matrix multiplication indicated in the above may be carried out by components along rows and columns. In the three equations, thus developed, if the first and second are divided by the third, one obtains the collinearity condition equations with respect to each photo coordinate (also considering $x_o = O = y_o$, i.e., photo coordinates referred to the principal point as the origin):

$$x = (x_p - x_o) = -f \frac{(X_p - X_o)m_{11} + (Y_p - Y_o)m_{12} + (Z_p - Z_o)m_{13}}{(X_p - X_o)m_{31} + (Y_p - Y_o)m_{32} + (Z_p - Z_o)m_{33}}$$

$$y = (y_p - y_o) = -f \frac{(X_p - X_o)m_{21} + (Y_p - Y_o)m_{22} + (Z_p - Z_o)m_{23}}{(X_p - X_o)m_{31} + (Y_p - Y_o)m_{32} + (Z_p - Z_o)m_{33}} \qquad (2)$$

These two equations imply that the two vectors r and $\vec{R}$ are collinear, i.e., the object point (*P*), the perspective centre (or, the exposure station, *O*), and the image point (*P*) lie on the same straight line. This assumed condition, however, does not hold true always in the physical reality. A variety of disturbances (e.g., lens distortion, atmospheric refraction, etc.) may cause the actual ray to deviate from the collinearity condition. However, by knowing such disturbances (determined in advance) one can apply appropriate corrections and use *refined* data before imposing the collinearity condition.

Each measure point on the photograph leads into the two condition equations (Eqs. 2), one for each of the *x* and *y* coordinates. These equations contain practically all elements of the problem (Fig. 18.3). They contain the image co-ordinates and the object space co-ordinates and are sufficient in themselves for the solution of any problem in conventional photogrammetry (Ghosh, 1988).

The basic projective equation (Eq. 1) or the collinearity equations (Eq. 2) may have to be modified or augmented in view of the specific system geometry. Ideas on this are presented in Ghosh (1988). There are, however, certain other problems. Firstly, the condition equations are non-linear. Secondly, in usual cases, more image points are observed than are necessary. To obtain a practical and statistically acceptable solution, therefore, the equations are linearised and then the method of Least Squares is applied to account for redundant data. Generally, the equations are linearised by using "Taylor" expansion and, in view of an iterative solution procedure, only the first order terms are retained. In case the camera position and/or its optical axis attitude are known or if such parameters are related to some function of time or certain other parameters, one may apply weight or parametric constraints to such solutions (Ghosh, 1988).

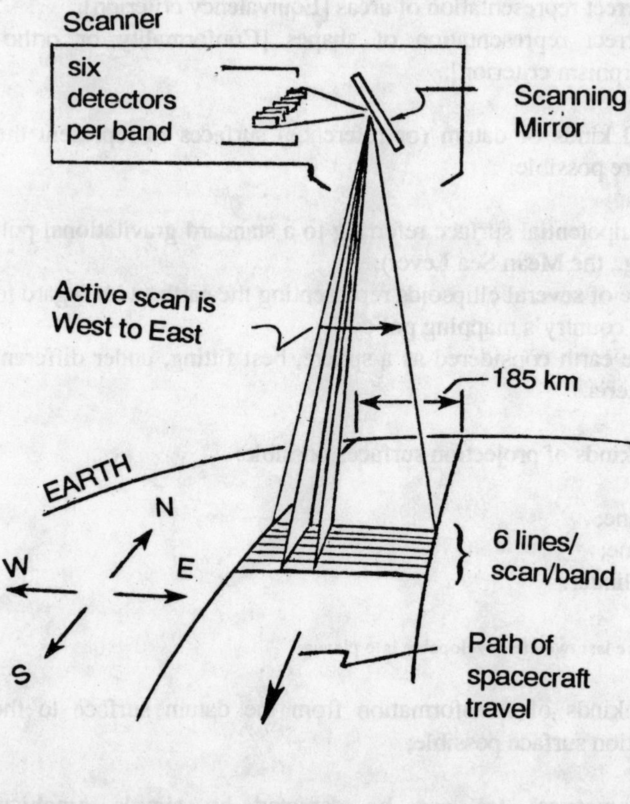

FIG. 18.3: MSS Scanning System

Ground — Map Relationship

Map projection or the representation of the earth on a plane surface (map) involves "stretching" or "shrinking" resulting in distortions. Certain properties of the distortions are favourable for the specific purpose at hand with regard to the involved area to be represented. Accordingly, one selects the specific "type" of map projection most suitable under the circumstances. There are several considerations in doing this:

A. Three principal cartographic criteria are possible:

1. Correct representation of distances [Equidistance criterion];
2. Correct representation of areas [Equivalency criterion];
3. Correct representation of shapes [Conformality or orthomorphism criterion].

B. Several kinds of datum (or reference) surfaces to represent the earth are possible:

1. Equipotential surface referring to a standard gravitational pull (e.g., the Mean Sea Level);
2. One of several ellipsoids representing the earth (with regard to the country's mapping policy);
3. The earth considered as a sphere, best fitting, under different criteria.

C. Three kinds of projection surfaces possible:

1. Plane;
2. Cone;
3. Cylinder.

***Note*: The last two are developable into planes.**

D. Three kinds of transformation from the datum surface to the projection surface possible;

1. Geometrical (as can be depicted by simple graphical illustrations);

2. Mathematical (obtainable only through mathematical computations based on certain criteria);
3. Semi-geometrical.

E. Three types of contact of coincidence of the projection surface with the datum surface possible:

1. Tangent (at a point — plane);
2. Secant (at a line — cone or cylinder);
3. Polysuperficial (polyconic or polycylindrical).

F. Numerous orientation of the projection surface with respect to the datum surface possible:

1. Normal (contact along one or more lines of latitude);
2. Transverse (contact along one or more selected meridians);
3. Oblique (contact along one or more selected great circles).

G. Two methods of projection possible:

1. Direct (directly from the ellipsoid to the surface of projection):
2. Indirect (first ellipsoid transformed into a sphere and then from the sphere to the projection surface).

In view of a worldwide need of uniformity in mapping, certain criteria stand out as primary requirements:

1. Conformality (for minimising directional errors);
2. Continuity over sizeable areas in one country and between countries;
3. Scale error to be within tolerable limits;
4. Uniformity and simplicity of transformation formulas between zones and countries;
5. Adaptability to various ellipsoids and earth sphere.

The Universal Transverse Mercator (UTM), which is a modification of the Gauss-Kruger projection has been developed in the recent years with regard to the above basic criteria. The basic features of the UTM (Snyder, 1984) are:

(a) A cylindrical surface assumed touching the ellipsoid along a specific meridian constitutes the projection for a "zone".
(b) The world is divided into 60 zones, each extending through 6° of longitude.
(c) A scale distortion (grid scale constant) of 0.9996 is applied along the central meridian. This limits the scale error to 1/2500 within each zone.
(d) A plane rectangular metric grid is superimposed on each zone. This permits easy referencing and transformation between zones, which can be easily programmed for working computers.
(e) The UTM is used between latitudes of 80°N and 80°S, most of the countries being covered thereby.

Approximately 65 countries use the UTM currently as the most general use projection within the country. Furthermore, the U.S.S.R. and many "Eastern" block countries use Gauss-Kruger (Transverse Mercator) also with 6° zones. Thus they are similar.

Most of the earth satellite imaging systems in view of their near polar sun-synchronous orbits, give images which are formed on an almost cylindrical surface in space. This can be considered as a specific map projection very nearly the same as the UTM. Colvocoresses (1974 and 1977) suggests that the "projection" inherent in such systems would be mathematically definable with the potential of developing into automated, computer cartographic systems of mapping. In this, the "pixels" (picture elements) can be discretely related to their positions on the surface of the earth. This projection, termed "Space Oblique Mercator", is somewhat skewed with regard to the regular UTM. However, the required augmentation is not complex and data handling is easily manageable with currently available technology.

The mapping problem can be viewed in terms of two distinct categories: (1) For better mapped areas of the world, like India or most European countries, photoidentifiable control is readily available. This permits geometric evaluation and fitting of the cartographic data to a recognised map projection. (2) In areas, like some African countries, where ground control is rare, mapping has to depend on orbital and sensor data. The curved Earth surface cannot be projected on a plane surface without some distortions. However, most satellite images have overall field angles of less than 30° and the curvature of the Earth across the scene is less than 6°. This generates each frame and each stereo-model with a near orthographic disposition giving tremendous cartographic fidelity which is certainly sufficient for most small scale regional mapping.

Conclusion

The Indian Space Research Organisation has already developed a sensing satellite to provide data for integrated surveys in India. Deekshatulu of the National Remote Sensing Administration (NRSA) as the Chief of the Indian delegation at the 1988 Kyoto Congress of the International Society for Photogrammetry and Remote Sensing (ISPRS) indicated, *inter alia*, the need in India for generating greater emphasis on regional remote sensing problems, on efforts to reduce ground truth data (for qualitative aspects) and on greater use of indigeneous regional low cost instruments. However, requirements for higher resolution and stereo coverage still remain the two most essential ingredients for quantitative tasks of remote sensing in India. We are already in the era of automated mapping based on earth-sensing space and aerial systems, wherein digital approaches by using computer technology have been specifically viable.

REFERENCES

American Society for Photogrammetry, 1983 : *Manual of Remote Sensing* (2nd Ed., 2 Vols).

Colvocoresses, A. P., 1974 : Space Oblique Mercator; *Photogrammetric Engineering*, Vol. 40, No. 8; August.

————,1977 : Proposed Parameters for Mapsat, *Photogrammetric Engineering and Remote Sensing*, Vol. 45, No. 4, April.

Ghosh S. K., 1988 : *Analytical Photogrammetry* (2nd Ed.), Pergamon Press.

ISPRS., 1988 : *International Archives for Kyoto Congress*; Vol. 27, Part A.

Munshi, M. K., 1983 : *A Study of the Determination of Wheat Crop Statistics in India Through the Utilisation of Landsat Data*. Ph.D. Thesis, Indian Institute of Technology, Delhi.

Snyder, John P., 1984 : *Map Projections Used by the US Geological Survey* (2nd Ed). Geological Survey Bulletin 1532.

United Nations, 1989 : *World Cartography, Vol. 20*; authored by A. J. Brandenberger and S. K. Ghosh; United Nations, New York.

[illegible] co-product will appear with greater dimension giving tremendous cartographic quality which is [illegible] required mapping.

Conclusion

The [illegible] Research Organisation has undertaken [illegible] satellite to provide data for [illegible] survey [illegible] India. The [illegible] of the National Remote Sensing Agency (NRSA) as the Chief of the Indian delegation at the 1988 Kyoto Congress of the International Society for Photogrammetry and Remote Sensing (ISPRS) indicated [illegible] cartographic [illegible] remote sensing products, [illegible] efforts to [illegible] ground truth data [illegible] greater use [illegible] low cost [illegible] requirements for higher resolution and [illegible] coverage [illegible] the two most [illegible] India. We [illegible] based on [illegible] satellite and aerial systems [illegible] approach by using computer [illegible].

[illegible] 2nd Ed. Penguin Press.

ISPRS [illegible] *Archives* [illegible] Vol 27 [illegible]

Hussain [illegible] A Study of the Geometric [illegible] Delft [illegible] Technology [illegible].

MAPPING AGRICULTURAL RESOURCES BY REMOTE-SENSING TECHNIQUE*

J. A. ALLAN

THE STUDY of agriculture and the disposition of the resources upon which agricultural development depends requires the deployment of a wide variety of methodologies of data acquisition and analyses. Rarely does one have enough data for a comprehensive review of renewable resources and very frequently, while some aspects are well covered, others are deficient. The reasons for the deficiencies vary, sometimes they arise because the acquisition of data by conventional methods would be too expensive to justify with respect to the likely returns from further development; on others there may be institutional impediments and a shortage of professional skills preventing data collection or there may be no local tradition for gathering such material. Finally, the deficiency may occur because the variables of concern are ephemeral and they change from season to season and from year to year.

From time to time a major new methodology becomes available through the development of a new technology. Such developments may fill a gap in the data required by resource managers and government, or they may merely provide something better than available to date. The launch of the LANDSAT series of satellites in 1972 was such an innovation. The first land resources survey satellite together with the second and third, launched respectively in 1975 and 1978, have provided a great deal of low resolution imagery which on occasions provides a very special perspective on rapidly changing environmental

---

*Parts of this article appeared in the Journal *Remote Sensing of Electromagnetic Spectrum*, Vol. 15, No. 2, 1978, pp. 40-51.

phenomena; phenomena which could not be monitored by any other means than the regularly imaging sensors on platforms imaging the same spot once every 18 days, or more frequently should there be more than one satellite in orbit (there have generally been two since 1975).

Agricultural studies in India can be implemented on the basis of very comprehensive data with respect to landuse and cropping, and information on water distribution is also readily available. In many areas, however, detailed soil maps are not available and in these circumstances the satellite now provides a very useful overview which makes it possible to map major terrain and related soil conditions extremely cheaply and rapidly. The imagery can be readily acquired from the United States, and in due course from an Indian receiving station, and despite its low resolution, the mapping potential of the imagery is enormous.

Remote-Sensing in Reconnaissance Surveys

The place of satellite imagery in reconnaissance surveys of renewable and non-renewable natural resources is firmly established. Reconnaissance surveys here refer to surveys of 1 : 30,000 or smaller, the purpose of which is to identify tracts which have a recognisable character. This character may be one of relative homogeneity in terms of landform, soils and vegetation, or of regular heterogeneity as reflected in the patterns and textures which appear on remotely sensed images. Reconnaissance survey is the first stage in a sequence which leads to detailed technical and socio-economic surveys (Fontaine, 1963 and Stewart, 1968) and they can, and often do provide the basis for the selection of areas in which detailed studies can be effectively deployed (Bawden, 1968). The object of the preliminary surveys is often to identify areas of differing landuse potential, leading to the provision of advice on the sequence of activity necessary to promote development (Spooner, 1958).

The development of photographic and scanning sensors (MSS) for satellite platforms has made available a new range of products for mapping terrain and natural resources. In the non-military sector these images have been on a small scale with nominal resolution for the LANDSAT MSS imagery of 80 metres and for Skylab of 10 metres in favourable circumstances, although many users would argue that effective resolutions are of the order of 200 metres and 30 metres, respectively. The relatively low resolution of the satellite system has reinforced the role of remotely sensed imagery in the level of reconnaissance surveys and has precluded it so far from any more detailed level of resource studies.

The literature is rich in its record of successful and partially successful identification on satellite images of vegetation and landuse boundaries as well as of related physiographic and soil phenomena (National Academy of Sciences, 1977). At the same time consultants and international agencies automatically order satellite imagery of their study area for many problems involving mapping, both for planning aerial photography and for the inventory of natural resources. This is strong confirmation that such imagery has a place in map-making at a reconnaissance level, especially in Asia and Africa. The purpose of the rest of this paper is to indicate some of the problems which can arise in organising surveys in developing countries and the contribution which LANDSAT imagery can make by making available an inexpensive (at least to date) image which is a very effective basis for the spatial stratification of a study area. Such spatial stratifications can be the basis for deciding upon areas for further study as outlined above, or as in the case study to be discussed here a basis for the selection of a properly structured sample of farms in a particular district of Western Uttar Pradesh in north-west India.

North-West India and its Maps of Renewable Natural Resources

The difficulty in gaining access to all types of large and intermediate-scale topographic maps makes the preliminary stages of any study of natural resources in northern India an exercise which demands much more detailed and extensive fieldwork than would be necessary if such documents were readily available. Air photographs are even more difficult to acquire, and since the northern plains are very flat with no vantage points, the absence of an overall view such as provided by an aerial image makes it very difficult to define the very important ecological boundaries and zones of transition, which minor changes of relief and groundwater circumstances bring about.

Air photographs and topographic maps (Survey of India at scales of 1 : 63,360 and smaller) do exist, and unpublished soil surveys have been completed but are not generally available to environmental scientists. The topographic maps are helpful in locating settlements and some parts of the irrigation system, but they lack contours which would be useful in infering the all-important groundwater situation.

The soil surveys which include maps are useful, but they are based on a small number of sample points and possess all the deficiencies which arise from such mapping. In Uttar Pradesh the publication of the district soils monographs has been completed to the point of profile description and soil sample analysis; the reports and maps have not been completed (Mehrotra *et al.*, 1972). The district of Bulandshahr which is the subject of this case study was one of the districts where no soil map existed (Fig. 19.1).

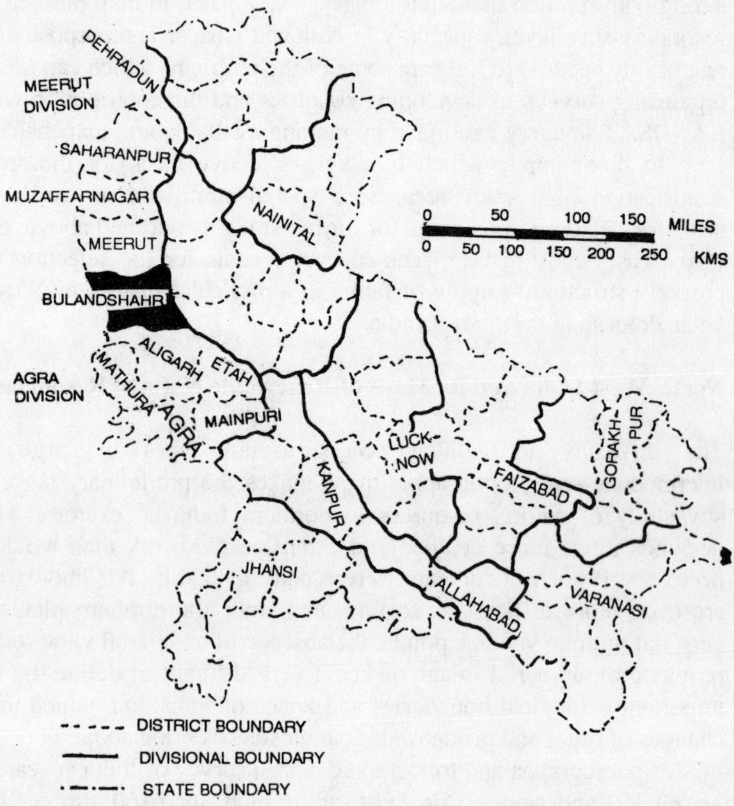

FIG. 19.1: Uttar Pradesh (India) showing position of Bulandshahr and surrounding districts in Meerut and Agra divisions.

One of the purposes of the survey carried out in the district of Bulandshahr was to sample agricultural practices and farm productivity in various parts of the district so that those areas which were well endowed in terms of soil and well provided with irrigation water could be compared and contrasted with areas which had poor soils which were often exposed to flooding and were affected by saline-alkali conditions. The problem which had to be solved before such a study could proceed was how the district could be sub-divided into land units (Fig. 19.2) so that both the well-favoured tracts and the poor land were correctly represented in a sample of farms.

The study in question was mounted in 1971-1972 and so it could not make use of the LANDSAT imagery which was being generated for the first time during the second half of 1972. In its absence it was necessary to sub-divide or stratify the area using the inadequate topographic maps of the area (inadequate in that they showed no relief features) supplemented by canal department maps of the irrigation network. Some additional information was taken from tube-well department maps and groundwater records. However, the main work of delineation was completed by a considerable input of fieldwork involving extensive travelling. The results of this preliminary study are shown in Fig. 19.2. The spatial stratification shown in Fig. 19.2 was the basis for the selection of a clustered random sample of farmers — the clusters being villages (Allan, 1973). The arrival of LANDSAT imagery served two purposes. First, it confirmed that the spatial stratification had been carried out effectively (Fig. 19.3). Secondly, it showed that such preliminary delineations as are necessary in studies of the sort being discussed here can be effected on the basis of satellite imagery much more economically, much more rapidly, and much more accurately in terms of the phenomena being classified than by any other method.

Reconnaissance Level Maps of Soils and Drainage from LANDSAT Images

North-West India has been particularly well covered by LANDSAT imagery. The coverage is good not only in the sense that the quality is good, but also in its comprehensive character in terms of recording the seasonal changes in precipitation, run-off and cropping. Coverage was particularly good in the 1972-1973 period, and there has been subsequent updating since 1975. The importance of the comprehensive seasonal cover is indicated in Fig. 19.4 which shows the relationship between precipitation and cropping as well as the timeliness of the 1972-73 imagery.

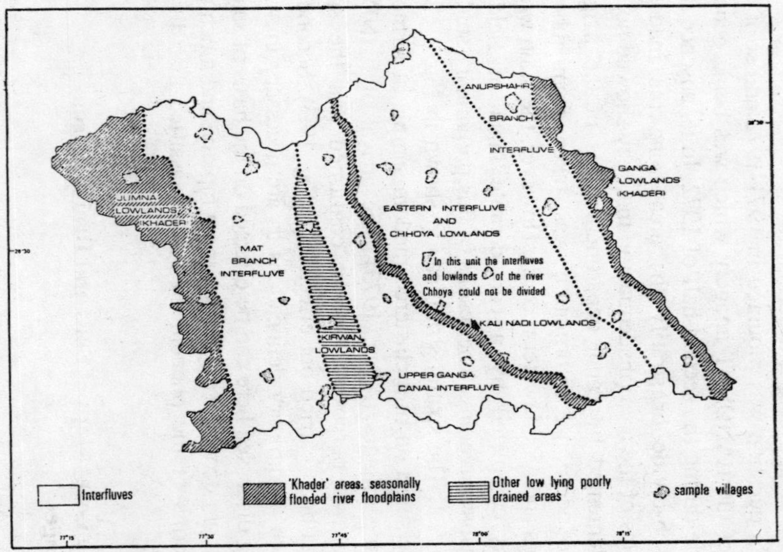

FIG. 19.2: Bulandshahr district — selected sample villages.

This division of the district on the basis of hydrological criteria was compiled on the basis of topographic maps, canal and tube-well department maps and some field survey.

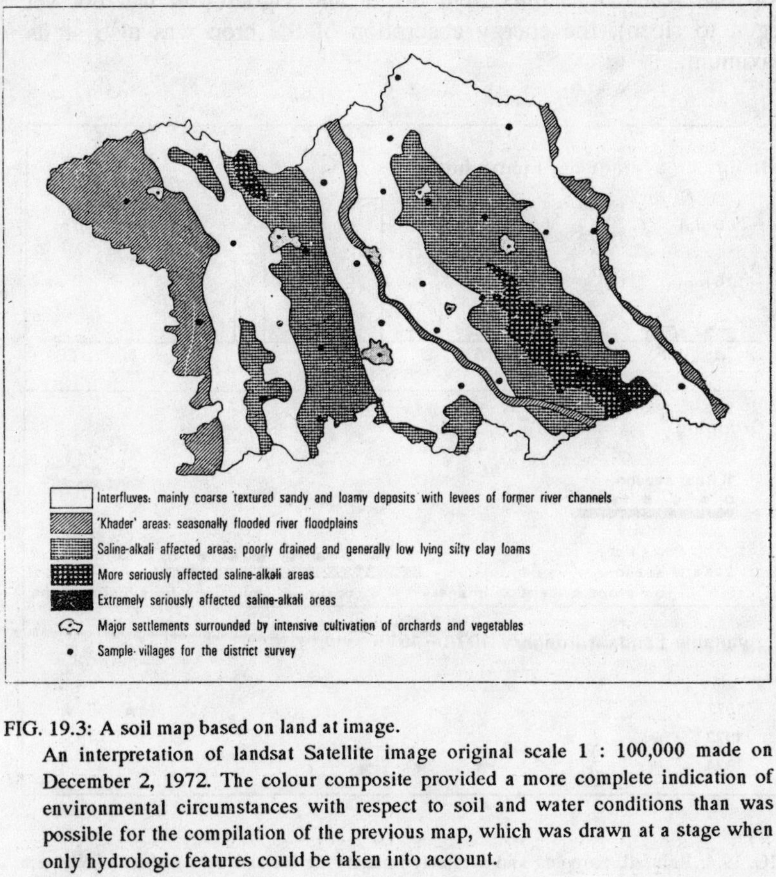

FIG. 19.3: A soil map based on land at image.

An interpretation of landsat Satellite image original scale 1 : 100,000 made on December 2, 1972. The colour composite provided a more complete indication of environmental circumstances with respect to soil and water conditions than was possible for the compilation of the previous map, which was drawn at a stage when only hydrologic features could be taken into account.

Wheat is the most important crop in the Bulandshahr District and in Western Uttar Pradesh as a whole. In 1971-1972 over 80 per cent of the cropped area was under wheat in the rabi (winter) season, and in many villages visited wheat was being raised on over 90 per cent of the cropped area. Such uniformity of cropping is unusual outside the plains of North America and the Soviet Union, but when it does occur it is a feature of great assistance to the user of the low resolution LANDSAT imagery. Fig. 19.4 shows that by December the crop cover should be sufficient to register on channels four and five of the LANDSAT system. Both the 1972-73 single-band imagery and derived colour composites confirmed the ability of the system to record the rabi (winter) season crop. Predictably by February at which time the wheat crop had reached its maximum height and vigour (but had not yet begun to ripen), the energy absorption of the crop was also at its maximum.

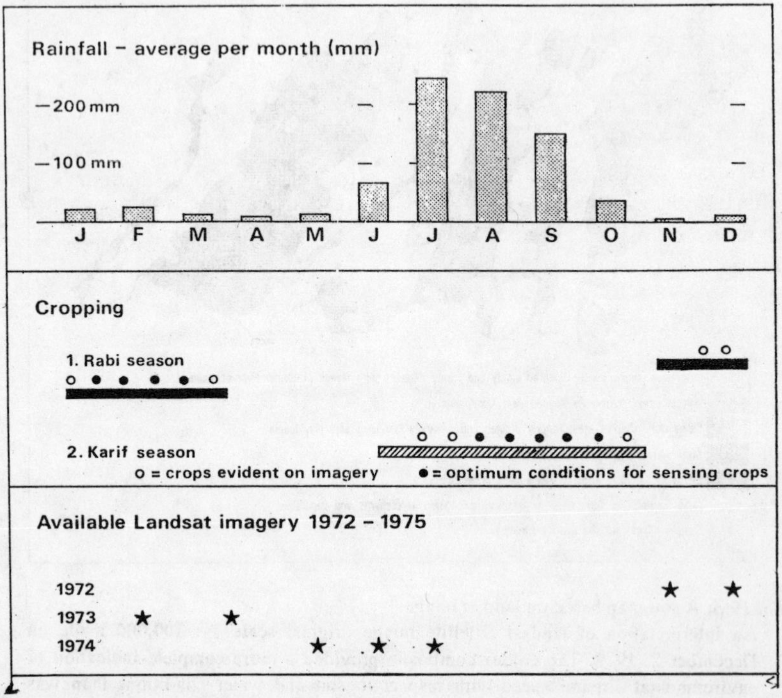

FIG. 19.4: Rainfall, cropping and available imagery for Bulandshahr district of Western Uttar Pradesh.

Source: *Indian Meteorological Department 1931-1960*, p. 361, *Field Work and World Bank*, 1976, IBRD, p. 1181.

Most evident, however, on February 1973 channel 5 image was the contrast between the areas on which wheat was being raised and the tracts without any crop. The crops were those which were generally low-lying and, therefore, liable to suffer from accumulations of naturally occurring and man-aggravated saline-alkali soil conditions. The landuse practices of the farmer were a clear reflection of soil quality and it was a simple task to delineate the areas partially and seriously affected by saline-alkali conditions (Fig. 19.3). It is useful to compare the results of the extensive field survey (Fig. 19.2) with those from the much quicker and more economical method based on the satellite images (Fig. 19.3). The exercise confirmed something which is by now common-place among those attempting to use remotely sensed imagery for the identification of landuse or vegetation boundaries and the associated disposition of soils, namely, that even at the low resolution of the satellite systems, timely imagery in favourable circumstances will reveal clear and significant boundaries.

Equally interesting and clear are the seasonal changes in the surface run-off on the comprehensive cover of northern India. In the study area being discussed here the rise and fall of the rivers following the summer monsoon are evident, with the river Jamuna almost ceasing to flow during the driest part of the year, towards the end of the dry season in May. The rise of the rivers is clear in July as is their subsequent decline through the winter. Associated with the changes in run-off are fluctuations in ground-water levels, and such changes are evident in the major low-lying areas of the river Ganges flood-plain where the presence or absence of groundwater flows is signalled by associated vigour or lack of vigour of vegetation at the edge of the levees which border the flood-plain.

The study of the use of LANDSAT imagery in an area where small changes of relief have great significance with respect to groundwater levels and related soil conditions shows that the overall view provided by such imagery can provide a valid basis for reconnaissance mapping. In the special circumstances of the study area where during one season a single crop dominated the landuse and where the regions of poor soil supported no crops, the satellite images allowed specialists with knowledge of the physiography of the district to recognise areas affected by saline-alkali conditions as well as those more seriously affected.

A second important feature revealed in this case study is in the contribution which satellite imagery can make in resources studies in providing a swift and economical means of effecting a spatial stratification of a study area such that tracts with good soil and water resources for agriculture can be distinguished from those with a poor agricultural potential.

REFERENCES

Allan, J. A., 1973 : *Bulandshahr District Study, Sample Design Sampling Procedures and Schedule of Questions.* ITC, Enschede, pp. 29.

Bawden, M. G., 1968 : "Methods Used to Assess the Land Resources of Eastern Bechuanaland", *Proceedings of the Toulouse Conference on Aerial Surveys and Integrated Studies.* Paris, UNESCO, pp. 409-410.

Fontaine, R. G., 1963 : "Ecological Studies as a Basis for Agricultural Development", *Ninth Technical Meeting of the International Union for the Conservation of Nature and Natural Resources.* Kenya. RT 9/iv/8.

India Meteorological Deptt., 1931-1960 : *Climatological Tables of Observatories in India.* Delhi, Government of India Press, p. 470.

Mehrotra, C. L., 1972 : *Soil Survey and Soil Work in Uttar Pradesh*, Vol. VIII. (Das, S. K. and Saksena, V. P.) Allahabad, Department of Agriculture, Uttar Pradesh, p. 360.

National Academy of Sciences, 1977 : *Resource Sensing from Space: Prospects for Developing Countries*, Washington, National Academy of Sciences.

Spooner, R. H., 1968 : "A Method Employed for Rapid Assessment of Landuse Prospects in an Extensive Area", *Proceedings of the Toulouse Conference on Aerial Surveys and Integrated Studies.* Paris, UNESCO, pp. 411-412.

Stewart, G. A., 1968 : *Land Evaluation.* Melbourne, MacMillans, p. 392.

World Bank, 1976 : *LANDSAT Index Atlas for the Developing Countries of the World*, Washington, The World Bank, p. 16.

INDEX